MILITARY PSYCHOLOGISTS' DESK REFERENCE

MILITARY PSYCHOLOGISTS' DESK REFERENCE

Editors

Bret A. Moore

Jeffrey E. Barnett

OXFORD
UNIVERSITY PRESS

OXFORD
UNIVERSITY PRESS

Oxford University Press is a department of the University of Oxford.
It furthers the University's objective of excellence in research, scholarship,
and education by publishing worldwide.

Oxford   New York
Auckland   Cape Town   Dar es Salaam   Hong Kong   Karachi
Kuala Lumpur   Madrid   Melbourne   Mexico City   Nairobi
New Delhi   Shanghai   Taipei   Toronto

With offices in
Argentina   Austria   Brazil   Chile   Czech Republic   France   Greece
Guatemala   Hungary   Italy   Japan   Poland   Portugal   Singapore
South Korea   Switzerland   Thailand   Turkey   Ukraine   Vietnam

Oxford is a registered trademark of Oxford University Press in the
UK and certain other countries.

Published in the United States of America by
Oxford University Press
198 Madison Avenue, New York, NY 10016

Library of Congress Cataloging-in-Publication Data
Moore, Bret A.
Military psychologists' desk reference / edited by Bret A. Moore, Jeffrey E. Barnett.
  p cm
Includes bibliographical references and index.
ISBN 978-0-19-992826-2
1. Psychology, Military—Handbooks, manuals, etc.   I. Barnett, Jeffrey E.   II. Title.
U22.3.M589 2013
355.0019—dc23
2013006571

Views expressed in this book are those of the authors and do not necessarily reflect official policy or position of the Department of the Army, Department of the Navy, Department of Air Force, Department of Veterans Affairs, Department of Defense, or the United States government.

In memory of Peter J. Linnerooth, Ph.D.; A great friend, dedicated father, courageous Army officer, and compassionate psychologist
—BAM

In memory of LTC Timothy B. Jeffrey, Ph.D., ABPP; A great leader, role model, mentor, and friend
—JEB

CONTENTS

FOREWORD

Psychologists in today's military wear multiple hats and are required to be well versed in numerous areas of the modern-day profession of psychology. Some of these varied roles include clinician, scientist, researcher, educator, consultant, expert witness, advocate, communicator, coach, mentor, and leader. The breadth of the profession's contributions is simply extraordinary. This comprehensive Desk Reference provides a convenient and visionary overview of many of these topics, as written by leading experts in their respective fields of military psychology. For those with an appreciation for the future, it provides an intriguing road map for where the civilian psychology community may very well evolve. The psychology of tomorrow will mature from what was not that long ago essentially "mom and pop" small private practices into integrated, multidisciplinary systems of care with an increasing emphasis on demonstrated mission-based, objective outcomes. Services will be patient-centered, holistic, and individually tailored, whether they are considered clinical, consultative, or operational in nature. Psychologists in today's military find themselves working in ever-expanding settings, delivering novel services, with varied populations that would have been entirely unthinkable to the forefathers of early military psychology who got their foot in the door with psychological screening and gained a solid foothold with clinical practice. From operational psychology, forensic psychology, and health psychology to neuropsychology, research psychology, and organizational psychology, the modern-day military psychologist is involved in nearly all aspects of the profession of arms.

The military and its sister federal agencies—the U.S. Department of Veterans Affairs (VA), the Federal Bureau of Prisons, and the U.S. Public Health Service—have long offered psychology unique and exciting opportunities to function to the fullest extent of its training and professional vision. Over the years the federal sector has increasingly become the employer of choice for new graduates and has been on the cutting edge in adapting to what must be considered unprecedented change in response to the advent of technology into the health care arena (e.g., telehealth, electronic health records, comparative effectiveness research, etc.). In the training arena, the military, along with psychology's leaders within the VA, have essentially defined the field of postdoctoral education for the profession. In so doing, the federal sector has become a catalyst for substantive policy discussions within the American Psychological Association (APA) governance, leading to extensive modifications within the national education (especially accreditation) and practice communities.

This thought-provoking Desk Reference provides an insightful overview of the depth and breadth of psychology's involvement over the years within the Department of Defense, as well as its willingness to address new and evolving challenges; for example, women in combat, suicide in the military, and most recently psychology's roles in interrogation and psychotropic management. The fundamental mission of the military has not changed—it remains to protect our national security. And yet, the military itself has rather dramatically changed over the years. Today's military is an all-volunteer force with many of those placing themselves "in harm's way" coming from the Reserve and National Guard. There are increasing numbers of women in leadership positions, with the US Army having selected its first female (and first nurse) as its Surgeon General. The landscape of warfare is changing (cyber war, remotely piloted aircraft, etc.), requiring psychologists to research and address the stressors unique to these new theaters of operation. The role and contributions of military families have become a significant priority for operational consideration. It is also the case that since 9/11 today's military is facing an entirely different type of enemy compared to previous conflicts, under very unusual if not unprecedented circumstances. The signature wounds of this conflict are heavily psychological in nature, for example, recovering from head trauma due to unexpected blasts, psychological stress (posttraumatic stress disorder, or PTSD, being an obvious example), and strategically addressing the beginning stages of reentry into civilian life for the Wounded Warriors and now equally important, their families.

What is perhaps the most significant contribution of this publication for the civilian reader is the manner in which the uniqueness of the military culture is systematically incorporated into each of the chapters, thereby providing that all-important underlying context for what is being discussed. For example, although for many clinicians absolute "patient confidentiality" is the bedrock

for the successful therapeutic relationship, one must appreciate that "confidentiality" must be conditional within a military context. Similarly, although in the private sector and civilian life concerns regarding "stigma" are unquestionably significant, just how significant may have a different consequence within the military, especially when seeing a psychotherapist may be perceived by superiors as jeopardizing a critical mission. The reader will also quickly come to appreciate that there are many subtle cultural nuances within the military—that each of the service branches is different—that assigned units and operational missions can make a real difference. Rank and years of experience—not to mention multiple deployments—may be seen as windows into what may perhaps be fundamentally different treatment populations.

Given the modern-day realities of limited financial and staffing resources, psychologists practicing in today's military and federal sectors must seek innovative ways to deliver their services in cost-effective, efficient, and evidence-supported ways. Community-based prevention, technology-enhanced interventions such as tele-mental health, group therapies, time-limited psychotherapies, embedded mental health, and primary care consultation are just a handful of examples of strategies currently being used to expand the reach of precious mental health resources. It is critical to the future of the profession that psychologists continue to conduct rigorous research studies and continually evaluate the effectiveness of programs and interventions to scientifically inform their decisions and guide the way to improved outcomes.

Military psychologists are uniquely trained and postured to lead the field in these multifront efforts and to share their "lessons learned" from experiences in the battlefield, home front, clinics, classrooms, laboratories, courtrooms, and offices. Whether the populations served are active duty, reserve, civil service, family members, veterans, or retirees, as a microcosm of society the evidence obtained from the military experience can often generalize to the larger community from which the military is drawn. Continued support from and partnerships with private industry, academia, civic leaders, professional organizations, and other influential stakeholders will be crucial to these efforts. Most important is maintaining the critical trust of those who serve our nation, including their families.

[The views expressed are personal and do not represent those of USUHS, the USAF, or the Department of Defense]

Patrick H. DeLeon
Uniformed Services University of the
Health Sciences
University of Hawaii
Former President, American Psychological
Association

Jay M. Stone
Uniformed Services University of the
Health Sciences, USAF

PREFACE

The psychological well-being of the men and women returning from the wars in Iraq and Afghanistan is one of the most discussed and contemplated mental health issues in our country today. Every week scores of articles on the topic are published in popular newspapers, magazines, and top scientific journals. Television and radio news programs fill much of their time debating the "epidemics" of PTSD and traumatic brain injury in our returning veterans and the potential fallout of a less than adequate military and Veterans Administration mental health system. However, this is only a small glimpse into the world of the service member and an even smaller one into the profession of military psychology.

Military psychology as a specialty within psychology has been around since the turn of the 20th century. It is likely one of the most diverse specialties within the field and includes numerous subspecialties, work settings, and career trajectories. In addition to addressing issues like the aforementioned PTSD and traumatic brain injury in service members and veterans, military psychology is positioned and equipped to influence such issues as psychological resilience, extended family stress, the role of technology in health care delivery, and ways to increase human performance under harsh conditions within a variety of settings

and contexts. A profession that was once seen as esoteric and mysterious has been normalized and integrated into the national health care discourse.

Because of the depth and breadth of military psychology and its far-reaching influence, it is imperative that not only the military clinician have access to a comprehensive resource covering this vast and expansive field, the nonmilitary clinician, researcher, educator, and policymaker should also have access to the most relevant and up-to-date information in the field. We believe *Military Psychologists' Desk Reference* meets this need.

The general format of *Military Psychologists' Desk Reference* may be familiar to the reader. It is based on the original and very successful *Psychologists' Desk Reference* edited by Koocher, Norcross, and Hill (2004) and also published by Oxford University Press. Consisting of nearly 70 brief, focused, and practical chapters, the *Military Psychologists' Desk Reference* highlights the most salient information in the field, which is summarized by leading experts within the military, Veterans Administration, and civilian sector.

The first section of this volume provides a brief overview of the history of military and VA psychology, basic military demographics, and invaluable information related to military cultural issues, both with clinical and non-clinical implications. Section two covers the major psychological specialties within the field to include military neuropsychology, aviation and operational psychology, combat and operational stress, human factors engineering, command and organizational consultation, and others. It also covers the unique roles psychologists play in supporting Special Operations Forces. Section three provides information on a number of professional issues in military psychology such as ethical challenges, scope of practice, professional education and training, challenges of women in combat, working with the media, professional burnout, and the controversial topic of psychologists' involvement in interrogations. Section four includes numerous chapters on clinical theory, research, and practice issues such as treating PTSD, suicide, resilience, violence, trauma assault, traumatic brain injury, sleep disorders, and many others. Section five closes out the volume with a chapter that includes the more commonly used military abbreviations and acronyms and a chapter that displays the various ranks used by the US Army, Navy, Air Force, Marine Corps, and Coast Guard.

It is our belief that the *Military Psychologists' Desk Reference* will become the authoritative guide within the field of military psychology. With over 100 of the field's leading experts in their respective areas, this volume addresses both broad and narrow aspects of military psychology. However, the book is by no means complete. Information relevant to those who serve and support military personnel is continually changing. And considering the vastness of the field, we have undoubtedly inadvertently neglected to include relevant information. As a remedy, we have created an e-mail account so that readers can share their thoughts and suggestions on how to make the next edition stronger.

Bret A. Moore and Jeffrey E. Barnett
MPDRfeedback@gmail.com

ACKNOWLEDGMENTS

There are many people who make a book like this a reality. We would like to thank Sarah Harrington and Andrea Zekus from Oxford University Press and Prasad Tangudu from Newgen Knowledge Works for all of their hard work and support during the publication process. We are grateful for the many experts in military psychology who agreed to contribute to this volume. An edited book is only as good as those who write the chapters. We are indebted to Gerald P. Koocher, John C. Norcross, and Sam S. Hill III for allowing us to adapt the format for our book from their very successful *Psychologists' Desk Reference*. Last, but certainly not least, we thank our families for enduring the many late (and early) hours in front of our computers. Editors are only as good as their loved ones.

ABOUT THE EDITORS

Dr. Bret A. Moore is the founder of Military Psychology Consulting and adjunct associate professor in psychiatry at University of Texas Health Science Center at San Antonio. He is licensed as a prescribing psychologist by the New Mexico Board of Psychologist Examiners and board-certified in clinical psychology by the American Board of Professional Psychology.

Dr. Moore is a former active duty Army psychologist with two tours of duty to Iraq totaling 27 months. He is the author and editor of nine other books, including *Treating PTSD in Military Personnel, Handbook of Counseling Military Couples, The Veterans and Active Duty Military Psychotherapy Treatment Planner, Living and Surviving in Harm's* Way, *Wheels Down: Adjusting to Life after Deployment, The Veterans and Active Duty Military Homework Planner, Pharmacotherapy for Psychologists: Prescribing and Collaborative Roles, Handbook of Clinical Psychopharmacology for Psychologists*, and *Anxiety Disorders: A Guide for Integrating Psychopharmacology and Psychotherapy*. He also writes a biweekly newspaper column titled *Kevlar for the Mind*, which is published by *Military Times*.

Dr. Moore is a Fellow of the American Psychological Association and has been awarded early career awards in military psychology and public service psychology from Divisions 18 and 19 of APA, respectively. His views and opinions on military and clinical psychology have been quoted in *USA Today*, the *New York Times, Boston Globe, TV Guide*, and on NPR, the BBC, CNN, CBS News, Fox News, and the CBC.

Dr. Jeffrey E. Barnett is a professor in the Department of Psychology at Loyola University Maryland and a licensed psychologist in independent practice in Annapolis, Maryland. He is board certified by the American Board of Professional Psychology in clinical psychology and in clinical child and adolescent psychology and is a distinguished practitioner of psychology in the National Academies of Practice.

Dr. Barnett is a former Army psychologist who was the first psychologist in the Army's Special Operations Command serving as the group psychologist for the 160th Special Operations Aviation Group (Airborne). He was a paratrooper, rappelle master, airborne pathfinder, and the first graduate of the Army's high risk survival, evasion, resistance, and escape (SERE) course.

Dr. Barnett is a past chair of the ethics committees of the Maryland Psychological Association and the American Psychological Association. At present, he is a member of the ethics committee of the American Board of Professional Psychology and serves on the Maryland Board of Examiners of Psychologists. Dr. Barnett has numerous publications and presentations to his credit that focus on ethics, legal, and professional practice issues for mental health professionals to include editing one book and coauthoring six. His recent books include *Ethics Desk Reference for Psychologists* (2008, with W. Brad Johnson) and *Ethics Desk Reference for Counselors* (2009, with W. Brad Johnson). He is a recent recipient of the American Psychological Association's Award for Outstanding Contributions to Ethics Education and its award for Distinguished Contributions to the Independent Practice of Psychology.

CONTRIBUTORS

Melissa M. Amick, Ph.D.
Psychologist, Spinal Cord Injury Service;
Investigator, Translational Research Center for
Traumatic Brain Injury and Stress; VA Boston
Healthcare System, Boston, MA; Assistant
Professor of Psychiatry, Boston University
School of Medicine, Boston, MA

Elizabeth H. Anderson, M.R.C.
VA North Texas Health Care System, Dallas, TX

Rodney R. Baker, Ph.D.
Department of Veterans Affairs (Retired),
San Antonio, TX

L. Morgan Banks, Ph.D.
Operational Psychology Support, LLC

Willie G. Barnes, D.Min, CFMFT
Chaplain, Colonel (ret.), Army National Guard,
Consultant, Defense Suicide Prevention Office,
Educator and Clinician

Paul T. Bartone, Ph.D.
Professor and Senior Research Fellow, Center
for Technology and National Security Policy,
National Defense University, Washington,
DC; Adjunct Research Professor, University of
Bergen, Norway

Charles C. Benight, Ph.D.
Director CU: Trauma, Health, and Hazards Center
and Professor of Psychology, University of
Colorado, Colorado Springs, CO

Elizabeth A. Bennett, Ph.D.
Professor of Psychology, Washington & Jefferson
College, Washington, PA

Robert M. Bray, Ph.D.
Senior Program Director, Military Behavioral Health,
RTI International, Research, Triangle Park, NC

Donna M. Brazil, Ph.D.
Psychology Program Director, Department
of Behavioral Sciences and Leadership,
United States Military Academy, West Point, NY

William L. Brim, Psy.D.
Deputy Director, Center for Deployment
Psychology; Associate Professor, Medical
and Clinical Psychology, Uniformed Services
University of the Health Sciences, Bethesda, MD

Craig J. Bryan, Psy.D., ABPP
Associate Director of the National Center for
Veterans Studies, Salt Lake City, UT; Assistant
Professor in the Department of Psychology, The
University of Utah, Salt Lake City, UT

Colette M. Candy, Ph.D.
Supervising Psychologist, Department of
Behavioral Health, Madigan Army Healthcare
System, Tacoma, WA

Vincent F. Capaldi, II, Sc.M., M.D., MAJ, MC
US Army, Assistant Professor, Department of
Psychiatry, Uniformed Services University of
the Health Sciences, Bethesda, MD

Melinda C. Capaldi, Psy.D., CPT, MSC
US Army, Clinical Psychologist, Bethesda, MD

Julie A. Cederbaum, Ph.D., MSW, MPH
Assistant Professor, University of Southern
    California School of Social Work, Los Angeles, CA

Wayne Chappelle, Psy.D., ABPP
Senior Aeromedical Clinical Psychologist, USAF
    School of Aerospace Medicine, Wright-Patterson
    AFB, OH

Roman Cieslak, Ph.D.
Senior Research Associate, CU: Trauma, Health,
    and Hazards Center, University of Colorado,
    Colorado Springs, CO; Associate Professor,
    Department of Psychology, University of Social
    Sciences and Humanities, Warsaw, Poland

Michael E. Clark, Ph.D.
Pain Section Leader, James A. Haley Veterans
    Hospital, Tampa, FL; Associate Professor,
    Department of Psychology, University of South
    Florida, Tampa, FL

Michael Crabtree, Ph.D.
Professor of Psychology, Washington & Jefferson
    College; Director and Chief Clinician, Washington
    Psychological Services, Washington, PA

Michael R. DeVries, Ph.D.
Operational Psychologist, US Army Special
    Operations Command, Fort Bragg, NC

Jean M. Dixon, M.S.N. Ed., RN, CCM
Active Duty Case Manager, Navy Region
    Mid-Atlantic, Reserve Component Command,
    Medical Hold Department, Norfolk, VA

Richard L. Dixon Jr., M.Ed., MMAS
Analyst, Lieutenant Colonel, US Army
    Reserve, Joint and Combined Operational
    Analysis, Suffolk, VA; Sergeant, Tucson Police
    Department, AZ

Kent D. Drescher, Ph.D.
Health Science Specialist, National Center for
    PTSD, VA Palo Alto Health Care System;
    Psychologist, The Pathway Home: California
    Transition Center for the Care of Combat
    Veterans, Yountville, CA

Eric B. Elbogen, Ph.D., ABPP-Forensic
University of North Carolina-Chapel Hill School
    of Medicine and Durham, VA, Medical Center

Brian Engdahl, Ph.D.
Counseling Psychologist and Faculty Member,
    Brain Sciences Center, Minneapolis Veterans
    Health Care System; Clinical Professor,
    Department of Psychology, University of
    Minnesota, Minneapolis, MN

Charles R. Figley, Ph.D.
Paul Henry Kurzweg Distinguished Chair, Tulane
    University, New Orleans, LA

Kimberly Finney, Psy.D., ABPP, ABMP
Clinical Associate Professor, University of
    Southern California, Los Angeles, CA

Jeanne M. Gabriele, Ph.D.
Local Evidence-Based Psychotherapy Coordinator,
    G.V. "Sonny" Montgomery Veterans Affairs
    Medical Center, Jackson, MS; Assistant Professor
    of Psychiatry and Human Behavior, University
    of Mississippi Medical Center, Jackson, MS

Dave Grossman, Lt. Col., USA (ret.)
Director, Killology Research Group

Veronica Gutierrez, Ph.D.
Counseling Psychologist in San Marcos and
    Oceanside, CA.

Lynn K. Hall, Ed.D., LPC, NCC, ACS
Dean, College of Social Sciences, University of
    Phoenix, Tempe, AZ

Jeanne S. Hoffman, Ph.D., ABPP
Chief, Pediatric Psychology Clinic, Department
    of Pediatrics, Tripler Army Medical Center,
    Honolulu, HI

Beeta Homaifar, Ph.D.
Clinical Psychologist, Boston VA Health Care
    System, Boston, MA; Assistant Professor,
    Department of Psychiatry, Boston University,
    Boston, MA

Pennie L. P. Hoofman, Ph.D.
Director, Aeromedical Psychology, US Army
    School of Aviation Medicine, Ft. Rucker, AL

C. Alan Hopewell, Ph.D., MP, ABPP,
    MAJ, US Army (ret.)
Director of Neuropsychology and Behavioral
    Health, Traumatic Brain Injury Clinic,
    CRDAMC, Ft. Hood, TX

Richard J. Hughbank, D.M., MAJ, US Army (ret.)
Assistant Professor in Criminal Justice and
  Homeland Security, Northwestern State
  University of Louisiana, Natchitoches, LA

Jared A. Jackson, Ph.D.
Clinical Psychologist, Captain, US Army

Larry C. James, Ph.D., ABPP
The School of Professional Psychology, Wright
  State University, Dayton, OH

W. Brad Johnson, Ph.D.
Professor of Psychology, US Naval Academy,
  Annapolis, MD; Faculty Associate, Johns
  Hopkins University, Baltimore, MD

Brian L. Jones, Ph.D.
Major, United States Air Force, 70th Intelligence,
  Surveillance & Reconnaissance Wing, Fort
  George G., Meade, MD

Michele J. Karel, Ph.D.
Psychogeriatrics Coordinator, Mental Health
  Services, VA Central Office, Washington, DC;
  Associate Professor, Department of Psychiatry,
  Harvard Medical School, Boston, MA

Terence M. Keane, Ph.D.
National Center for PTSD at VA Boston Healthcare
  System and Boston University School of
  Medicine, Boston, MA

Mark P. Kelly, Ph.D., ABPP-CN
Program Director, Postdoctoral Fellowship in
  Clinical Neuropsychology, Walter Reed National
  Military Medical Center, Bethesda, MD

William D. S. Killgore, Ph.D., LTC, MS, USAR
Director, SCAN Laboratory, McLean Hospital,
  Belmont, MA; Associate Professor of Psychology,
  Harvard Medical School, Boston, MA

Nathan A. Kimbrel, Ph.D.
Clinical Research Psychologist, VISN 17 Center
  of Excellence for Research on Returning War
  Veterans, Waco, TX; Assistant Professor, Texas
  A&M Health Science Center, Temple, TX

Daniel W. King, Ph.D.
Research Professor of Psychology and Psychiatry,
  Boston University; VA Boston Healthcare
  System, Boston, MA

Lynda A. King, Ph.D.
Research Professor of Psychology and Psychiatry,
  Boston University and VA Boston Healthcare
  System, Boston, MA

Heidi S. Kraft, Ph.D.
Clinical Psychologist, San Diego, CA

Gerald P. Krueger, Ph.D., CPE
Adjunct Assistant Professor of Military
  Psychology, Uniformed Services University of
  the Health Sciences, Bethesda, MD

William Sean Lee, D.Min., B.C.C.
Chaplain, Colonel, Joint Force Headquarters
  Chaplain, Maryland Army National Guard,
  Baltimore, MD

Peter J. N. Linnerooth[†], Ph.D.
Independent Practice, Mankato, MN

Dolores Little, Ph.D.
Psychologist and VA Medical Center
  Administration Department of Veterans
  Affairs (ret.)

Brett T. Litz, Ph.D.
Director, Mental Health Core, Massachusetts
  Veterans Epidemiological Research and
  Information Center (MAVERIC), VA Boston
  Healthcare System, Boston, MA; Professor,
  Boston University, Boston, MA

Judith A. Lyons, Ph.D.
Team Leader, Trauma Recovery Program, G.V.
  "Sonny" Montgomery Veterans Affairs
  Medical Center, Jackson, MS;
  Associate Professor of Psychiatry and Human
  Behavior, University of Mississippi Medical
  Center, Jackson, MS

A. David Mangelsdorff, Ph.D., M.P.H., FAPA,
  FAPS, FAAAS
Professor, Army-Baylor University Graduate
  Program in Health and Business Administration,
  Fort Sam Houston, TX

Brian P. Marx, Ph.D.
National Center for PTSD at VA Boston
  Healthcare System and Boston University
  School of Medicine, Boston, MA

Michael D. Matthews, Ph.D.
Professor of Engineering Psychology,
  Department of Behavioral Sciences and
  Leadership, U.S. Military Academy,
  West Point, NY

Nancy A. McGarrah, Ph.D.
Cliff Valley Psychologists, Atlanta, GA

[†]Peter J.N. Linnerooth unfortunately passed away
before this book was completed.

Cindy McGeary, Ph.D., ABPP
Associate Faculty–Research, Department of
    Psychology, University of Texas at Arlington,
    Arlington, TX

Don McGeary, Ph.D., ABPP
Assistant Professor in Psychiatry, University
    of Texas Health Science Center San Antonio,
    San Antonio, TX

Brock A. McNabb, MSW
Department of Veterans Affairs, Honolulu, HI

Donald Meichenbaum, Ph.D.
Distinguished Professor Emeritus, University of
    Waterloo, Ontario, Canada; Research Director
    of the Melissa Institute for Violence Prevention,
    Miami, FL

Eric C. Meyer, Ph.D.
Clinical Research Psychologist, VISN 17 Center
    of Excellence for Research on Returning War
    Veterans, Waco, TX; Assistant Professor, Texas
    A&M Health Science Center, Temple, TX

Laurence Miller, Ph.D.
Independent Practice, Boca Raton, FL; Adjunct
    Professor of Psychology, Florida Atlantic
    University, Boca Raton, FL

Paul Montalbano, Ph.D., ABPP (Forensic),
Deputy Director, Postdoctoral Fellowship Training
    Program in Forensic Psychology, Walter Reed
    National Military Medical Center, Bethesda, MD

Jennifer L. Murphy, Ph.D.
Clinical Director, Chronic Pain Rehabilitation
    Program, James A. Haley Veterans' Hospital,
    Tampa, FL; Assistant Professor,
    Department of Neurology, University of South
    Florida, Tampa, FL

Walter Erich Penk, Ph.D., ABPP
Professor, Psychiatry and Behavioral Sciences,
    Texas A&M College of Medicine; Consultant,
    Department of Veterans Affairs, VA
    Rehabilitation Research and Development

Alan L. Peterson, Ph.D., ABPP
Professor, Department of Psychiatry, Chief,
    Division of Behavioral Medicine, Director,
    STRONG STAR Multidisciplinary PTSD
    Research Consortium, University of Texas
    Health Science Center at San Antonio, San
    Antonio, TX

Matthew C. Porter, Ph.D.
Assistant Professor, California School of
    Professional Psychology, Alliant International
    University, San Diego, CA

Lewis Pulley, M.S., PsyM.
The School of Professional Psychology,
    Wright State University, Dayton, OH; Captain,
    US Air Force

Michael G. Rank, Ph.D.
Clinical Associate Professor, Director,
    San Diego Academic Center, School of Social
    Work, University of Southern California,
    San Diego, CA

Greg M. Reger, Ph.D.
Deputy Director, Emerging Technologies Program,
    National Center for Telehealth and Technology,
    Joint Base Lewis-McChord, WA

David S. Riggs, Ph.D.
Director, Center for Deployment Psychology,
    Research Associate Professor, Uniformed
    Services University of the Health Sciences,
    Bethesda, MD

M. David Rudd, Ph.D., ABPP
Co-Founder and Scientific Director, National
    Center for Veterans Studies; Dean, College
    of Social and Behavioral Science, Professor of
    Psychology, University of Utah

Mark C. Russell, Ph.D., ABPP
Chair, Psy.D. Program, Antioch University
    Seattle, WA

Morgan T. Sammons, Ph.D., ABPP
Dean and Professor, California School
    of Professional Psychology,
    San Francisco, CA

Mary E. Schaffer, Ph.D.
Chief, Training and Research Division,
    Department of Behavioral Medicine,
    Brooke Army Medical Center, San Antonio, TX

Richard Schobitz, Ph.D., CDR, USPHS
Chief, Training and Research Division, Department
    of Behavioral Medicine, Brooke Army Medical
    Center, San Antonio, TX

David S. Shearer, Ph.D.
Clinical and Prescribing Psychologist, Director of
    Behavioral Sciences, Family Medicine Residency,
    Dept of Family Medicine, Madigan Army
    Healthcare System, Tacoma, WA

Michelle D. Sherman, Ph.D.
Director, Family Mental Health Program, Oklahoma
    City VA Medical Center; Core Investigator, South
    Central Mental Illness Research, Education and
    Clinical Center (MIRECC); Clinical Professor,
    Department of Psychiatry and Behavioral Sciences,
    University of Oklahoma Health Sciences Center

Avron Spiro III, Ph.D.
Research Professor, Departments of Epidemiology and Psychiatry, Boston University Schools of Public Health and Medicine, Boston MA; Research Career Scientist, VA Boston Healthcare System, Boston, MA

Maria M. Steenkamp, Ph.D.
Clinical Research Psychologist, VA Boston Healthcare System, Boston, MA

Amy E. Street, Ph.D.
Women's Health Sciences Division, National Center for PTSD, VA Boston Healthcare System; Associate Professor, Division of Psychiatry, Boston University School of Medicine, Boston, MA

Diana L. Struski
San Antonio, TX

Connor Sullivan
Research Assistant, Department of Psychology, University of North Carolina, Chapel Hill, NC

Alina Surís, Ph.D., ABPP
Chief of Psychology, Mental Health Service, VA North Texas Health Care System; Associate Professor, Psychiatry, University of Texas Southwestern Medical Center, Dallas, TX

Michael G. Sweda, Ph.D., ABPP (Forensic)
Board-Certified Forensic Psychologist; Director, WRNMMC Forensic Psychology Fellowship; Deputy Director, WRNMMC Center for Forensic Behavioral Sciences, Fremont Bldg, Bethesda, MD

Richard G. Tedeschi, Ph.D.
Professor of Psychology, University of North Carolina Charlotte, Charlotte, NC

David F. Tharp, Psy.D.
VISN 17 Center of Excellence for Research on Returning War Veterans, Waco, TX; Associate Professor, Texas A&M Health Science Center, Temple, TX; Lieutenant Colonel (Dr.), United States Air Force Reserves

Jennifer J. Vasterling, Ph.D.
Chief of Psychology, VA Boston Healthcare System, Boston, MA; VA National Center for PTSD, Boston, MA; Professor of Psychiatry, Boston University School of Medicine, Boston, MA; Lecturer, Harvard Medical School

Dawne Vogt, Ph.D.
Women's Health Sciences Division, National Center for PTSD, VA Boston Healthcare System; Associate Professor, Division of Psychiatry, Boston University School of Medicine, Boston, MA

Joseph Westermeyer, M.D., M.P.H., Ph.D.
Staff Psychiatrist, Minneapolis VA Medical Center; Professor of Psychiatry, University of Minnesota, Minneapolis, MN

Emile K. Wijnans, Ph.D.
Clinical Psychologist, United States Army Drill Sergeant School, Ft. Jackson, SC

Sherrie L. Wilcox, Ph.D., CHES
Research Assistant Professor, University of Southern California, Los Angeles, CA

Thomas J. Williams, Ph.D.
Professor, US Army War College, Carlisle, PA

Kristin N. Williams-Washington, Psy.D.
Clinical Psychologist, Upper Marlboro, MD

Blair E. Wisco, Ph.D.
National Center for PTSD at VA Boston Healthcare System and Boston University School of Medicine, Boston, MA

Stacey Young-McCaughan, RN, Ph.D.
Professor, Department of Psychiatry, Division of Behavioral Medicine, University of Texas Health Science Center, San Antonio, TX

PART I
History and Culture

1 EARLY HISTORY OF MILITARY MENTAL HEALTH CARE

Brian L. Jones

Mental health care providers in the military follow an uncommon charge in comparison to their civilian counterparts. Issues such as differing arenas of practice and ethical quandaries only scratch the surface of the complexities found in being a military officer and a mental health care provider. Despite the efforts of military mental health providers and the current federal budgetary emphasis placed on the mental health of military forces and their family members, the state of military mental health care has not always been this robust (Laurence & Matthews, 2012). It is difficult, if not impossible, to discuss the development of mental health care in the military without simultaneously noting that each step was paved during a particular time in this nation's history of war.

TRAUMATIC STRESS IN WAR

Posttraumatic stress disorder (PTSD) was added to the formal diagnostic nomenclature in 1980, though psychiatric symptoms stemming from stress-related combat have always existed. History is replete with indications of war-related difficulties (uncontrollable shaking, heart palpitations, going blind on the battlefield) all the way back to ancient Greece. Epizelus is recorded as going blind on the battlefield after a man next to him was killed in a war between the Greeks and Persians (Jones, 1995). While symptoms of PTSD were likely

exhibited in all previous military conflicts, they were addressed by other names. The changes in terms used to denote combat-related stress, while interesting from a historical perspective, also enrich our understanding of how viewing these symptoms differently and more accurately over time spawned the development of military mental health care.

The term "nostalgia" was coined by Swiss physician Johannes Hofer during the 17th century and referred to homesickness with the belief that symptoms derived from a soldier's desire to return home. In the 18th century an Austrian physician, Josef Leopold Auenbrugger, wrote about nostalgia, listing the symptoms as sadness and being taciturn, listless, and isolating (Jones, 1995). During the Napoleonic wars PTSD symptoms were termed "exhaustion." During the American Civil War, terms like "soldier's heart" and "effort syndrome" were coined. The symptoms during World War I (WWI) and World War II (WWII) were identified as "shell shock" and "battle fatigue," respectively (Kennedy, Boake, & Moore, 2010). As the culture slowly began to develop an understanding of people returning home from war with psychological wounds, during the Vietnam War service members with such wounds were said to be suffering from "post-Vietnam syndrome." The gradual emergence of mental health services in America evolved over time hand in hand with the sociocultural political climate in terms of understanding the psychological effects one might experience subsequent to the brutality of war.

PRE–WORLD WAR I

Although operational psychology practices (often referred to as PSYOPs) were employed during the American revolutionary war, there was little to no attention given to the possible mental health difficulties attributable to the exposure to trauma during war (Kennedy & McNeil, 2006). It wasn't until the American Civil War that documentation of mental health disorders was initiated. A great deal of documentation considered substance abuse problems, which were rampant due to the management of pain from amputation by way of narcotics (Watanabe, Harig, Rock, & Koshes, 1994). The brutality of the Civil War contributed to significant psychiatric trauma. Nostalgia was the second most common diagnosis made by Union doctors. New terms were coined, including "soldier's heart" and "exhausted heart." Like "nostalgia," these new terms explicated the symptoms exhibited by emotionally distraught soldiers, particularly paralysis, tremors, sudden changes in mood, and a deep desire to return home. The advent of neurosurgery during the Civil War was seminal in distinguishing maladies that had an organic basis from those more psychological in nature (Jones, 1995).

WORLD WAR I (WWI)

Although the "official" birth of military psychology occurred during WWI and psychologists were utilized at the time, their duties mostly centered on activities such as testing, assessment, and selection. However, in terms of military mental health care, it was during WWI that physicians in the military began to notice the traumatic reactions of soldiers. Initial thought centered on the idea that an actual shock to the nervous system had occurred. Hence, the phrase "shell shock" was used to account for a soldier exposed to shelling and subsequently developing blindness or a peculiar gait, detaching from activities of daily living, and/or suffering from amnesia. However, this notion was quickly abandoned when it was discovered that soldiers never exposed to shelling experienced the same symptoms (Jones, 1995). Suddenly, this constellation of symptoms was viewed as a psychiatric problem, and applicable psychiatric care was offered near the front. The intervention of PIE (proximity, immediacy, expectation of recovery) was developed and utilized to decrease the number of shell shock cases unable to return to fighting the war (Kennedy & McNeil, 2006). The concept of PIE is understood as a foundational ✳ intervention in combat-related stress, and at least some variation remains in use by all branches of the military today.

WORLD WAR II

The advent of WWII saw another increase in the use of military psychologists utilized in formalized screening (testing, assessment, etc.). Unfortunately, the emphasis placed on screening meant that there was little emphasis on forward deployed mental health care workers. Failing to capture any lessons learned during WWI about combat-related stress reactions, the belief at the time was that they could screen out those individuals predisposed to such reactions. The terms "combat fatigue" and "combat exhaustion" both underscore the thinking at the time, which was that these symptoms were largely due to long deployments. While there was a significant increase in the number of early discharges due to combat-related stress, there was finally an appreciation of the importance of mental health intervention on the battlefield and preparing military personnel for the psychological consequences of engaging in combat (Kennedy & McNeil, 2006).

This was also the period of time during which ✳ military psychologists were first assigned to hospitals. As WWII came to an end, it was clear that physicians could not adequately manage the overwhelming numbers of service members needing mental health care. Psychologists were able to fill this void and proved to be propitious in delivering quality mental health care (i.e., individual and group psychotherapy), especially in Veterans Administration (VA) treatment settings after the conclusion of the war (Ball & Peake, 2006).

Just as after WWI, the end of WWII saw the demobilization of psychologists. However, the growing consensus among decision makers was that a benefit to having a psychologist in the military is the power of influence, which might not be available from a civilian psychologist working within the military system. Consequently, in 1947 psychologists were given permanent active duty status as military members (Kennedy & McNeil, 2006). Similarly, although utilized in a civilian capacity prior to this time, by the end of WWII social workers were granted active duty status as military officers. Also of particular note during this time period, the addition of "gross stress reaction" to the formal diagnostic nomenclature provided clinicians a common frame of reference for service members suffering from the stressors of combat.

KOREAN WAR

By the time the Korean War started in 1950, the prior half-century had witnessed the incremental development of the military mental health care provider from civilian to active duty during wartime to regular active duty. Beginning in the Korean War active duty mental health providers found themselves in positions not encountered before (stationed overseas, in combat zones, on hospital ships, etc.). Unfortunately, because of the hasty beginning of the war, there were not the appropriate support units in place. This meant that the lessons learned from WWI and WWII in terms of forward deployed mental health intervention were not available at the beginning of the war (Kennedy & McNeil, 2006). The immediate impact of this lack of intervention was significant, but as the practices of combat stress intervention were gradually employed there

was an increase in service members returned to duty. The end of the Korean War saw the Army focus its attention on organizational principles (motivation, morale, leadership) and psychological warfare, while the Navy and Air Force began to focus on performance enhancement, specifically through the study of human factors (Kennedy & McNeil).

VIETNAM WAR

During the Vietnam War, military psychologists continued to serve in combat zones, applying the well-established principles of combat stress intervention practiced during WWI, WWII, and the Korean War. Compared to these previous wars, there appeared to be a reduced amount of traditional combat-related stress in Vietnam. These symptoms were again described as "combat fatigue." However, there was more attention given to problematic behavioral issues than to mental health diagnosis, as service members were seen as exhibiting character disorders (Kennedy & McNeil, 2006).

Problems with abusing and being dependent on alcohol, narcotics, and other substances have existed in most militaries worldwide since historical records have been kept. However, the war in Vietnam was characterized by it. Part of this can be attributed to the zeitgeist of the 1960s and early 1970s in the United States, which was much more indulgent and lenient regarding the use and abuse of substances. Likewise, there was a concomitant increase in alcohol and drug rehabilitation. Prior to the 1970s, attempts to solve these problems in both military and nonmilitary settings were woefully inadequate because of the belief that substance abuse and dependence emerged from a lack of discipline. Not understanding the "disease" component of substance use disorders, treatment options gave way to a variety of other mechanisms to address this apparent dereliction of duty. This changed in 1971, when the treatment of substance use disorders became a reality with the assistance of a congressional mandate (Watanabe et al., 1994).

The Vietnam War was unique in multiple ways. The combination of jungle warfare, cruel and inhumane experiences upon capture, poor unit cohesion due to staggered deployment rotations, and a largely nonsupportive public created an atmosphere ripe for the development and sustainment of PTSD. It is no surprise that there is an increase in rates of PTSD experienced by veterans of that war. In addition to the rising rates of PTSD, the ending of the war in Vietnam brought with it an understanding that a more systematic approach was necessary in responding to critical incidents that were not combat-related, namely training accidents and suicide (Kennedy & McNeil, 2006). Responding to critical incidents in this manner continues today with each service forming its own practices: Special Psychiatric Rapid Intervention Team (SPRINT) in the Navy; Traumatic Event Management (TEM) in the Army; and Traumatic Stress Response (TSR) in the Air Force.

THE FIRST GULF WAR

Though the First Gulf War lasted just under seven months, both operations Desert Shield and Desert Storm contained significant combat stressors not encountered in previous wars. Exposure to chemical and biological weapons, extreme desert conditions such as sandstorms, and greater numbers of enemy forces threatened to increase the possibility of psychological casualties due to combat-related stress. Forward deployed mental health care was once again utilized as well as a psychologist serving for the first time aboard a Navy aircraft carrier (Kennedy & McNeil, 2006). The availability of these services, as well as the brevity of the war and small number of American casualties, is likely the reason that there were a reduced number of service members unable to return to fighting due to combat-related stress.

Unfortunately, the availability and good response of mental health care could not account for the delayed incidence of PTSD, which has continued to increase over time in veterans of both Operations Desert Shield and Desert Storm. Likewise, a phenomenon known as "Gulf War Illness" or "Gulf War Syndrome" has plagued veterans from this war (Kennedy & McNeil, 2006). An enigmatic constellation of medically inexplicable physical and psychological symptoms, this condition continues to persist in complicating the lives of veterans and baffling researchers and clinicians in terms of determining the etiology and best course of treatment.

GLOBAL WAR ON TERROR (GWOT), OVERSEAS CONTINGENCY OPERATION, AND BEYOND

For almost the entirety of the first decade of the 21st century, service men and women spent time supporting Operations Enduring Freedom (OEF) and Iraqi Freedom (OIF). Both operations were part of the GWOT, which officially became known as Overseas Contingency Operation under the administration of US President Barack Obama. At the time of this writing OIF has concluded, while operations in OEF continue. For both operations the principles of forward deployed mental health care have been and continue to be implemented, reducing the number of psychological casualties associated with combat. Despite the thriving practice of mental health care in combat environments, veterans of both OEF and OIF suffer from PTSD. Also, because of the type of weaponry used, traumatic brain injury (TBI) is one of the signature wounds of both operations. This has presented another difficulty for clinicians and researchers in terms of developing the appropriate treatments to target and manage the sequelae of complicated physical and psychological symptoms, which are often comorbid with PTSD (Kennedy et al., 2010).

Though this chapter has focused primarily on the emergence of military mental health care in its relation to wars over the last century, it is worth noting the current state of affairs. What started as rudimentary mental health principles in forward deployed locations designed to get service members back to combat (i.e., PIE) has blossomed into a thriving and robust panoply of available services to military members, their families, and retirees. There is no question that mental health care in deployed locations requires creativity and adaptability to address the relevant needs of service members (Ball & Peake, 2006). However, the majority of military mental health care providers find themselves in other settings. Providing direct care to active duty members and their families, consulting with leaders about unit cohesion or other organizational concerns, recommending whether or not a service member is fit for duty, and giving a briefing about suicide prevention are all activities that one military mental health care provider might engage in over the course of a single day. Moreover, he or she has to be willing and able to engage in these activities in an outpatient clinic, on an inpatient psychiatry ward, as part of a primary care staff, on a ship, as a member of a disaster team, or in a classroom.

Historically speaking, the inherent difficulties in our nation's wars have challenged mental health care providers to search for innovative ways to manage the stressors of those in combat and learn the invaluable lessons of previous wars (Ball & Peake, 2006). From each conflict in our nation's history has emerged the ingenuity to learn from past oversights and capture the essence of how to better manage the mental health needs of those serving our country. It is now imperative for those in the field to consider the changing landscape of warfare (cyber war, remotely piloted aircraft, etc.) and carefully think through the stressors unique to these new theaters of combat. Simultaneously, there must be concentrated attention given to the development of more advanced methods of treatment delivery (e.g., telepsychology)

while continuing to provide the appropriate treatments for active-duty members and their families.

References

Ball, J. D., & Peake, T. H. (2006). Brief psychotherapy in the U.S. military: Principles and applications. In C. H. Kennedy & E. A. Zilmer (Eds.), *Military psychology: Clinical and operational applications* (pp. 61–73). New York, NY: Guilford Press.

Jones, F. D. (1995). Psychiatry lessons of war. In R. Zajtchuk & R. F. Bellamy (Eds.), *Textbook of military medicine: War psychiatry* (pp. 1–33). Washington, DC: Office of the Surgeon General, US Department of the Army.

Kennedy, C. H., Boake, C., & Moore, J. L. (2010). A history and introduction to military neuropsychology. In C. H. Kennedy & J. L. Moore (Eds.), *Military neuropsychology* (pp. 1–28). New York, NY: Springer.

Kennedy, C. H., & McNeil, J. A. (2006). A history of military psychology. In C. H. Kennedy & E. A. Zilmer (Eds.), *Military psychology: Clinical and operational applications* (pp. 1–17). New York, NY: Guilford Press.

Laurence, J. H., & Matthews, M. D. (2012). The handbook of military psychology: An introduction. In J. H. Laurence & M. D. Matthews (Eds.), *The Oxford handbook of military psychology* (pp. 1–3). New York, NY: Oxford University Press.

Watanabe, H. K., Harig, P. T., Rock, N. L., & Koshes, R. J. (1994). Alcohol and drug abuse and dependence. In R. Zajtchuk & R. F. Bellamy (Eds.), *Textbook of military medicine: Military psychiatry: Preparing in peace for war* (pp. 61–90). Washington, DC: Office of the Surgeon General, US Department of the Army.

2 HISTORY OF MILITARY PSYCHOLOGY

C. Alan Hopewell

PSYCHOLOGISTS JOIN THE WAR TO END ALL
WARS

In an abridged addition to his 1890 seminal
work *The Principles of Psychology*, William
James described his hope that by treating psychology as a natural science, he could help
"her" become one (James, 1890). It was at that
point that psychology as a formal discipline
stood on the verge of changing military operations forever.

The outbreak of World War I saw the United
States Army grow from an undertrained peacetime reserve force of 190,000 to one of 3,665,000
in only 20 months. In 1917 a committee from
the National Research Council proposed that
Surgeon Major General Gorgas commission
psychologists as active duty officers in order
to implement the newly devised mental testing techniques to address the problems of rapid
military induction, the need to screen for mental defects or psychiatric problems, and to make
assignments. General Order 74 commissioned
the president of the American Psychological
Association, (APA), Robert Yerkes, as a Major,
and commissioned the first 16 active duty psychologists as first lieutenants. A somewhat parallel effort was also developed by Walter Scott
and Walter Bingham with the Committee on
Classification of Personnel, as they were uncertain of the "theoretical" nature of Yerkes's
project and wanted to implement practical, busi-
ness-oriented programs. The Navy showed no

interest, but psychologists were soon involved
in the infancy of aviation personnel selection
and training.

After a successful trial program, Camp
Greenleaf was established at Fort Oglethorpe,
Georgia, for centralizing and standardizing
the training of psychology officers and technicians, using. The Army Alpha tests were used
for group assessments and the Army Beta tests
for illiterates. By May of 1918, there were 24
induction camps with psychological companies.
Eventually, 1,750,000 soldiers were examined,
an astounding 47% of the entire Army. As
part of this rapid mobilization, a young corporal with a newly minted Master's degree was
trained by Captain Edwin Boring. Put to work
screening soldiers at Camp Logan, near Corpus
Christi, Texas, David Wechsler later adapted
the Army Alpha Test, transforming it into the
Wechsler-Bellevue Intelligence Test.

By war's end, the psychology companies
were being asked to do more and more, to
include forensics, clinical consultation, solve
problems in training and morale, and developing strategies for forward psychiatry and
interventions for combat stress. In addition,
the Committee on Classification of Personnel
was formally incorporated into the military at
the initiative of the General Staff. By late 1918,
remaining psychological staffs were working
in collaboration with the Division of Physical
Reconstruction of the Surgeon General's
Office. This established psychology's role in

areas of physical disability, to include many of
the brain injuries suffered in the war. In this
regard, psychology as a whole succeeded much
more than Yerkes probably could have imagined, setting the stage for its resurgence during
World War II.

MILITARY PSYCHOLOGY "REBOOTS" FOR WORLD WAR II

As war once again broke out in Europe, many
of the lessons learned from World War I were
quickly "rebooted" for World War II. One hun-
dred and forty psychology officers were initially
commissioned as First and Second Lieutenants
assigned under the Personnel Research Section
to induction stations. By the spring of 1942,
six clinical psychologists had been directly
commissioned as first lieutenants to the then
existing Army General Hospitals. Colonel W.
C. Menninger eventually implemented more
uniform procedures with an ultimate allotment
total of 346 officers, to include five enlisted
Women's Army Corps (WAC) candidates who
were commissioned as psychologists.

The Air Force Aviation Psychology Program
of 1941 began to accept Army enlisted personnel and to test and to train them for the Army
Air Corps. The Army Research Institute (ARI)
for the Behavioral and Social Sciences had been
established somewhat earlier, in 1939, with its
historical roots going back to World War I.
These two groups, the "clinical" officers and
the "behavioral/applied scientist" psychologists, therefore, began to shape what we today
recognize as Army and eventually military
psychology as a whole.

The Army General Classification Test, the
Wechsler-Bellevue Intelligence Scale, and the
Stanford Binet Intelligence Test for the first time
became authorized tests of intelligence under
the guidance of TB MED 115. TB MED 155
similarly addressed brain-damaged and aphasic
patients and authorized the use of the Goldstein-
Scheerer Test of Abstract and Concrete Behavior.
Personality assessment emphasis, however, once
again was placed upon the attempted prescreening the attempted prescreening of those who
might develop stress disorder. In addition, it

should be recalled that troops were only in combat on the ground in Europe for ten months,
thus providing little actual "ground time" for
a conflict, which otherwise lasted four years for
US forces so most psychologists were found in
the United States, England, or further behind
the combat lines in the Pacific.

Clinical interrogations of the senior Nazi
leadership interned at war's conclusion were
extensive, this being the discipline's first
encounter with "detainee ops." The Wechsler
and Rorschach were first administered by psychology technicians to 56 senior surviving Nazi
leaders during their top-secret incarceration at
the Palace Hotel in Mondorf, Luxembourg, a
mission known as "Operation Ashcan." After
they were transferred to Nürnberg for their
final trials, psychologist G. Gilbert used the
data collected along with his subsequent interviews for a comprehensive study of Nazi personalities, although he had first been assigned
as an intelligence officer (Dolibois, 1989).

Those most in need of psychological treatment at the conclusion of hostilities were
the prisoners of war, either captured aircrews
in Europe or mostly the survivors of the
Philippine assaults. Underappreciated at the
time, it is now estimated that half of those
captured in Germany and Japan during World
War II developed posttraumatic stress disorder. Most received little in the way of formal
treatment. The majority of captured aircrews
in Europe were held at "Stammlager," where
conditions were at least partly tolerable. But
hundreds were also incarcerated at Buchenwald
as well as other extermination camps. Many of
these latter troops, along with the POWs held
by the Japanese, were more severely tortured
and were specifically targeted for liquidation,
especially after the Dresden air raid (Edwards,
2012). Most of the treatment that could be
conducted with these more seriously damaged
POWs was done back in US hospitals.

However, whether returning from combat,
repatriated, or freed at the end of the war, almost
all troops from either the European or Pacific
Theaters spent 2–4 weeks aboard ships returning to CONUS, where they invariably "talked"
to other soldiers or sailors about their experiences (Settles, 2012, personal communication).

Although many World War II military person-nel are known after the war for "not talking" about their combat experiences, the informal "talking therapy" conducted aboard ship did contribute to an eventual understanding of how important such cathartic therapy could be.

Army psychology training was eventually authorized at the Adjutant General's School at Fort Sam Houston with 24 officers starting in October 1944, and a total of 281 in the end graduating. With the current interest concern-ing traumatic brain injury due to blast injuries experienced in the Wars on Terror, it is of inter-est to note that of the 88 hours of instruction, four hours were devoted specifically to "Brain-Injured Patients." Many of these students later became prominent neuropsychologists.

At war's end, Edward Boring, having served as a Captain under Major Yerkes, edited an influential text dividing the "psychological business of the Army and Navy" into seven major categories, many of which still underlie modern military psychology (1945, p. 3):

- **Observation**—accuracy in perception
- **Performance**—action and movement; the acquisition of skills; efficiency in work and action
- **Selection**—classification; the choice of the right man for the right job
- **Training**—teaching and learning and the transformation of attitudes into accom-plished skills
- **Personal adjustment**—the individual's adjustment to military life, his motivation, his morale, and his reaction to stress and fear
- **Social relations**—leadership; the nature of panic; the relations with peoples of different races and customs
- **Opinion and propaganda**—assessment of public opinion and attitudes—psychological warfare

In February 1946 the School of Military Neuropsychiatry was moved to Brooke General Hospital, a Chief Clinical Psychologist was established to oversee operations, and Division 19 of the APA, Military Psychology, was in part spawned from the American Association for Applied Psychology. The fol-lowing year Army psychologists obtained permanent active duty status, and in 1949 the first internship programs were established. In 1946 Congress established the Office of Naval Research, which included behavioral science, and as the Air Force became a separate service it created the Human Resources Research Center in 1948 to carry on the work of the Army Air Force Aviation Psychology Program.

KOREA

The exploding conflict on the Korean penin-sula soon saw psychologists in new positions in service overseas, in the combat zones them-selves, and on hospital ships. With the benefit of experience, this time psychologists pushed for the application of combat stress princi-ples. The new emphasis upon these theorems saw the return to duty rate increase from the World War II levels of only 40% to rates of 80% to 90%. Combat stress operations were carried out by the 212th Psychiatric Battalion, which led to the award of the first Bronze Star Medal to a combat operational psychologist, Richard H. Blum, although the award was not made until 2005.

Partly due to the nature of the war, other significant advances now began to occur in the areas of Operational Psychology. New tech-niques of communist "brainwashing," con-tinuous propaganda, reeducation, and types of torture which even the Japanese had not inflicted, also resulted in marked psychological revisions to the burgeoning survival schools. Rates of having a mental health condition rose to 88% to 96% among the surviving American POWs from Korea. This, along with later experiences from the torture suffered in the Vietnamese "hotels" led to extraordi-nary changes in survival, evasion, resistance, and escape (SERE) training. Established by the United States Air Force at the end of the Korean War, SERE was extended during the Vietnam War to the Army, Navy, and Marines. Most higher level SERE students are military aircrew and special operations personnel con-sidered to be at high risk of capture.

Psychological warfare (PSYOPS) also began to be used with more effect, and in 1950 the 1st Loudspeaker and Leaflet Company arrived in South Korea to begin operations. Following the war, the Army finally began to devote significant resources to the areas of motivation, leadership, PSYOPS, and human/ecological systems.

VIETNAM

As the Vietnam War lengthened and personnel issues became critical, the Armed Services Vocational Aptitude Battery (ASVAB) was implemented in 1968. This has provided a consistent aptitude tool that has come to be heavily relied on by military psychologists, especially in studies of brain injury. Also building on the principles of forward mental health lessons of Korea, a number of psychologists served in the combat zones of Vietnam, and psychologists were stationed aboard naval ships. This era saw the formal adoption of the diagnosis of Posttraumatic Stress Disorder (PTSD), as well as a better understanding of the subclinical concepts of Combat Operational Stress (COS). However, since most veterans were not career soldiers and were discharged after their tour of duty, the majority of treatment for both PTSD and substance abuse was done by the Veterans Administration and not by military psychologists themselves. Although the initial estimated rates of PTSD were proven to be too high (Frueh et al., 2005), the lack of treatment, the public shunning of the veterans, and the deliberate mistreatment of PTSD survivors for political purposes left thousands of Vietnam Veterans scarred for years after the conflict ended (Spinrad, 1993).

AFTER VIETNAM

Following the war, a number of line officers were able to attend civilian psychology programs and were then able to transition to the Medical Service Corps and serve as fully qualified psychologists. These officers, along with both ROTC graduates and officers directly commissioned from graduate programs, brought a unique combination of military and psychological expertise to the work of rebuilding the Army from the standpoint of the provision of mental health services and to meet the Medical Corps motto "to conserve the fighting strength."

By now, Army psychologists generally were assigned mental health treatment duties at either Army Medical Centers or Army Community Hospitals under the auspices of MEDCOM, or to FORSCOM units. Earlier, psychologists tended to spend most, if not all, of their entire career in one or the other capacity, but starting in the 1990s, it became more common for these officers to switch between MEDCOM and FORSCOM assignments, thus broadening career experience and opportunities. These years saw the growing numbers of trained neuropsychologists either commissioned or trained as officers, along with the establishment of the long-term studies of traumatic brain injury (TBI) through the Defense Veterans Brain Injury Center with military psychologists as critical contributors to this long-standing project.

The Uniformed Services University of the Health Sciences (USUHS) was established by Congress in 1972 and by 1978 offered a Clinical Psychology Ph.D. This now also includes a Ph.D. in Medical Psychology for military students. Walter Reed, Brooke, and William Beaumont Army Medical Centers began to train interns. Naval internships were established at Bethesda and the Naval Medical Center San Diego, and the Air Force internship was begun at Wilford Hall, Lackland Air Force Base.

The eruption of regional conflicts such as the Baader-Meinhoff bombing of USAFE headquarters at Ramstein, the conflicts in Bosnia, Panama, the first Gulf War, and so forth, saw military psychologists as being integral to the development and implementation of forward combat mental health provision and counterterrorism techniques. The Psychology at Sea program has seen the assignment of psychologists to carriers to help reduce the previously alarming number and very difficult to accomplish medical evacuations of naval personnel at sea to an astonishing 1–2% level.

THE WAR ON TERROR AND BEYOND

Although few realized it at the time, the 1979 embassy hostage crisis in Iran (partly handled by military psychologists), the crises in Lebanon, and the 1990 Gulf War with Iraq proved to be among the opening steps to the Global War on Terror (GWOT), which engulfed the United States on September 11, 2001. During the Gulf War, despite the low number of combat stress casualties, after-action analysis indicated that if significant casualties had occurred, mental health teams would have found it very difficult to carry out their mission. These teams were not adequately staffed, equipped, or trained in peacetime to perform their wartime role. The full outbreak of the GWOT proved this analysis to be prescient, and subsequently produced unprecedented strains on the provision of mental health services of the military from 9/11/2001 until the present.

Expanded military training programs helped to produce officers to meet these challenges. Neuropsychological training had been done for years at Madigan Army Medical Center, and formal neuropsychology fellowships were eventually established at Walter Reed Army Medical Center in 1991 and at Tripler Army Medical Center (TAMC) in 1995. The same year also saw the establishment of the Psychopharmacology Fellowship at TAMC. Navy and Air Force officers are more often sent to civilian training institutions for fellowship and advanced training. Advanced fellowships now generally include Neuropsychology, Child Psychology, Behavioral Medicine, and Forensics. The mid-1990s also saw the initiation of the Department of Defense Psychopharmacology Demonstration Project (PDP), in which ten initial military psychologists were trained and credentialed for prescriptive privileges. By 2006, other officers, now bolstered with state medication licenses, had begun to enter active duty. These officers have been assigned both to combat areas and stateside, serving as very effective force multipliers.

With combat and counterterrorism operations still continuing over 11 years later, much of the effort of military psychology has been in terms of forward combat operations, with heavy reliance on reserve units and their mental health assets.

Neuropsychologists set up increasingly sophisticated assessment and management strategies for TBI patients. Innovative research in TBI continues to expand, such as with computer programs and hand-held devices.

The nature and continuation of combat operations brought several thousand detainees into military custody during operations, and numbers of psychologists were, therefore, also needed for detainee operations, both in Iraq and Afghanistan (Kennedy, Malone, & Franks, 2009). Kennedy et al. (2009) noted that this is the first wartime scenario in which detained enemy combatants have been provided unfettered access to mental health evaluation and treatment services during their detention, much of that being provided by psychologists.

With the continuing innovation and modernization of 21st century military psychology, Robert Yerkes's vision of psychology as James's "natural science" which could change the composition and function of the entire military, has come to fruition in ways that the original 17 active duty military psychologists could scarcely have imagined.

References

Boring, E. G. (Ed.). (1945). *Psychology for the armed services*. Washington, DC: Infantry Journal.

Dolibois, J. E. (1989). *Pattern of circles: An ambassador's story*. Kent, OH: Kent State University Press.

Edwards, E. C. (2012). The lost airmen of Buchenwald. Stalag Luft III Reunion, 17 April, Dayton, Ohio.

Frueh, B. C., Elhai, J. D., Grubaugh, J. M., Kasdan, T. B., Sauvageot, J. A., Hamner, M. B., ...Arana, G. W. (2005). Documented combat exposure of US veterans seeking treatment for combat-related post-traumatic stress disorder. *British Journal of Psychiatry, 186*, 467–472.

James, W. (1890). *The principles of psychology*. New York: Holt.

Kennedy, C. H., Malone, R. C., & Franks, M. J. (2009). Provision of mental health services at the detention hospital in Guantanamo Bay. *Psychological Services, 6*(1), 1–10.

Spinrad, P. S. (1993). Patriotism as pathology: Anti-veteran activism and the VA. *Journal of the Vietnam Veterans Institute, 2*(1), 42–70.

3 HISTORY OF PSYCHOLOGY IN THE DEPARTMENT OF VETERANS AFFAIRS

Rodney R. Baker

Psychologists in the Department of Veterans Affairs (VA) have joined their military psychologist colleagues in providing transition mental health care for those leaving the military and becoming veterans ever since the end of World War II (WWII). As the end of that war was nearing, President Harry S. Truman knew that millions of active military personnel would be coming home as veterans. He asked General Omar N. Bradley, the popular and successful commander of the 12th Army forces in Europe in WWII, to use his organization and people skills to assume leadership of the then Veterans Administration to prepare the VA to provide the best possible medical care promised its veterans.

Bradley accepted the challenge knowing that it would not be easy. The VA health care system was poorly organized, and extreme shortages existed in personnel, as many doctors, nurses, and other health care providers had enlisted in the military. As of June 30, 1945, 74% of the 2,300 doctors in the VA were actually on active military duty assigned to provide health care in veterans' hospitals. Many of those doctors would end their service to the VA with discharge from the military (Baker, 2012).

Bradley knew he would need others to help him and turned to some of those he served with and trusted during his command in the WWII European theater. One of his first appointments was that of General Paul R. Hawley to serve as the VA's first chief medical director. A highly respected physician, Hawley had served as chief surgeon in Europe during WWII and had been recognized as helping the military offer exceptional care to soldiers on the battlefield. Bradley and Hawley were successful in getting Congress to pass legislation that they needed to revitalize the VA, and on January 3, 1946, President Truman signed Public Law 293, which made major changes to the organization of medical care in the VA (Baker, 2012). The legislation provided new health care departments in the VA's Central Office in Washington, DC, that were responsible for establishing and monitoring the quality of care in VA hospitals in medicine, surgery, rehabilitation, psychiatry and neurology, and other health care disciplines. The legislation also authorized the VA to enter into training affiliations with medical schools to train and supervise interns and residents assigned to the VA and help ensure the quality of care for veterans. By the end of the year, the VA had established affiliations with 63 of the nation's 77 medical schools.

In the spring of 1946 James Grier Miller, a Harvard-trained psychiatrist and psychologist, accepted an appointment to head clinical psychology. Miller successfully argued that the VA needed doctoral-trained clinical psychologists with experience in providing treatment and assessment services for veterans and received authority to hire 500 clinical psychologists (Baker & Pickren, 2007).

THE VA PSYCHOLOGY TRAINING PROGRAM

At the end of WWII, psychology was primarily an academic and research discipline and few members of the profession were being trained to provide clinical services. The directory of the American Association for Applied Psychology listed 650 members in applied settings in the entire country (Baker & Pickren, 2007). Miller would need to recruit three-fourths of that number to fill the 500 doctoral psychologist positions he had been authorized to hire.

Miller knew that the VA would have to train the psychologists needed to fill these positions. He convinced Bradley that the legislation authorizing the VA to establish training affiliations with medical schools would also allow affiliation agreements with universities who were training clinical psychologists. He proposed that psychology graduate students would be employed as part-time staff with a training assignment of delivering clinical services to patients under university faculty supervision. His plan was approved, and in 1946, the VA funded 225 training positions in psychology that paid students an hourly salary. Miller recognized that not all graduate programs in clinical psychology were providing training in psychological service delivery. He asked the American Psychological Association (APA) to identify universities that provided that training, a list that he would use to select students for the new training program. In the fall of 1946, 22 universities on the resulting list had proposed 215 students for training that the VA accepted and hired. The following year, APA formalized a process of helping the VA identify universities for the VA psychology training program and 470 students were accepted and funded for the 1947 training year. Baker and Pickren (2007) noted that because of these actions the VA is generally acknowledged to be responsible for APA developing its professional psychology accreditation program for universities and, later, for accreditation of sites providing internship training for clinical and counseling psychologists.

The number of part-time VA psychology training positions grew to 650 in 1950. In 1952 the VA began developing vocational counseling programs in its hospital, using doctoral counseling psychology staff, and the following year added 55 training positions for graduate students in counseling psychology. For fiscal year 1956, 771 clinical and counseling psychology graduate students were appointed to part-time training positions in the VA (Baker & Pickren, 2011).

In a 10-year review of the training program, 80% of the graduate students in the program went to work for the VA after their training, even without any required payback work obligations (Baker & Pickren, 2007). The program clearly succeeded in helping the VA meet its recruitment goals for treatment-experienced psychologists. It can be noted that the vast majority of all students in universities after WWII were veterans themselves, most receiving their education with G.I. Bill of Rights legislation, and many in psychology graduate training eagerly sought acceptance into the VA training program hoping to help fellow veterans. The wish to help their fellow veterans in a training capacity turned into a later desire to work with veterans as VA staff psychologists.

THE GROWTH OF VA PSYCHOLOGY

As VA psychology trainees and staff entered the mental health programs in the VA, the influence and importance of psychology steadily grew. The immediate post-WWII era demands made on the VA were to provide treatment to a large and growing hospitalized veteran population in which patients with serious mental illness occupied 58% of available beds (Baker & Pickren, 2007). From 1946 to 1988, psychology responded to these demands in three major areas. First, psychologists increased the number of health services for patients, especially noted in the promotion and use of group psychotherapy. Second, they played critical roles in moving the VA from an exclusive use of inpatient treatment to starting outpatient mental health clinics. Last, they helped develop nontraditional treatment approaches for the mental health care of veterans.

The development of nontraditional treatment approaches defined the reputation for innovation that VA psychology enjoyed in the 1960s that has continued to the present. The need to treat large numbers of veterans in the post-WWII era, for example, led psychologists in mental health clinics to explore the potential for use of group psychotherapy. By 1960, group therapy had been found effective for treating veterans, even surpassing the effectiveness of individual psychotherapy in many cases, and the VA published a "Manual of Group Therapy" written by two VA psychologists and a psychology consultant. Described by Baker and Pickren (2007), the manual not only reviewed the theoretical bases for group psychotherapy but also was one of the first publications to give practical advice in conducting effective group psychotherapy sessions. Its chapters discussed topics ranging from different kinds of groups and desired outcomes, time and frequency of group meetings, preparing the patient for group therapy, and handling hostile, dependent, silent, and talkative patients. The manual helped establish a sound theoretical and therapeutic basis for group psychotherapy, and in the 1980s, the number of patients receiving group psychotherapy services in the VA continued to grow and exceeded the number receiving individual psychotherapy by a factor of three (Baker & Pickren, 2007).

The nontraditional treatment approaches being utilized by VA psychologists in part emerged from their awareness of the limitations of the reliance of psychiatry on its historical roots in psychoanalytic theory and practice, limitations especially noted with the serious psychiatric problems of the veteran population. Psychologists joined their non-VA colleagues in looking at behavioral and other therapeutic applications for care. In 1965 the VA sponsored a conference to examine the latest treatment approaches psychologists used with their patients. The papers presented included work on attitude therapy, token economy programs, day treatment centers, and therapeutic milieu programs. Also highlighted were the activities of psychologists working with different patient populations in such treatment areas as renal dialysis, open-heart surgery, automated retraining of patients with aphasia, and other medical programs of the general VA hospital (Baker & Pickren, 2007).

From 1970 to 1981, the number of specialized treatment programs in the VA continued to increase. Mental health outpatient clinics almost doubled, as did day treatment centers, and a fourfold increase in new day hospitals served to meet the more intensive care of acute psychiatry outpatients. Specialized inpatient and outpatient alcohol abuse and drug dependency programs also saw substantial growth (Baker & Pickren, 2007).

In the 1970s, the large number of Vietnam-era veterans seeking treatment for post-traumatic stress disorder (PTSD) presented a unique problem for psychologists and other mental health professionals. Other than in the military, the PTSD treatment experience in the non-VA sector was essentially limited to treatment for trauma resulting from sexual assault and natural disasters, and military and VA psychologists had to draw on their own resources in starting treatment programs to meet the needs of veterans with combat-related PTSD.

The care of Vietnam veterans brought several other problems to the VA. Improvements in military care on the battlefield resulted in many more survivors with serious physical disabilities, which prompted the VA to increase the number of rehabilitation and spinal cord injury programs. Vietnam veterans also felt alienated from mainstream society and believed, correctly in many cases, that VA employees shared the public's ambivalence, even anger, with the US government's support of that war that carried over to its veterans. Even the existing WWII veterans in the VA's care were not overly friendly with the new veteran population.

In 1971 a psychologist leader in VA's Central Office in Washington, Charles A. Stenger, served as Chair of the VA's Vietnam Veterans Committee and was called on to organize a series of conferences on treatment issues for Vietnam-era veterans. The conferences highlighted the unique problems of Vietnam veterans, and participants were challenged to generate initiatives and create programs to address these problems. Out of these conferences, over 30 inpatient PTSD treatment units and almost 100 PTSD outpatient clinics were started. Each of these programs almost universally included psychologists as staff in developing these programs.

The VA was designated a cabinet level department in 1989 and renamed the Department of Veterans Affairs. The 1990s saw a continued growth of mental health treatment programs for veterans that created important roles for psychologists in new programs for treating homeless veterans, in psychosocial rehabilitation programs with work therapy and residential rehabilitation care, and in traumatic brain injury centers. The number of women veterans seeking care in the VA had been steadily increasing along with the increased role of women in the military. Over 60,000 female veterans received health care in the VA in fiscal year 2004, many receiving that care in new women's health or sexual trauma treatment clinics (Baker & Pickren, 2007).

In addition to their treatment activities, psychologists were major participants in VA research programs directly related to improving patient care. The list would include early research projects in the VA's pioneering use of cooperative research programs involving multiple hospitals following the same research protocol. Because of their training in research, psychologists assumed national administrative leadership in these projects and participated at the local hospital level in cooperative research topics ranging from evaluating the effectiveness of psychotropic medications to cooperative research in tuberculosis. Psychologists followed participation in these early research programs with key research in suicide, life-span topics and problems of the elderly, neuropsychological assessment and brain functioning, and PTSD. During the 1950s, VA psychologists participated in as many as 500 studies a year, many emerging from the dissertation research of psychology students. In 1956 psychologists and their trainees were involved in 409 of 653 mental health research projects and, in fact, were conducting one-third of all research in the VA, both in mental health and non-mental-health treatment areas (Baker & Pickren, 2007).

The successes and resulting support of the VA training program led the VA to assume major roles in internship training and accreditation. By 1985, 84 VA training programs were accredited by APA, and in 1991, one-third of all APA accredited internship program were VA based. The VA started funding psychology postdoctoral training programs in 1991 in substance abuse followed by postdoctoral training in geriatrics, PTSD, and psychosocial rehabilitation. When APA began accrediting postdoctoral training programs, the VA assumed a major role in promoting that level of training and accreditation. In 2005 the VA represented almost half of the APA accredited postdoctoral training programs for adults in the country.

At the end of 1988 the VA employed over 1,400 psychologists in its 172 medical centers and associated outpatient clinics. That year, the annual mental health treatment survey listed 1,241 programs in its hospitals, almost all of which included psychologists in a full- or part-time staff capacity (Baker & Pickren, 2007). In addition to general psychiatry programs, the list included inpatient and outpatient programs in the treatment of alcohol and drug dependence and inpatient and outpatient treatment of PTSD. Also included were mental health outpatient clinics, day treatment centers, day hospitals, vocational assessment and compensated work therapy programs, services for homeless veterans, biofeedback and pain clinics, neuropsychology evaluation clinics, sexual dysfunction clinics, and sleep disorder clinics.

VA PSYCHOLOGY TODAY

The 1990s saw the growth of traumatic brain injury centers and psychosocial residential rehabilitation treatment programs. Ten Mental Illness Research, Evaluation, and Clinical Centers (MIRECCs) were funded from 1997 to 2004 in regional areas across the country to conduct specific research in the mental health problems of veterans most needing attention (Baker & Pickren, 2007). The focus of these centers ranged from treatment of PTSD to psychosocial rehabilitation to problems of the elderly veteran. The MIRECCs also had additional internship training positions for psychology graduate students as well as training positions for students in other mental health training professions.

The contributions of VA psychologists to the health care of veterans today include some unique challenges not encountered in other wars. The improvements in health care by the military for those wounded in combat, already noted for Vietnam veterans, continued to improve. In the prolonged second Iraq war and the war in Afghanistan, however, the improvement in military care had increased the numbers of severely wounded veterans with multiple and complex trauma. The complexity of injury led the VA to create in 2005 four regional polytrauma centers, where psychologists joined other health care specialists to coordinate needed rehabilitation care for these veterans. Treatment of pain and the sequelae of head injury, irreversible physical disability, and the resulting emotional problems of patients treated by psychologists in these centers complemented their efforts in providing needed counseling to the wives, husbands, children, and parents of these veterans. The VA continued to build polytrauma centers, opening the latest in 2011. Psychologists were additionally providing similar rehabilitation services to veterans in traumatic brain injury sites and spinal cord injury units closer to the veterans' home after discharge from the regional polytrauma centers.

Over the last several years, Congress has recognized the debt owed to veterans with mental health problems and has added significant funding for VA mental health programs. At the end of the fiscal year 2005, the VA employed 1,685 doctoral psychologists. By the end of the 2011 fiscal year, 3,741 doctoral psychologists were providing mental health services to veterans. In addition to staff psychologists, 437 predoctoral psychology interns and 245 postdoctoral interns were receiving training and adding services to veterans (A. Zeiss and R. A. Zeiss, personal communication, February 2, 2012).

The service of VA psychologists to veterans in clinical treatment programs, research, and training can only be introduced in this brief chapter. If history serves as prologue, the past service of VA psychologists for care of our nation's veterans promises a future with similar excellence of care for veterans. Military mental health providers aware of this rich tradition can, without hesitancy, refer and encourage their patients leaving the military to seek care in the VA.

References

Baker, R. R. (2012). Historical contributions to veterans' healthcare. In T. W. Miller (Ed.), *The Praeger handbook of veterans' health: Vol. 1: History, veterans eras & global healthcare* (pp. 3–23). Westport, CT: Praeger Security International.

Baker, R. R., & Pickren, W. E. (2007). *Psychology and the Department of Veterans Affairs: A historical analysis of training, research, practice, and advocacy.* Washington, DC: American Psychological Association.

Baker, R. R., & Pickren, W. E. (2011). Department of Veterans Affairs. In J. C. Norcross, G. E. Vandenbos, & D. K. Freedheim (Eds.), *History of psychotherapy: Continuity and change* (2nd ed., pp. 673–683). Washington, DC: American Psychological Association.

4 DEMOGRAPHICS OF THE US MILITARY

Richard L. Dixon Jr. and Jean M. Dixon

The US Military is made up of six branches. However, only four branches are routinely identified and tasked with the mission of defending the United States' interests abroad. The largest and oldest branch of service is the Army, comprising Active Duty, Reserve, and National Guard soldiers. The next-largest branch is the Air Force, comprising Active Duty, Reserve, and National Guard airmen. The third-largest branch is the Navy, followed by the Marines. Both of these services comprise Active Duty and Reserve members. The remaining two branches are the Coast Guard and the Uniformed Public Health Service, which also comprise Active Duty and Reserve members. To be an effective clinician, it is imperative that you have a working knowledge of the demographic make-up of the combined US Military.

The first concept to be understood is the difference between the Active Duty, Reserve, and National Guard components. This can be very confusing, even to those who wear the uniform. The common denominator for all the Services is that the individual has voluntarily sworn an oath to defend and protect the Constitution of the United States. The differences relate to the amount of personal time the individual has agreed to give and to the general area (land, air, or sea) that the individual has agreed to protect and defend. Additional variables include the benefits that the individual has been promised

as well as the rank and responsibility that an individual has agreed to.

Active Duty service members make up the largest component of the Navy, Marines, Coast Guard, and Uniformed Public Health and approximately half the service numbers in the Army and Air Force. The remainder of each component is composed of reserve members. Reservists are discussed below. Active Duty members are those who have agreed to be available 24 hours a day, 7 days a week. The Active Duty component of each branch of service is tasked with being the primary responder and protector of the United States' interests. The people serving on active duty are stationed both domestically and abroad. They live in communities composed of other Active Duty military members and their families. The usual time span of active duty service ranges from 3 to 8 years depending on one's specialty. Most, but not all, service members have spent at least 1 year on active duty prior to serving in the reserve component.

The reserve component is made up of service members who have agreed to be available to augment the active duty component as needed. The size of the reserve component varies with the different branches. The Army and Air Force Reserve components are further divided into three entities: Reserve, Reserve National Guard, and Active National Guard. The Reserve entity branches are further divided into multiple categories based on the amount and location of training that the service member participates in such as: (1) Those assigned to a specific unit (Troop Program Unit-TPU), which drills one weekend a month and has a 2-week period of active duty for training once a year; (2) Those assigned to a specific unit (IMA-Individual Mobilization Augmentee or IA-Individual Augmentee) but only drill with that unit for 2 weeks a year; and (3) Those service members (Individual Ready Reserve-IRR) who are not assigned to a unit where they drill but can be called to active duty to augment units (active duty or reserve) during times of need. For the remainder of this chapter, only the four main branches of service will be discussed.

THE MILITARY AT WAR

A recent report by ABC News titled *U.S. Veterans by the Numbers*, which aired 9/11/11, reported that 22,658,000 active and reserve personnel have served in the military since 9/11/01. As of Veteran's Day, 2011, the current number of personnel on active duty was 2,317,761. Comparing these numbers to the 2010 US Census of 308,745,538 people, less than 1% of Americans are currently serving in uniform, while the total number of veterans who have served since 9/11/01 makes up only 7% of the population. It is also reported that of this total number of veterans who have served, only 8% have been female.

ABC also reported figures related to the hostilities in Iraq and Afghanistan, noting that 1,348,405 veterans have been deployed to either Iraq or Afghanistan. Of these, 977,542 have been deployed more than one time. In addition, this article reported that 1,286 service members have become amputees as a result of these two wars. The 2010 Department of Defense Demographics Profile of the Military Community (DoD, 2010) is the most recent set of data available and will serve as the basis for data in this chapter except where noted.

AGE

The youngest age at which one can enlist in the service is 17 years with parental permission and 18 years without parental permission. The mandatory retirement age differs by service and career track but is either 60 for nondemand or 62 for demand careers such as medical and nursing officers. The average age of service members varies by rank and component. In general, the Active Duty (AD) component of the military is younger (average 28.5 years) than the Reserve component (average 32.2 years). Officers are older than enlisted, with Reserve component officers averaging 40.1 years of age while reserve enlisted average 30.8 years. The Marine Corps is made up of the youngest service members in both the Active Duty and Reserve components while the Air Force Reserve is made up of the oldest service members followed closely by the Army Reserve.

GENDER

While the military is mostly composed of males, women do make up 14.4% of the total active duty force and 17.9% of the total reserve forces. Active duty women are most commonly found in the Air Force and Navy and are least likely to be in the Marine Corps. In comparison, Reserve and National Guard women are most commonly found in the Air Force and Army followed by the Navy and Marines.

RACE/ETHNICITY AND REGION OF ORIGIN

Overall, 30% of Active Duty service members report minority status, versus 24.1% of the Reserves. Five percent of the minorities on active duty are foreign-born. The Navy has the highest ratio of foreign-born troops at 8%, followed by the Army, Air Force, and Marines. The majority of foreign-born service members come from the Philippines and Mexico (Batalova, 2008). Approximately two-thirds of the foreign-born troops serving in the Armed Forces have become naturalized US citizens. Since September 2001, almost 75,000 service members have become naturalized US citizens. Statistics for fiscal year 2011 showed a total of 10,334 naturalized service members were currently serving in the military (US Citizenship and Immigration Services, 2011).

Watkins and Sherk (2008) found that American Indians and Alaskan natives are two and a half times (2.68) more likely to serve in the military based on 2007 statistics released by the Defense Manpower Data Center. These researchers calculated a troop-to-population ratio using this data and population numbers from the US Census Bureau for males ages 18 to 24. A ratio of 1.0 is equal to the general population for each race. Blacks were found to be only slightly more represented in the military at 1.08 and were virtually equal to the ratio of whites at 1.03. Those of Asian descent were found to be less represented in the military at a ratio of 0.94 while those of Hispanic ethnicity were found to be least represented at a ratio of 0.65.

Another category assessed was the region of origin for the troops, enlisted and officers, who joined the military. The statistics are broken down by state and region and are based on the troop to population ratio. Montana had the highest number of troops entering the military with a ratio of 1.67 followed by Nevada at 1.50, Oregon at 1.39, Maine at 1.35, and Arkansas at 1.32. When broken into regions, people from the South were most represented at a ratio of 1.19, while the Northeast is the least represented area at a ratio of 0.73. The West and Midwest were virtually equal in representation at ratios of 0.94 and 0.98 respectively.

RANK AND EDUCATION LEVEL

Each branch of the military is made up of enlisted personnel and officers. In general, the ratio of officer to enlisted is one officer to every five enlisted personnel in the Active Duty corps and one officer to every 5.7 enlisted personnel in the Reserves. This ratio varies depending on services, with the Marine Corps having the least number of officers and the Air Force having the largest number of officers.

Education levels vary greatly within the various services. On average, 82.8% of all AD officers have a bachelor's degree or higher versus 4.9% for the enlisted. Further, 93.6% of the enlisted have a high school diploma and/or some college. In the RC, 84.5% of officers have a bachelor's degree or higher versus 8% for the

enlisted, with 88.2% of RC enlisted having a high school diploma and/or some college. Since 1995 both officers and enlisted in the RC have increased their level of education, AD officers have decreased the number of bachelor's and advanced degrees, and AD enlisted have continued to increase their education levels. While rare at less than 1%, some enlisted service members do not have their high school diplomas or GED.

The pay scale for officers and enlisted is determined by the Department of Defense and is the same for all services. Pay is based on the service member's rank and years in service. There is one pay scale for service members on active duty and another pay scale for service members participating in Guard and Reserve drill. Pay scales are updated each year and can be found on the Defense Finance and Accounting Service website found at www.dfas.mil.

MARITAL STATUS AND DEPENDENTS

The most recent data show that 56.4% of all AD service members are married, while the overall marriage rate for RC service members is 48.2%. In both AD and RC, officers have a higher marriage rate than enlisted. Both components have shown a decrease in the marriage ratio since 1995. Attempts to match the divorce rate to deployments have not been conclusive. Troop divorce rates have gone from 2.6% in 2001 to 3.6% in 2009 with no changes in 2010 (Bushatz, 2010). A Rand study found that the military divorce rate was 3% in 1996 and validated previous studies that there is no correlation between deployments and divorce (Karney & Crown, 2007).

Of interest is the comparison of marital status to the different services. The highest percent of marriage for AD and RC was found in the Air Force (59.2% to 58.1% with 57.4% in the Air Guard) while the lowest percent was in the Marines (48.8% to 32.1%). The Navy also showed a higher percent of marriage between AD and RC at 54.3% and 58.2%. The Army showed the largest variance between AD and RC marriage rates at 58.7% AD, 45.9% RC, and 44.7% National Guard. The military has followed the same trend as the civilian sector in regard to the

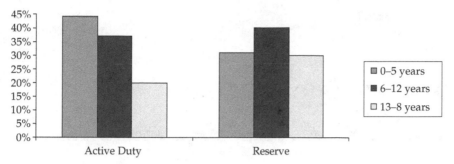

FIGURE 4.1 Active Duty/Reserve Age Comparison Chart

increase of dual income marriages. One variation that is unique to the military is the "dual military couple" where both spouses are also members of the armed forces. This means that both partners are subject to frequent changes of station as well as deployments. Though the military tries to keep dual service couples stationed together, it depends on the individuals' specialties and ultimately the needs of the service.

Of all AD service members, 43.7% have children, whereas 42.8% of the RC service members have children. The average for both components is two children per family. Of the AD component, 35.6% are married to a civilian, 2.8% are dual military couples, and 5.3% are single parents. In the RC, 32.3% are married to a civilian, 1.4% are dual military couples, and 9.1% are single parents. As in the civilian world, the majority of single parents are female service members.

The AD component tends to have younger children compared to the RC. Among AD service members with children, the highest percentage of children is aged birth to five years. In the RC the highest percentage of children are aged 6 to 12 years (see Figure 4.1).

The DoD maintains many programs to help the 1.8 million dependent children of service members. Many of the programs are run out of military installations, which means that RC families may have a much harder time accessing these resources as they tend to have little to no access to military facilities. Examples of available resources are the Family Advocacy Program (FAP) and the Child and Youth Behavioral/Military and Family Life Consultant (CYB-MFLAC) Program, which specifically helps RC families with nonmedical counseling services.

References

Batalova, J. (2008). Immigrants in the US Armed Forces. *Migration Policy Institute*. Retrieved from http://www.migrationinformation.org/feature/display.cfm?ID=683

Bushatz, A. (2010). Troop divorce rates level in 2010. *Military.com News*. Retrieved from http://www.military.com/news/article/troop-divorce-rates-level-in-2010.html

Department of Defense (DoD). (2010). Demographics 2010 profile of the military community. *Military Community and Family Policy*. Retrieved from http://www.militaryhomefront.dod.mil//12038/Project%20Documents/MilitaryHOMEFRONT/Reports/2010_Demographics_Report.pdf

Karney, B. R., & Crown, J. S. (2007). Families under stress: An assessment of data, theory, and research on marriage and divorce in the military. *RAND Corporation*. Retrieved from http://www.rand.org/pubs/monographs/MG599.html

Martinez, L., & Bingham, A. (2011). U.S. veterans by the numbers. *ABC News*. Retrieved from http://abcnews.go.com/Politics/us-veterans-numbers/story?id=14928136#

US Citizenship and Immigration Services. (2011). *Naturalization through military service: Fact sheet*. Retrieved from http://www.uscis.gov/portal/site/uscis/menuitem.5af9bb95919f35e66f614176543f6d1a/?vgnextoid=26d805a25c4c4210VgnVCM100000082ca60aRCRD&vgnextchannel=ce613e4d77d73210VgnVCM100000082ca60aRCRD

Watkins, S., & Sherk, J. (2008). Who serves in the U.S. military? The demographics of enlisted troops and officers. *Heritage Foundation*. Retrieved from http://www.heritage.org/research/reports/2008/08/who-serves-in-the-us-military-the-demographics-of-enlisted-troops-and-officers

5   MILITARY CULTURE

Lynn K. Hall

The members of the US military are, indeed, a diverse group of people in American society that must be understood as uniquely different from the civilian world. As Reger, Etherage, Reger, and Gahm (2008) state, "to the extent that a culture includes a language, a code of manners, norms of behavior, belief systems, dress, and rituals, it is clear that the Army represents a unique cultural group" (p. 22). Virtually every author or expert on military life reviews the characteristics of military culture as a foundation for understanding the military; for this chapter the most commonly discussed characteristics will be consolidated into three elements. While each is multifaceted, the three elements that will be presented are the hierarchical nature of the military, the military imperative to focus first on the mission, and the internal, or inward-facing, focus of the military. Hopefully, the majority of the unique dynamics of the military can be captured by presenting these three overarching elements.

HIERARCHICAL STRUCTURE

The consideration of the rank and pay grade structure of the military is perhaps the best place to start when considering its hierarchical structure. Service members who enter the military as enlisted, as opposed to as an officer, and progress only through the lowest 3 or 4 pay grades have much less influence or control than those who continue on beyond these lower grades to eventually become Non-Commissioned Officers (NCOs), for instance, as a Senior Master Sgt at pay grade E-8. Even as an NCO, however, the enlisted service member never has quite the power or authority of an officer. Someone going into the military as an officer (most often with a college degree) even at the lowest pay grade will have more power and authority than an NCO at a pay grade of 5, 6, or 7 of the enlisted rank. As Mary Wertsch wrote in 1991, the reasons that the military is organized in this manner are very obvious, but it seems that "the only equality among officers and enlisted is in dying on the battlefield" (p. 288).

In addition to the rank and pay grade structure of the military, another important dynamic related to the hierarchical structure has to do with discipline and etiquette. Military discipline is "the orderly conduct of military personnel perfected through repetitive drill that makes the desired action a matter of habit" ("Military Culture," 2008, p. 5). The goal of discipline is two-fold: (1) to impose order to "minimize the confusion and disintegrative consequences of battle" (p. 5) and (2) to "ritualize the violence of war, to set it apart from ordinary life" (p. 5). Military discipline and etiquette is obviously internal with the rituals of boot camp, training exercises, and protocol detailed for every military event or maneuver, but also external, visible to the greater society. In fact, the discipline and etiquette of the military must even be sanctioned by the larger society where it is honored and even copied (Hall, 2012). The outward evidence of discipline and etiquette is often demonstrated in many military ceremonies with "bright colored uniforms and unfurled flags ..., drum rolls and bugle calls ..., foot parades and more contemporary air shows" ("Military Culture," 2008, p. 6). An important goal of these military ceremonies is to remind the civilian society of the importance of the military to their personal well-being and attempt to share the burden of the commitment of the military with the larger society (Hall, 2012).

The hierarchical structure as well as the discipline and etiquette enforced within the military creates an obvious authoritarian structure. All of these characteristics create the tools for the military to produce a service in which order is maintained, confusion is minimized, necessary action is a matter of habit, and violence is ritualized.

There are many resulting consequences of the hierarchical or authoritarian structure of the military, both to individual service members and their dependents or families. Some of these include the jargon of the military and the difficulty in understanding the common acronyms, the rituals that are routinely performed on installations, the etiquette around who addresses whom and how that is done, and how families should act—including often understood but unspoken patterns of interaction between spouses and children of the military, depending on whether they are families of enlisted or officers. These patterns often also extend into the family structure, where autocratic systems are maintained and enforced, sometimes to the detriment of the individual development of the family members. In some families where the authoritarian structure is carried over into authoritarian parenting, it is fairly common to find rigid rules for behavior, a lack of tolerance for activities that hint at individuation, unwillingness to allow questioning of any authority, and even inappropriate violations of privacy (Hall, 2008).

While military families live within the military culture, they also move back and forth between the military and the civilian worlds, often creating a considerable amount of confusion for some family members. "The great paradox of the military is that its members, the self-appointed front-line guardians of our cherished American democratic values, do not live in a democracy themselves" (Wertsch, 1991, p. 15).

THE IMPORTANCE OF THE MISSION

After interviewing hundreds of adults who grew up as military dependents (brats), Wertsch (1991) defined the importance of the mission as one of the major characteristics of the military. Martin and McClure (2000) explained that historically the demands of the military require a total commitment that is the "very essence of the concept of military unit cohesion" (p. 15).

One of the elements of military cultures outlined in an article in the *Encyclopedia of Violence, Peace and Conflict* ("Military Culture," 2008) is the element of professional ethos, defined as "a corporate identity based on expert knowledge of and control over the means of violence" (p. 5). Even if it is uncomfortable for our civilian society to acknowledge there must be a commitment in military service that "presumes personal willingness to kill and accepts the risk of being killed, for oneself and for those one commands" ("Military Culture," 2008, p. 5). This focus on the mission demands a corporate, or collectivistic, mentality that requires absolute military cohesion or the "feelings of identity and comradeship that soldiers hold for those in their immediate military unit ....[and] the commitment and pride soldiers take in the larger military establishment to which their immediate unit belongs" ("Military Culture," 2008, p. 8).

The impact of this element of military culture is strongly felt by the military family, including absences of the service member from the family, the psychological conflict between the value of the family versus the importance of the mission, and constant preparation for disaster. These consequences often lead to boundary ambiguity, or "a state in which family members are uncertain in their perception about who is in or out of the family and who is performing which roles and tasks within the family" (Faber, Willerton, Clymer, MacDermid, & Weiss, 2008, p. 222). Wertsch (1991) discovered in her interviews with adult military brats that the real "determining factor" for most families was not their families, but rather the all-powerful military mission, "without which their lives would have no meaning" (p. 292). The ability to constantly make the changing alliances from the family of the military to the personal family, and back again, is the basic foundation of the military. While research has shown that solid families do indeed assist in creating better performing service members, it is also clear that "it is still a difficult balancing act for service members to be a part of both of these families who are so integral to the success of the mission and to their personal career" (Hall, 2008, p. 53).

Families are also aware that the importance of the mission means the possibility of disaster. From the beginning of the all-volunteer service in the 1970s until the Gulf War in 1991, young people entered the military with the idea that they would never be in harm's way. That scenario has changed; the focus of military training in the last two decades has been to plan for disaster. Martin and McClure (2000) acknowledge that military service is now an unlimited commitment and service members may be asked to sacrifice their lives. It may be a misunderstanding in the civilian world, but often it is overlooked that "a central truth [of military service is] that at any moment they may be called upon to give their lives—or lose a loved one—to serve the ends of government" (Wertsch, 1991, p. 16). This constant burden faced by service members and their families leads to a level of stress unknown by most civilian families.

INWARD FOCUS OF THE MILITARY

As noted in the previous element, the first priority of the military is the importance of the mission. As a result, service members and families have a continued and necessary inward focus to the military rather than an outward more community-focused perspective. The characteristics of this element include paradoxically an external locus of control, frequent loss and transition issues such as changing schools for children, loss of friends and community, deployment and other reasons for parent absence, and isolation issues. In a world of external locus of control, families (and service members) have almost no control over where they go and how often they move. Everyone associated with the military is subject to the control of the system that determines the best use of its personnel based on how the mission can best be served. Service members "live with the expectation of deployment, the inability to quit their job, a loss of control over significant life decisions…and the requirement to respond to others" (Reger et al., 2008, p. 29).

The isolation of living in the military leads families to focus solely on what lies ahead and can lead to a "lack of concern for the wider community in which they live" (Hall, 2008, p. 48). For instance, for those who spend time living abroad, the housing areas are usually isolated from the local communities leading to an "oddly isolated life, one in which it is possible to delude oneself that one is still on American soil" (Wertsch, 1991, p. 30). Even those children who attend public schools in the United States often reflect that there is a sense of "us versus them" (Hall, 2008). We know that on average, military families move every 2 to 3 years and secondary students move three times more often than civilian high school students. "Even the experience of moving (or PCSing, ironically called Permanent Change of Station) comes with rituals, taking up to 2 or 3 weeks to get all the permissions, requisite signatures, school records, household goods packed and shipped…which reinforce the needs of the military over the family" (Hall, 2012, p. 143).

These characteristics of military life often lead to attachment issues for service members and their families. "As the military service member experiences a series of separations and reunions, the attachment systems of each partner [and dependents] is activated" (Hall, 2012, p. 145). Attachment theory can provide a valuable tool for psychotherapists in working with military families as they face these many loss and transition issues. As noted by Basham (2008), attachments that transform from security to insecurity can result in relationships becoming chaotic and disorganized. "In these families, we see the insidious effects of affect dysregulation on the parents and children, disrupted attachments, and erratic parenting, which fuel disorganized attachments and increased behavioral problems in children" (Basham, 2008, p. 90). Understanding the constructs of attachment theory can lead to better understanding of issues faced by our military families.

The inward focus of the military leads to three psychological traits outlined in 1991 by Mary Wertsch. These traits are important as they often capture many of the defining issues that contribute to family dysfunction. The traits are secrecy, stoicism, and denial. The work of the military and certainly many occupations within the military demand secrecy, not only from the outside civilian world but even from family members and other members of the military.

The constant preparation for disaster and the focus on the importance of the mission leads to a level of stoicism, or sense of the appearance of being able to handle any stress or burden for both service member and family. This trait of stoicism is at the heart of the commitment to the military made by each service member but, at its extreme, also leads to the stigma against seeking help or even acknowledging any weakness that may render the service member dysfunctional. "Stoic behavior is rewarded, whereas emotionality is not only discouraged but often punished; often the first casualties are family relationships" (Hall, 2008, p. 57).

Denial is a constant; families and service members are asked to deny personal freedoms, individual preferences, and sometimes even offenses such as domestic violence or child abuse or neglect. Denial is made even more subtle by expecting families to keep their emotions in check and their fears unexpressed. "If these families had to constantly be conscious of these fears, it would be unbearable, so the possibility of disaster is repressed; in the process, most other feelings are denied also. Warriors cannot do their duty without denial, and the spouses and families need the denial to not feel so vulnerable" (Hall, 2008, p. 57). This constant level of denial does not allow access to the grief work that would lead to healing during the many emotional and physical transitions required of military members and their families.

These traits, which could arguably be defended as crucial to the success of the mission and the military, can also determine whether military members, and even their families, access necessary treatment or assistance when faced with family or personal concerns. "To the extent that seeking psychological treatment is defined as 'weakness,' soldiers may be slow to pursue services" (Reger et al., 2008, p. 27). These traits are the norms of the military, and our goal in psychotherapy is to acknowledge and work with the consequences of the traits, not necessarily to eliminate them.

Military service is important to the lives of many members of any society. Appreciating the reasons why someone joins the military also helps us understand some of its cultural aspects. Being able to serve the greater good, following a family tradition, gaining skills for a future career, identifying with the warrior mentality, and finding a way to better one's life, all lead different people to join the military. These, and many others, are places to start in understanding military culture. But for whatever reason people join, they are joining a culture very different from our civilian culture. As mental health professionals, we must understand and value the culture in order to best serve those who live and function within it.

References

Basham, K. (2008). Homecoming as a safe haven or the new front: Attachment and detachment in military couples. *Clinical Social Work Journal, 36,* 83–96.

Faber, A. J., Willerton, E., Clymer, S. R., MacDermid, S. M., & Weiss, H. M. (2008). Ambiguous absence, ambiguous presence: A qualitative study of military reserve families in wartime. *Journal of Family Psychology, 2,* 222–230.

Hall, L. K. (2008). *Counseling military families: What mental health professionals need to know.* New York, NY: Routledge, Taylor, and Francis.

Hall, L. K. (2012). The military lifestyle and the relationship. In B. A. Moore (Ed.), *Handbook of counseling military couples* (pp. 137–156). New York, NY: Routledge, Taylor, and Francis.

Martin, J. A., & McClure, P. (2000). Today's active duty military family: The evolving challenges of military family life. In J. A. Martin, L. N. Rosen, & L. R. Sparacino (Eds.), *The military family: A practice guide for human service providers* (pp. 3–24), Westport, CT: Praeger.

Military culture. (2008). In *Encyclopedia of violence, peace, and conflict.* Retrieved from http://www.credoreference.com/entry/estpeace/military_culture

Reger, M. A., Etherage, J. R., Reger, G. M., & Gahm, G. A. (2008). Civilian psychologists in an Army culture: The ethical challenge of cultural competence. *Military Psychology, 20,* 21–35.

Wertsch, M. E. (1991). *Military brats: Legacies of childhood inside the fortress.* St. Louis, MO: Brightwell. (Originally published by Harmony Books).

6 PERSONALITY AND MILITARY SERVICE

Michael R. DeVries and Emile K. Wijnans

In order to effectively treat individuals in psychotherapy, it is necessary to know something about their personality. One's personality tells us how they see themselves, the world, and their relationships with others. While there is not a single military personality style, it is likely that people who join the military have certain common personality traits. Additionally, differences between the personalities of those in the military and the general population are likely to increase over time due to the influence of the military culture and the winnowing process that occurs within the military ranks.

Arguably, the military is a strong contextual situation that moderates the expression of personality variables (Darr, 2011). Likewise, service members enlist for distinct periods of time and must continually reaffirm their interest in the military throughout their career by reenlisting. It is likely that those who stay in the military for their entire career (some 17%) have similar traits, leading them to find satisfaction in the service. The effect of the military context and factors that lead people to remain in the military for an entire career likely contribute to homogenizing the military population over time. Therefore, psychotherapists working with military service members will benefit from an understanding of how certain personality traits, or the lack there of, may influence their client in the distinct culture of the military.

As we discuss personality in the context of the military we must be cautious not to oversimplify. Despite the perhaps common perception, the military is not a collection of completely homogeneous individuals. Consequently, individuals vary in how much they resemble any prototype. A common misconception is that all service members are similar. For example, people unassociated with the military may know intellectually that there are many different jobs in the military, but they often do not truly appreciate the various roles service members play and skills they possess. Picano, Williams, and Roland (2006) discuss personality traits which are most salient in personnel performing high risk military specialties such as aviation or explosive ordinance disposal, and which distinguish them from the general military population. It may be possible then to make some useful hypotheses about individuals in certain specialties, and more generally, about all military personnel. Despite variation across the military, the unifying factor that makes a discussion of the military personality worthwhile is the similarities that exist between individual cultures of the Armed Forces that are, collectively, very different from civilian culture.

THE FIVE FACTOR MODEL OF PERSONALITY

Currently, personality is generally characterized in terms of the five factor model (FFM) of personality (Costa & McCrae, 1988). The five factors generally agreed on to describe personality are extraversion, agreeableness, neuroticism/emotional stability, conscientiousness, and openness. The following discussion of personality in the military uses the FFM as defined by Barrick, Mount, and Judge (2001) to understand normal personality and facilitate discussion.

Extraversion: This dimension of personality consists of sociability, dominance, ambition, positive emotionality, and excitement-seeking.

Agreeableness: Agreeableness is defined as cooperation, trustfulness, compliance, and affability.

Emotional stability: Sometimes referred to by its opposite, emotional stability is the lack of anxiety, hostility, depression, and personal insecurity.

Conscientiousness: Closely related to work ethic, conscientiousness is associated with dependability, achievement striving, and planfulness.

Openness to experience: Openness is defined as intellectual interest, creativity, unconventionality, and broad-mindedness.

PERSONALITY AND JOINING THE MILITARY

The military attracts people with a wide variety of personalities and cultural backgrounds, and, as stated earlier, the military is made up of a wide variety of job types. The personality drawn to spend months on board a ship or submarine may vary significantly from the personality drawn to the infantry, aviation, medical services, or mortuary affairs. Despite differences, people join the military for many similar reasons: college money, escape from a "dead end lifestyle" or job, service to their country, honoring family tradition, seeking excitement, and so forth. Service members who join in order to secure money for college, receive job training, or to escape financial hardship may not have personality traits in common, but those who join out of a sense of duty or adventure may be displaying core personality traits.

In general, service members are more willing to leave behind the safety and security of home. This is particularly true over the last 10 years, as the United States has been continuously at war since 2001. Individuals who have joined since 2001 have committed to service in the military in the time of war with the expectation that they will be deployed around the world and in war zones. The typical service member seems to be willing to sacrifice all that is valuable to him or her for a greater cause. In his book on the American soldier, Peter Kindsvatter (2003) discusses the reasons soldiers "rally to the flag" and choose to serve. He notes that motivations range from "enthusiastic volunteer to resentful draftee" (p. 1). Though the draft ended in 1973, our current military is likely still characterized by the same "mix of enthusiasm, resignation, and resentment" (p. 4) described by Kindsvatter. How service members view their service is likely to be related to why they joined. Did they join out of a perceived need to escape something or a financial need, or did they join to serve and fulfill a sense of duty? The latter is one facet of conscientiousness. So, while we can't assume all military members have the same level of motivation and reasons for service, a desire to serve and sense of duty may be one commonality.

Kindsvatter (2003) states that World War I provided an opportunity for young men to prove their courage and manhood. The young soldier was "fascinated by the prospect of adventure and heroism" (p. 6). Excitement-seeking may also be part of the "typical" military personality style. The cliché is that people join the military to see the world. This facet of extraversion may draw people to consider joining the military and testing themselves amid what may be novel, real-life danger. Additionally, military jobs often involve travel, working with new technology, varied and fast-changing responsibilities, meeting people from around the world, and doing adventurous things (flying in helicopters, repelling from towers, training in hand-to-hand combat, etc.) that are less often available to the general public. Many service men and women may well share a greater sense of adventure and desire for excitement, on average, than age-matched civilian peers.

While Kindsvatter (2003) studied the reasons soldiers join the service from a qualitative perspective, there is some quantitative evidence that those who join the service are different from their civilian counterparts. Jackson, Thoemmes, Jonkmann, Lüdke, and Trautwein (2012) completed a longitudinal study of German soldiers and their civilian counterparts. They studied cohorts of German citizens who left high school and completed either 9 months of military service or 9 months of community service. German laws required that eligible students serve in the military or choose to complete community service. They found that high school students who chose the military were less agreeable, less neurotic, and less open than their peers who did not join the service.

One cannot infer too much about personality from a service member's initial job choice, as not everyone lands their first choice. Often, new recruits must choose from a few options rather than the field that most holds their interest. New service members enlist for a defined period of time, typically 2 to 6 years, and likely display a greater cross-section of the general population in their personality styles. After this initial enlistment, when they begin to grasp what the military entails, a winnowing occurs. Service members then choose to leave military service, stay in their current job, or reclassify into another specialty. It is likely that those who choose to reenlist (with officers and warrant officers, this process is not technically reenlisting but they similarly choose to stay or resign their commission or warrant) have more in common than those serving their initial term of service.

PERSONALITY OVER TIME IN THE MILITARY

The military is a powerful environmental factor that likely shapes, to some degree, the personalities of the young men and women who serve. Some argue that personalities tend to be stable over a lifetime, specifically after age 30 (Costa & McCrae 1988). However, 66% of our military is under the age of 30, and it is likely that the vast majority joined when they were younger than 30 years old (Pew Social and Demographic Trends, 2011). Roberts, Walton, and Viechtbauer (2006) found that younger individuals showed the greatest degree of change in longitudinal studies of personality change. It is likely that young service members are in the process of solidifying their adult personality styles when exposed to the military.

Darr (2011) investigated the role of the military in moderating personality. She postulated that while personality tends to be consistent over time, strong situations can influence the expression of personality traits. She argues that the military is a significantly strong situation that influences the expression of certain personality characteristics of service members. In the military, the service member faces punishment, financial and otherwise, if they do not fulfill their duties. Regardless of which service one joins, the institution has incredible power over the individual. The chain of command has ultimate authority and responsibility for the service member to the point that they can impose fines, confinement, loss of privileges, loss of rank, and even removal from military service. Any environmental effects the military may have on personality traits are likely to be more significant the longer the service member is exposed to the culture. Matthews (2009) reported that over 47 months of training at West Point, character strengths were relatively stable, so any adaptation that takes place may take many years. So what then can we say about personality traits of military service members over time?

Certain personality traits such as conformity, emotional stability, and conscientiousness are reinforced by military culture, resulting in the expression of certain personality traits. There is some evidence that agreeableness may be discouraged by military training and experience. Furthermore, those with certain complementary traits may be encouraged to remain in the military. From the early stages of one's military career, a pruning process occurs. McCraw (1990) studied Air Force personnel in technical training. He found that there were significant personality differences between those service members who presented for treatment with a desire for discharge from the military and those who were identified as being well adjusted, with no desire to separate. Well-adjusted service members were significantly different from those seeking discharge on the California Personality Inventory scale of Achievement via Conformance. This scale is described as tapping into factors that facilitate achievement through conformity. It is not a surprise that well-adjusted service members are more willing to conform given the requirements for conformity (e.g., dress code, rank structure and submission to authority, military courtesies and traditions, etc.) in the military.

In their study of German service members and matched civilian counterparts, Jackson et al. (2012) found that those who chose the military were less agreeable after initial training than their peers in the cohort who completed community service. Furthermore, the group differences in this sample persisted five years later, after individuals from both cohorts had attended college or entered the work force.

This study suggests that military service may produce lasting changes in personality in those who serve.

Because the military has distinct criteria for performance and routine formal personnel evaluations, poor job performance can lead to failure to progress in rank or even removal from military service. Individuals who do not progress in rank will eventually be separated from the service, but they may choose to leave the service earlier if it is no longer rewarding. This competitive, performance-based culture means that personality traits that are associated with job performance are likely to be encouraged. Barrick et al. (2001) argue that most meta-analyses of personality measures have shown that conscientiousness and emotional stability are positively correlated with job performance across nearly all jobs. This is consistent with Picano et al. (2006), who found that across various high-risk military jobs, many of which use personality assessment in the selection process, personnel score higher in conscientiousness and emotional stability. Presumably these selection programs are selecting top performers, though, as these programs are typically filled with volunteers and not everyone in the service is eligible, they do not obtain all the military's top performers. To the degree that the military reinforces good job performance through effective selection and/or accurate, routine performance evaluation, it is likely that conscientiousness and emotional stability are encouraged.

ADAPTIVE ANXIETY

While the military is likely to select for and encourage emotional stability, there are some aspects of anxiety that are prevalent and adaptive in the military culture. Military service appears to widely and regularly reinforce traits that echo symptoms of obsessive-compulsive personality disorder (OCPD). Structure, orderliness, attention to detail, drilling to perfection, and precision are values that permeate the military environment. Nearly everything in the military comes with a checklist. Even the most routine maintenance procedures are outlined in a manual so that everyone is trained in the exact same procedure. The culture is such that if there is a problem or accident, the first step in the investigation of the cause will be to determine which manuals, regulations, and standard operating procedures apply and whether they were followed. Regularly, service members are expected to check and recheck their own and others' gear, prior to a training event or mission. When these checks are complete, a commander or supervisor may check the gear again. This behavior clearly makes sense when one considers that the safety of a soldier and his or her unit-mates may depend on the individual service member having all his or her gear, having it properly maintained, and having it in the designated location.

Leaders have responsibility for the behavior of everyone who falls under them. Experience has shown that this culture of regulations and accountability can lead even senior leaders to display anxious behaviors out of fear of the consequences of missing a detail. The next higher level of command or the real dangers of hazardous training and combat may impose consequences. Such behaviors do not typically lead to clinically significant impairment; however, they may look like symptoms of OCPD such as preoccupation with lists and details, lack of flexibility, and devotion to work. Outside the military, the intensity and preponderance of such traits may appear unusual and be problematic in social or occupational contexts.

References

Barrick, M. R., Mount, M. K., & Judge, T A. (2001) Personality and performance at the beginning of the new millennium: What do we know and where do we go next? *Personality and Performance, 9*(1/2), 9–30.

Costa, P. T., & McCrae, R. R. (1988). Personality in adulthood: A six-year longitudinal study of self-reports and spouse ratings on the NEO Personality Inventory. *Journal of Personality and Social Psychology, 54*(5), 853–863.

Darr, W. (2011). Military personality research: A meta-analysis of the Self Description Inventory. *Military Psychology, 23,* 272–296.

Jackson, J. J., Thoemmes, F., Jonkmann, K., Lüdke, O., & Trautwein, U. (2012). Military training and personality trait development: Does the military make the man or does the man make the military. *Psychological Science, 23*(3), 270–277.

Kindsvatter, P. S. (2003). *American soldiers: Ground combat in the world wars, Korea, and Vietnam.* Lawrence: University Press of Kansas.

Matthews, M. D. (2009). The soldier's mind: Motivation, mindset, and attitude. In S. M. Freeman, B. A. Moore, & A. Freeman (Eds.), *Living and surviving in harm's way* (pp. 27–49). New York, NY: Taylor & Francis.

McCraw, R. K., & Bearden, D. L. (1990). Personality factors in failure to adapt to the military. *Military Medicine, 155,* 127–130.

Pew Social and Demographic Trends. (2011). *The military-civilian gap: War and sacrifice in the post 9/11 era.* Washington, DC: Pew Research Center.

Picano, J. J., Williams, T. J., & Roland, R. R. (2006). Assessment and selection of high-risk operational personnel. In C. H. Kennedy & E. A. Zillmer (Eds.), *Military psychology: Clinical and operational applications* (pp. 353–370). New York, NY: Guilford.

Roberts, B. W., Walton, K. E., & Viechtbauer, W. (2006). Patterns of mean-level change in personality traits across the life course: A meta-analysis of longitudinal studies. *Psychological Bulletin, 132*(1), 1–25.

7 IMPACT OF MILITARY CULTURE ON THE CLINICIAN AND CLINICAL PRACTICE

William L. Brim

The ideal of cultural awareness and competence for health care providers has traditionally been related to work with ethnic and racial minorities; however, it is also an essential component in working with service members, veterans, and their family members. For the military and veteran population there is a significant stigma associated not only with seeking behavioral health care but also with seeking any medical care. This stigma, which is born out of tenets of military culture, will not be overcome by clinically competent, well-meaning providers who are not aware of and sensitive to the nuances and impact of the military culture.

Military culture can be defined as the total of all knowledge, beliefs, morals, customs, habits, and capabilities acquired by service members and their families through membership in military organizations. Military culture includes both explicit elements such as clearly defined organizations, roles, and relationships, and implicit elements such as the warrior ethos, a set of universal values, and guiding ideals. Like all cultures, military culture is defined by its values, ideals, and codes of conduct; physical objects such as uniforms and technologies; and behaviors. Acquisition and assimilation of this culture and the underlying warrior's ethos falls on a continuum with some subcultures and members more strictly adhering to the ideal (e.g., pilots or special forces) and with

adherence to the ideal varying throughout the career and life span of the service member or veteran (e.g., during deployment in a combat role versus in garrison in an administrative role versus after retirement or separation). The development of a warrior's ethos begins in the earliest stages of enlistment with the oath of office and continues through basic training and throughout the individual's career. Family members will also vary in their degree of adoption of and assimilation into the military culture.

This warrior culture provides the member with the strength, resilience, and ability to push forward in combat even in the face of overwhelming odds. It is a collectivistic, strength-based culture that often places it at odds with the medical and behavioral health culture that is individualistic and pathology focused. A member who has adopted a warrior ethos finds strength and purpose in self-sacrifice, in suppressing emotion and in learning to tolerate pain, whereas the medical culture will encourage seeking help, expressing emotions, and reducing pain. Providers who do not recognize these differences in cultural orientation are at risk of pathologizing the culturally derived thoughts, emotions, and behaviors of members; making inaccurate diagnoses; and providing inappropriate services, thus reinforcing stigma.

EXPLICIT CULTURAL FACTORS

The explicit cultural components of the military are the concepts, behaviors, and objects that are observable from the outside and are the aspects of the culture that most people are familiar with. These components include the organizations, roles, and relationships that can have a profound impact on how service members and veterans view themselves and others. Compared to the implicit elements of military culture, these concepts are more diverse because the various missions performed by the different services require different structures and roles and because each branch has its own unique heritage and traditions. Health care professionals may feel overwhelmed by the idea of trying to learn all of the nuances of military organizational structures, occupational specialties, rank hierarchies, and military jargon. You do not need to memorize rank charts and know all occupational specialties in order to be effective and reflect cultural sensitivity; you only need to know how these explicit military organizations, roles, and relationships might affect the health and well-being of your patient and have a willingness to understand it from the point of view of the patient.

For military leaders, military structure, missions, and roles are conceptualized in terms of organizational charts and capabilities. To the individual service member, the organizational levels, missions, and roles are about belonging to a team and serving others in pursuit of common goals. The first decision often faced by the individual member is which branch to serve in, a decision that will likely impact nearly every aspect of their experience in military life. It is often very interesting and enlightening to ask a member why they chose to join the service that they did. To become a member of a team an individual must relinquish some portion of their independent identity and self-determination, find a balance between personal goals and shared goals of the team, and agree to adopt the team's goal as their own. This giving up of some individualism can often be a difficult transition and a recurring theme when challenges arise. Likewise, experiences that impact the ability of the member to remain a part of the team—promotion, changes in assignment, illness, injury, separation, and retirement—can also precipitate difficulties.

Another explicit component of the military culture is the contract between the member and the government. This contract from the standpoint of the individual member can be both protective and restrictive, a source of security and a necessary evil. The contract between service member and service branch takes the form of the written enlistment or commissioning contract and the spoken oath. Both forms state that the member is required to obey all lawful orders and perform assigned duties and imply or state that the member will be subject to military justice and may be required to serve in combat. In exchange for good order and discipline, becoming a member of the team, and having a willingness to endure hardships and face possible death, members are provided with food, shelter, clothes, health care benefits, social opportunities, and pay.

Perhaps the most significant contractual issue impacting the health care provider is the fact that the Department of Defense is both the employer and the provider of the health care benefit. Because of this, and the fact that physical and mental readiness is a job requirement, service members do not have full confidentiality of their health records. Members, and their health care providers, have an obligation to inform the command about health conditions that may affect military readiness and fitness for duty. Health care professionals providing treatment to service members should become aware of the regulations addressing these limitations to confidentiality and privilege as well as the reporting requirements and duty status levels determined by physical and behavioral fitness. For the civilian provider working outside of the DoD, State confidentiality rules apply. However, the Member maintains an obligation to report medical and psychological health issues to their chain of command and often should be encouraged to do so in order to make use of resources that can assist in recovery and might not otherwise be available.

IMPLICIT CULTURAL FACTORS

The implicit cultural components of the military are the foundational and often unconscious intellectual and emotional content, the guiding ideals and values, which might be thought of as the "why" behind the "what" of the more explicit military culture components. These components are introduced in the initial military training of both enlisted members and officers and are expressed in writing and in symbols, ceremonies, behaviors, and stories. While being able to identify some of the overt, explicit components of military culture described above is helpful, having an understanding of the underlying reasons behind these components is an especially helpful tool for the health care provider seeking to understand their patient. While all members, when on active duty, will wear the proscribed uniform and rank and will comply with the military courtesies overtly, the extent to which they embrace the underlying core values and guiding ideals on the inside will have much more of an impact their worldview and sense of self. The degree to which a service member embraces the core values varies from member to member and may change over the course of the individual life span from active service to veteran. While it is difficult to clearly articulate and comprehensively define these guiding ideals and values, we can catch glimpses of them through the oath that all members take and the core values that they strive to follow.

The oath that all military members take is perhaps their first introduction to the military culture. The oath can be traced back thousands of years to Greek and Roman times, and while it varies from country to country and over the course of time, those who serve in uniform have always made a promise to defend their social order, maintain allegiance to their homeland, and obey the lawful orders of their superiors.

The core values described by each service are another aspect of the implicit component of military culture. The core values are central to military culture, they are the underlying rules for living that all military organizations are committed to being guided by and that members strive to embody every day and in every situation. While each service articulates the core values in slightly different words, the common themes are:

- Honor and Integrity—an adherence to what is right,
- Courage—the willingness to face physical and mortal danger without retreating,
- Service—sacrificing for others without complaint or expectation and,
- Duty and Commitment—the keeping of a moral promise to oneself and others.

Service members will often express these values in statements such as, "I wanted to be a part of something bigger than myself."

Other characteristics of the warrior ethos appear to include the following:

- a dedication to live every day by a moral code
- a commitment to defend the social order
- the finding of meaning through selfless service to others
- pride
- a willingness to suffer and face death
- competitiveness
- a will to win

Most service members, veterans, and their families may not have given these foundational concepts a lot of conscious thought and may not be able to articulate them and their impact on their daily life, but the degree to which they have adopted and assimilated the military culture will impact how they view the world and handle distress. For the health care provider working with this culture; the life, health, and functioning of your patient may only make sense when viewed in the context of these guiding ideals and principals. Looking for evidence that your patient feels they have lived up to, or failed to live up to these guiding principles and values, may give insight to their presenting concerns.

STRESSORS ASSOCIATED WITH THE MILITARY CULTURE

Cultural competence for health care professionals working with the military population is increased by an awareness of the routine stressors faced by service members and their families in addition to the challenges and possible experience of life threat, loss, and moral injury associated with deployment. Service members and their families report many of the same stressors that other patients report: stress at work such as problems with coworkers or supervisors, financial problems, family health problems, marriage, divorce, having a baby, or a death in the family. However, military members also routinely experience somewhat unique stressors as well: for example, frequent moves with their inherent disruptions in employment and schooling and conflicts between military and family responsibilities. Additionally, in many ways, military members are always preparing for deployment and spend a significant amount of work and family time getting ready for, or reintegrating from, operational missions.

The deployment process is often conceptualized in terms of phases each with a unique, though often overlapping set of stressors. The three most basic phases are predeployment, deployment, and redeployment and reintegration. Each phase may include common and often escalating routine stressors as well as the less common life threat, loss, and moral injury events that may occur. The deployment cycle has an impact on the unit, the member, and their family as well as on the community as a whole.

Given the range of stressors and events that may affect the member or their family, the culturally aware health care professional will be prepared to inquire into the impact of the deployment cycle, stress and stressors associated with each phase, and the resources available to the member. The member should always be asked if the reason for the visit might be related to any type of deployment or extended family separation. It should also be noted that members and their families will also report some positive and rewarding growth experiences related to deployment.

MILITARY HISTORY ASSESSMENT

With a heightened sense of the heterogeneity of the military culture, the culturally aware provider understands that it is no longer sufficient to just ask, "Are you a veteran or military member?" The following are some suggested questions that might constitute a military history portion of your assessment. The culturally sensitive provider will recognize that not all of these questions will be relevant in every interview; they are meant as an extensive, though not exhaustive guide, and a potential starting point for understanding the member's worldview. The member's response may generate more inquiry along a particular topic or may indicate that this is not a relevant area.

- *Have you or someone in your immediately family served in the military?* This is the generally the best way to ask because it is inclusive of family members and captures both active and veteran status.
- *What branch of service were/are you in?* Recognizing which service the patient is/was in can provide insight to the types of missions that the service member may have been involved.
- *Were you ever in the Guard or Reserve?* Following up a positive response to this question with the questions below about duties and deployments is important.
- *What years did you serve?* This question gives you a time frame of their service and lets you know how long they were in the military. You can use this information to inform later questions. For example, if service dates were between 1965 and 1973 they may have served in Vietnam, if between 1969 and 1972 they may have been selected to join in the draft lottery.
- *What was your rank?* Be aware of the rank structure. A basic understanding of the potential responsibilities associated with different rank is helpful.
- *What is/was your occupation(s) in the military?* The career field tells what a person does within the military and can include fields such as Armor, Infantry, Aviation, Hospital Corpsman, or Dental Technician.
- *How many duty assignments did you have and where were they?* Often, people are surprised with the number of moves that a military member may have had over the

course of their time in service. It might also be relevant if they had only a few assignments over the course of a 20-year career. In general, military members can expect to have a move (called a Permanent Change of Station or PCS) every 3–4 years, but many will have more or less.

- *What were some of the reasons you decided to join the military originally? Were they different from the reasons you stayed in the military (if they served more than 1–4 years).* Service members and veterans will sometimes give a superficial answer such as, "for the college money," but as you build rapport through your sincere interest in their experience you will get deeper answers. Few people would risk going to war and potential serious injury for "money for college." Service members and veterans often have a sense of patriotism and service to the community that is a significant driver for their choices and a powerful tool in therapy.

- *What were the major milestones in your career?* The answers may be related to promotions, selection for leadership positions, or the earning of awards. This information may also be valuable as a part of your conceptualization and treatment planning as it will give you some insight into the types of things they value.

- *What was the impact of military service on your family?* Military members clearly indicate that family concerns are a top stressor. Understanding the member's perspective of how military service in general affected the family is an important acknowledgment of their whole military service.

- *Were you ever deployed?* Deployed is a relatively new term in the general public. You may also ask if they ever served in combat, "went to war," or served during a war or conflict.

- *What was the most rewarding part of deployment?* Do not assume that every aspect of the deployment was negative, service members will often say that being deployed was the most important growth experience of their life. Members will describe learning things about themselves and finding strengths they never knew they had.

- *What was most difficult part of deployment?* Never assume that the answer to this question will be seeing combat or other war related trauma.

- *When did you deploy? Where?* Understanding where and when a member was deployed may give significant insight to what their experiences may have been.

- *How many times did you deploy and how long were the deployments?* In Operation Iraqi Freedom/New Dawn and Operation Enduring Freedom members and veterans have experienced multiple deployments. You may have to inquire about the experience of each deployment separately.

- *What were your duties in theater?* Many times the duties assigned to a member in the deployed environment differ from what they do as a regular duty when in garrison.

- *Did you see combat?/How often were you "outside the wire"?* A Service member does not need to have been designated a "combatant" to have seen combat or to have been affected by the results of combat. One population of service members with the highest percentage of PTSD are health care professionals who never went out on patrol and were designated "noncombatants."

- *Did you deploy with your unit or were you an individual augmentee?* Members who deploy as an augmentee to a larger already intact unit often report more psychological health issues and report they felt less unit cohesiveness and support.

- *If Guard or Reserve, what was the impact on your life of being deployed versus impact when you came home and returned to civilian life?* Guard and Reserve members are at higher risk for psychological health issues.

- *Did you feel supported by the unit?* Several recent studies and surveys have indicated that unit cohesion and the impression of unit leadership can have an impact on psychological health.

- *Were you exposed to blasts while deployed?* This is an opportunity for you to assess the possible presence of traumatic brain injury and add to your differential diagnosis and may lead to referral for neurological evaluation.

- *What was your exposure to death of unit members, enemy combatants, or civilians?*

This can be a difficult topic to broach with a new client and should be addressed carefully.

- *Do you feel like there are any lasting physical or psychological effects of your exposure to these potentially traumatic events?* This is a way to transition from discussing an event to a person's reaction to the event.
- *What is the possibility that you will get deployed again?* Military members may report that they wish to deploy again as soon as possible. Often the deployed setting is one that is more compatible for their "new normal." They may be reluctant to deploy again.
- *How was coming home from deployment, was it different than you expected?* There are many stressors associated with homecoming, often members report feeling overwhelmed by the immediate emersion back into their in-garrison life. Often members come home with high expectations for the homecoming. There are roles and rules to relearn, and many of the same issues that were present

prior to deployment are still there or have worsened with time and separation.

- *What was impact of deployment on family?* Many members will report that the deployment was hardest on the spouse or family members. However, often you will hear from family members that they learned they had strengths and capabilities that they did not know they had when the member was deployed.

Health care providers who work with service members, veterans, and their families have an obligation to understand the unique cultural aspects of the military and how these cultural factors inform the worldview, thoughts, emotions, and behaviors of their clients. Making use of a few well-thought-out assessment questions and understanding the rationale underlying the questions can reflect cultural awareness and enhance the therapy experience for the client and provider.

PART II

Military Psychology Specialties and Programs

8   AEROMEDICAL PSYCHOLOGY

Pennie L. P. Hoofman and Wayne Chappelle

Aeromedical psychology applies clinical psychology principles, methods, and techniques within the aviation population, focusing on the overall behavioral health and safety of the individuals and the effects on the crew and the unit. Aeromedical psychologists address all aircrew, not just pilots. "Aircrew" and "crewmembers" are all-inclusive terms and refer to anyone on flight status involved in flight duties, including but not limited to pilots, crew chiefs, flight engineers, flight medics, navigators, weapons systems operators, flight surgeons, and operators of unmanned aircraft systems (UAS)/ remotely piloted aircraft (RPA).

Safety is the military aeromedical psychologist's primary concern—safety of the aviation unit, the flight crews, the passengers, and each individual on flight status. The guiding principle is to keep crewmembers flying safely. Many individuals in aviation do not fully trust this principle as it is common in the aviation culture to avoid disclosing symptoms to the flight surgeon, who serves as the aviation primary care provider, or to avoid visiting the mental health office. This goes beyond the stigma of being diagnosed with a mental health disorder. In addition to the concern that others will perceive one as being unstable and weak, there is the perceived threat to one's career, livelihood, self-identification, and self-worth, all due to loss of flight status. Conversely, the crewmember's well-being and career can also be "saved" when evaluations result in accurate diagnoses and proper treatment, allowing a return to full flying duties—and flying safely.

Due to the nature of military flight and the medical standards being more stringent than for general military service, recommendations from the results of a Fitness for Duty evaluation for a nonaviation service member do not necessarily equate with fitness for Full Flying Duty (FFD) for aircrew. Being aware of the aeromedical implications of evaluations can minimize psychologists recommending a return to flight for someone who is a risk to aviation safety or make a recommendation that will ground someone who is capable of safely flying.

ENVIRONMENTAL DEMANDS AND HUMAN FACTORS IN AVIATION

Any psychologist who evaluates and treats aircrew and provides recommendations concerning fitness for flying should have a clear understanding of the crewmember's duties, the unique features of the aviation culture, and the rigors of flying where the demands and risks to personal safety are high. Military aviation operations strain crewmembers' physical and psychological disposition and present additional demands and risks beyond the traditional stressors associated with commercial flying and other military occupational groups.

Military aviation operations may occur at night and in situations with a reduction in visual cues that are often relied on for flying safety. Night vision goggles add demands due to the restricted field of view and the physiological impact of wearing the devices for extended periods of time. Flying at extremely high speeds requires a high degree of alertness and concentration over an extended period of time, often leading to physical and psychological fatigue. Flying over extreme terrains increases the hazards that can interfere with safe flight, such as mountains and hills that are not visible during difficult weather conditions. "Brown-outs" and "white-outs" decrease visibility and can cause spatial disorientation. Spatial disorientation can occur due to other visual or vestibular illusions caused by the environment. Flying in extreme weather conditions decreases visibility and increases reliance on instruments, thereby increasing stress and fatigue due to required sustained alertness. Flying at high altitudes increases the risk of hypoxia, which impairs physical and cognitive abilities and is often not readily recognizable by aircrew. Conversely, flying at very low altitudes increases the risk of unknown or unseen obstacles that quickly become hazards if the crewmembers' alertness diminishes. Birds present a hazard at all altitudes. Flying in combat or other hostile conditions creates additional physiological, cognitive, and emotional demands. Extended operations can lead to fatigue, which impairs concentration, communication, attention, and judgment.

Aircraft design and airframe characteristics can also increase cognitive and physiological stress and divert aircrew attention from operational duties. These design challenges include cockpit and instrument illumination, the large amount of data to monitor, seat discomfort (seats are built for safety, not comfort), visibility, noise, and vibrations. The noise and vibrations inherent in aircraft, especially rotary-wing, can impair concentration and have short-term and long-term health effects, such as increased fatigue, muscular tension, increased blood pressure and heart rate, and chronic back and neck pain.

Consider the extreme demands on crewmembers' cognitive, emotional, and physiological status with each of these factors, many of which occur simultaneously. Attention, concentration, memory, information processing speed and accuracy, judgment, and communication must be optimal to decrease the risks of human factors in aviation. Additional stressors stemming from one's health, family, or any other source can diminish crewmembers' abilities and increase the risks of flying. Although crewmembers tend to compartmentalize very well, each person has his/her own threshold, of which he/she may not be aware or may deny. Extremely stressful family situations impact the entire crew when one's attention and concentration are compromised. Thus, the crewmember might be grounded by the flight surgeon until the issues resolve, depending on the situation.

PSYCHOLOGICAL ATTRIBUTES OF MILITARY AIRCREW AND ASSESSMENT

Experience with aviation commanders, crewmembers, and the literature suggests that those who pursue a career as military aviators possess high levels of courage, self-discipline, competitiveness, self-confidence, stress tolerance, impulse control, perseverance under adversity, desire to succeed, and a strong attraction to high-risk activities. These personality traits typically accompany a high-average to superior level of intelligence, visual-spatial aptitude, dexterity, coordination, and reflexes that are combined with a strong motivation to pursue a career in aviation. Awareness of the cognitive aptitudes (ability), personality traits (stability), and motivation of military air crewmembers is critical for psychologists tasked with evaluating suitability for flying. Although most of the literature focuses on the personality and cognitive aptitudes of pilots of manned aircraft, more recent research revealed attributes of successful RPA operators: high stress tolerance, comfort working in a confined space with others, positive social interpersonal exchanges, willingness to take risks, high levels of adaptability, and resilience to stress. Cognitive abilities that are key to successful performance include situational awareness, vigilance, spatial analyses (i.e., ability to mentally manipulate two-dimensional objects into a three-dimensional mental image), reasoning, rapid speed of information processing, and visual tracking, searching, and scanning, as well as complex and divided attention.

Motivation, defined as the inherent drive, desire, and sense of reward a person experiences from pursuing a profession, is critical to performance, particularly in the military, where the threats to safety are substantial and perseverance through adversity is essential. Assessment of motivation is a core piece of the adaptability rating for military aviation and medical flight screening for military crewmembers in manned or unmanned airframes. The specific cognitive aptitudes and personality traits essential for performing and adapting to the rigors of military aviation may reveal who has the ability and stability, but motivational attributes may reveal who will succeed and remain in the field. (See the following for details and additional references about attributes: Chappelle, McDonald, & McMillan, 2011; Kratz, Poppen & Burroughs, 2007; Paullin, Katz, Bruskiewicz, Houston, & Damos, 2006; and Picano, Williams, & Roland, 2006.)

Awareness of the aforementioned attributes allows the psychologist to better prepare to interact with crewmembers. A psychologist unfamiliar with the aviation culture and the degree of crewmembers' distrust of the behavioral health process may misinterpret the crewmember's behavioral presentation and might erroneously assume pathology or lack of pathology. The crewmember may present as highly defended and uncooperative, appear very arrogant, minimize or deny having any problems, or behave in a passive-aggressive manner. Most are very cooperative, but few openly discuss their problems without considerable reassurance that their flying career will not be affected. Developing rapport can lead to a more cooperative crewmember who provides more accurate data during objective psychological testing and comprehensive clinical interview. If unfamiliar with the crewmember's exact job, it is important to inquire about his/her work environment and specific duties. In addition to building rapport, this inquiry also provides a framework for making appropriate recommendations. It is highly recommended to present information succinctly without the use of psychological jargon and to take time up-front to have a frank discussion with the crewmember about the nature of the assessment and the impact on aviation safety as well as the importance of the crewmember's own health and well-being.

For psychological evaluations of aircrew, providing recommendations based solely on a brief clinical interview does not provide a comprehensive assessment of vital areas of psychological functioning. Subtle impairments in cognitive performance may represent a risk to safety and mission completion. These changes can often be discovered via objective testing of intellectual and emotional disposition that are not clearly evident or revealed by self-report. Aeromedical policy requires cognitive assessments for aircrew with a history of cognitive difficulties stemming from a head injury, medical illness, developmental disorder, emotional problems, or subtle cognitive degradation due to medication use. When conducting such an evaluation, the psychologist should obtain objective testing that focuses on general intellectual functioning (e.g., attention/concentration, memory, spatial judgment, reasoning) as well as emotional-social disposition (e.g., presence of depression, anxiety, and/or irritability, social discomfort). Assessment instruments that assess for the degree of guardedness or defensiveness regarding self-disclosure of personal problems are useful. When interpreting results it is best to use available norms for an aviation-specific population to enhance the ability to make effective recommendations.

Assessment of applicants for selection to flight duties and assessment of existing aircrew for a return to flight duties requires the psychologist to consider specific criteria set forth in the service-specific regulations and policies. In general, the criteria to consider when recommending flight duties include the following: (1) the condition must not pose a risk of sudden incapacitation; (2) the condition must not pose any potential risk for subtle incapacitation that might not be detected by the individual, but would negatively affect higher order senses (e.g., alertness, situational awareness, information processing) relevant to performance; (3) the condition must be resolved or nonprogressive and expected to remain so under the unique stresses and demands of one's aviation duties; (4) if the possibility of progression or recurrence exists, the first symptoms or signs must be easily detectable and not pose a risk to the individual or the safety of others; (5) the condition cannot require exotic tests, regular invasive procedures, or frequent absences to monitor for stability or progression; and (6) the condition must be compatible with the performance of sustained flying operations and not jeopardize the successful completion of a mission. These criteria clearly indicate that the presence of a psychological disorder is inadequate for determining a crewmember's suitability for flying. The aeromedical criteria for many aircrew positions requires functioning beyond the absence of pathology. The key is any sort of change that leads to subtle performance decrement that compromises performance of aviation duties and increases risk to safety. For example, an aviator can have a history of an adjustment disorder diagnosis and no longer meet the diagnostic criteria. However, if the risk of recurrence is considered moderate to high when exposed to the demanding conditions of military aviation, then the person's psychological disposition could reasonably be considered unsuited for aviation, irrespective of the person's general fitness for military duty. An aeromedical evaluation should be approached with specialized questions, testing, and interview techniques that assess for a high level of ability, stability, and motivation to fly.

POLICIES AND REGULATIONS

When considering recommendations for treatment or duty, the psychologist should base professional decisions regarding suitability for flying on guidance described in the service-specific aeromedical standards for flying and the aeromedical waiver guide. Army Regulation (AR) 40–501 Chapters 4 and 6 and the associated Aeromedical Policy Letters (APLs), Air Force Instruction (AFI) 48–123 and the associated waiver guide, and MANMED Chapter 15–62 through 15–69, which includes the associated Aeromedical Reference and Waiver Guide (ARWG), provide guidance for crewmembers of their respective military branch. Interpreting the various waiver guides can be confusing. Contacting an aeromedical psychologist or flight surgeon for guidance minimizes improper evaluations and inappropriate recommendations.

Although there is considerable overlap among the branches regarding aeromedical policy, there are differences the psychologist should consider. One major difference among the services is the use of selective serotonin reuptake inhibitors (SSRIs) and selective norepinephrine reuptake inhibitors (SNRIs) to treat behavioral health disorders. Psychotropic medications of any kind are disqualifying for anyone in any service on flight status. Until 2006, use of these medications was not waiverable in any service. Since 2006, individuals on flight status in the Army can request consideration for waivers, through their flight surgeons, for the use of SSRI/SNRI treatment for psychiatric diagnoses as well as a waiver for the disorder being treated. A cognitive evaluation by a psychologist is required before the flight surgeon submits a waiver request. Although a waiver may be requested, it is not necessarily granted.

CONSULTATION WITH FLIGHT SURGEONS

The psychologist will invariably consult with the crewmember's flight surgeon to the extent needed to serve the interests of the individual aviator as well as preserve the integrity of aviation-related operations. It is important to discuss with the crewmember the content and purpose of the consultation ahead of time and when circumstances arise to offset unrealistic expectations and overcome obstacles related to disclosure. Frequent, open, and responsive communication between the psychologist and flight surgeon is necessary, especially when there are noticeable changes or concerns in a crewmember's psychological disposition that can affect safety. Flight surgeons find it useful when psychologists address the following issues: (1) specific changes in the symptoms and diagnosis that may or may not be suitable for flying; (2) specific recommendations for type and length of psychological treatment and recurrence rates based on professional literature; (3) potential restrictions in duties as related to preservation of occupational safety; and (4) specific recommendations regarding additional tests or evaluations that may be useful for diagnostic clarification and treatment.

Psychologists must remain cognizant of a crewmember's privacy and confidentiality. Only discuss information relevant to the issues at hand, such as the safety of military operations and the well-being of the aviator. However, be careful not to go too far in the direction of sharing too little information. The flight surgeon is responsible for making recommendations about flying to the commander who makes the ultimate decision about flying status; the commander needs to have all of the relevant information to ensure the best decision is made. In communications, reports, and recommendations, avoid psychological jargon and technical terms that may be misunderstood—and be succinct. As long as flight surgeons and commanders have the impression a military or civilian psychologist functions to preserve the integrity and safety of military operations and personnel, they are typically respectful of the recommendations and boundaries of confidentiality.

FUTURE CONSIDERATIONS

Although the literature regarding psychological attributes affecting performance is growing, there remains a dearth of research surrounding the performance of crewmembers other than pilots. For aeromedical evaluations of applicants to various crewmember positions, occupationally specific normative data for the various service-specific military aircrew positions and data that distinguish those who complete training from those who do not is critical. Research regarding the cognitive and personality attributes of other aircrew would lay the foundation for finding what treatments, if needed, might be most effective for other crewmembers. Additionally, aviation is expanding into remotely piloted aircraft at a rapid rate, and it is widely perceived that such aircraft will take over the missions traditionally associated with manned airframes. Having a clear understanding of the psychological profiles of those engaged in military aviation of unmanned airframes is essential to keeping up with progress.

References

Air Force Instruction 48–123. (2011). *Medical examinations and standards*. Available at http://www.e-publishing.af.mil.

Army Regulation 40–501. (2011). *Standards of medical fitness*. Available at http://www.apd.army.mil.

Chappelle, W., McDonald, K., & McMillan, K. (2011). *Important and critical psychological attributes of USAF MQ-1 Predator and MQ-9 Reaper pilots according to subject matter experts* (USAF Technical Report: AFRL-SA-WP-2011–0002).

Kratz, K., Poppen, B., & Burroughs, L. (2007). The estimated full-scale intellectual abilities of U.S. Army aviators. *Aviation, Space, and Environmental Medicine, 78,* 261–267.

MANMED. United States Navy Manual of the Medical Department, NAVMED P-117. (2005). Available at http://www.med.navy.mil/directives/Pages/NAVMEDP-MANMED.aspex.

Paullin, C., Katz, L., Bruskiewicz, K. T., Houston, J., & Damos, D. (2006). *Review of aviator selection* (Technical Report 1183). Arlington, VA: US Army Research Institute for the Behavioral and Social Sciences.

Picano, J. J., Williams, T. J., & Roland, R. R. (2006). Assessment and selection of high-risk operational personnel. In C. H. Kennedy & E. A. Zillmer (Eds.), *Military psychology: Clinical and operational applications* (pp. 535–570). New York, NY: Guildford Press.

9 ASSESSMENT OF AVIATORS

Pennie L. P. Hoofman and Wayne Chappelle

Applicants to military flight school complete a service-specific paper-based or computer-based assessment that measures aptitude for specific abilities related to flight (Wiener, 2005). The purpose of these assessments is selection and primarily addresses the *ability* of prospective aviators. Even with very recent changes in the selection instruments that include indirect and direct measures of *motivation* (Bruskiewicz et al., 2007), these batteries do not measure cognitive, emotional, or psychological domains in a way that clinical psychologists do in order to assess emotional *stability*. Assessment of applicants for aviation assignments in special operational settings is handled differently and will not be mentioned in this chapter, as the topic is addressed elsewhere (Picano, Williams, & Roland, 2006).

Military flight school applicants and trained pilots do not undergo a formal psychological assessment unless they have a history of behavioral health concerns, including substance use related issues, head trauma/cognitive difficulties, or use of psychotropic medications. Since neurological and psychiatric disorders are disqualifying for aviation, many pilots or prospective flight students do not seek behavioral health care when needed. Likewise, they may not readily report symptoms they have had or treatment they have sought. For those who have sought treatment, they may be unaware of any diagnoses actually made by the behavioral health provider. The role of the psychologist is to assess the aviator or flight school applicant and make recommendations to the flight surgeon for suitability for current and future flight duties based on history, current functioning, and prognosis for recurrence of symptoms. The retrospective aspect of the evaluations can be a challenge, making the clinical interview and collection of data extremely important in these evaluations.

Aviators and applicants tend to present themselves in a very positive light with few, if any, weaknesses. As a group, aviators do tend to be a very healthy population. When inquiring about one's history during the clinical interview, the psychologist may need to ask questions in different ways and urge the individual to be forthright since discrepancies in information will delay the waiver process. Waiver requests are reviewed thoroughly, and any discrepancies result in disqualification or the aeromedical summary being sent back to the flight surgeon for clarification. Depending on the presenting issue, the evaluating psychologist may also request documentation from previous providers or collateral information from peers or family with the proper written releases of information. For example, if a pilot or applicant has a history of delirium, a brief psychotic disorder, or generalized anxiety disorder, records of previous treatment and information from the commander combined with the results of the current evaluation provide a more comprehensive picture for the psychologist to make recommendations for aviation duties.

Referrals for aeromedical psychological evaluations originate from two primary sources: flight surgeons and commanders. When the flight surgeon learns that an applicant or trained aviator has experienced a neurological or psychiatric disorder that may have affected cognitive abilities, a neuropsychological evaluation must be conducted to assess *ability* beyond what the entrance examination measures. It is important for the psychologist to be aware of service-specific requirements for exceptions to policy or waivers for flight school applicants and trained aviators. In general, military aeromedical policies require an evaluation of intellectual functioning when there is a history of cognitive difficulties stemming from a head injury, medical illness (e.g., bacterial meningitis, obstructive sleep apnea, multiple sclerosis), developmental disorder (e.g., attention deficit disorder, learning disorder), alcohol/substance abuse, or emotional difficulties (e.g. anxiety or depression). It is critical to have a clear understanding of how changes in cognitive functioning, whether obvious or subtle, may negatively impact performance and adaptation to the rigors and demands of military flying. The following domains must be assessed when cognitive ability is in question: memory, attention, concentration, reasoning, verbal and visual information processing (speed and accuracy), motor skills, reaction time, and visual-spatial abilities.

A meta-analysis of military pilot selection literature over the past twenty years concluded that inherent cognitive aptitudes relevant to pilot performance include general intelligence, general verbal and quantitative abilities, dexterity, perceptual speed and information processing, reaction time, and visual-spatial abilities (Paullin, Katz, Bruskiewicz, Houston, & Damos, 2006). The breadth and depth of cognitive assessment depend on the reasons for the evaluation. According to aeromedical policies, a training applicant or trained aviator with low general cognitive ability and borderline functioning in the aptitudes mentioned above should likely not engage in aviation duties. It stands to reason that high levels of intelligence and inherent cognitive aptitudes are critical to training and adapting to the operational demands of military flying.

Although the most recognized intelligence test used for the evaluation of cognition is the Wechsler Adult Intelligence Scale-4th Edition (WAIS-IV), there are other instruments that are useful and reasonable alternatives: the Multiple Aptitude Battery-II, the MicroCog, or the CogScreen. These instruments are computer based time-efficient measures that provide a level of sensitivity and specificity for identifying problematic areas of cognitive functioning. However, if time is limited, simple measures such as the Wonderlic Personnel Test (WPT) and the Digit Symbol Coding subtest from the WAIS-IV may also be utilized to obtain an effective estimate of general intellectual functioning. The Wechsler Abbreviated Scale of Intelligence (WASI) may also be considered. The psychologist may choose to utilize other measures based on experience, keeping in mind that the instrument selected must be reliable and valid for the domains to be evaluated.

In addition to the discretion a psychologist exercises regarding selection of cognitive assessment instruments, it is important he or she utilize occupationally specific normative data to ensure effective interpretation of test results. In general, aviators are prescreened for this position resulting in the selection of individuals with cognitive aptitudes that are generally in the high average to very superior range of functioning. Scores that may be considered within normal limits for the general population may be well below normal and representative of significant weaknesses when compared with aviator specific normative data. Some assessment batteries include norms for aviation populations but this does not presume that the battery is better suited to assess the domains in question. The evaluator should select the battery or test that best assesses the issues in question.

It is also important to have some form of baseline testing as a comparison for an aviator's cognitive assessment scores. If no testing had been conducted previously, general intellectual functioning can be estimated from various demographic, academic, and achievement-oriented variables. General intellectual functioning can also be estimated from an applicants' scores on the Armed Services Vocational Aptitude Battery (Kratz, Poppen, & Burroughs, 2007; Orme, Brehm, & Ree, 2001).

Because military psychologists are called on to assess the stability of military personnel in high-risk jobs, they regularly assist flight surgeons in making recommendations to commanders about whether a pilot is aeromedically fit to continue his or her flying duties when there are concerns about the person's emotional or interpersonal disposition, namely, *stability*. For example, highly anxious, hostile, depressed, isolative, or impulsive persons are considered incompatible for the rigorous and inherently dangerous nature of military flying. Such traits can conceivably elevate the risk for an aviation mishap. If a military or civilian psychologist discovers an aviator or applicant is perceived to have problematic personality traits or behavioral patterns that interfere with flight safety, crew resource management, or ability to effectively perform aviation duties, then the psychologist can recommend to the pilot's commander administrative action that may involve restriction or removal from flying and aircrew duties in general. It is important to note that a diagnosed psychiatric personality disorder is not necessary. Rather, the reasonable perception that a pattern of behavior or specific traits interfere with occupational performance and adaptation is often enough to consider a person disqualified from aviation-related duties.

An extensive meta-analysis of the literature over the past 20 years regarding military aviator selection conducted by Paullin et al. (2006) reported that personality traits relevant to aviator performance include conscientiousness, integrity, achievement orientation, emotional stability, resilience, openness, self-confidence, self-esteem, and risk tolerance. Furthermore, a meta-analysis of personality data from assessment and selection programs of high-risk, high-operational military professions that included aviators reported that additional personality traits relevant to performance include initiative, motivation, drive, self-discipline, dependability, and cooperation

(Picano et al., 2006). Such traits are considered important to adapting to the rigors of highly demanding and dangerous conditions and job tasks. For instance, personality traits related to crew resource management may affect a pilot's performance differently in a multicrew aircraft than in a single- or two-seater jet aircraft. Regardless of one's view regarding the pattern of specific characteristics that constitute the right stuff, personality is considered to have a key role in succeeding as an aviator.

Particularly useful instruments in the assessment of personality include measures of pathology, such as the Minnesota Multiphasic Personality Inventory-Revised Clinical scales and the Personality Assessment Inventory, as well as measures of normal functioning (NEO Personality Inventory- 3rd Edition and the 16 Personality Factor test). Having a thorough assessment of an aviator's emotional and social disposition is key to understanding areas of strengths and weaknesses when determining a person's level of risk for adaptation difficulties. Such assessments should be included when evaluating aviators (or applicants) with a history of emotional (e.g., anxiety or depression) or behavioral difficulties (e.g., conduct related incidences), alcohol or substance use problems, and relational difficulties (e.g., partner relational problems, avoidant or dependent traits).

Assessment of problematic alcohol use and other substance use typically occurs by the service-specific alcohol treatment programs and most often relies on self-report questionnaires. However, if the psychologist is asked to evaluate an aviator or applicant with a question of problematic alcohol use or substance use, interpreting the results and making recommendations in the aviation context is vital. Although the *Diagnostic and Statistical Manual, 4th edition, Text Revision* (DSM-IV-TR) provides time-frame guidelines of 12 months for abuse and dependence, it is concerning when use of alcohol and other substances results in repeated incidents of domestic violence, fights with strangers, missing duty, disorderly conduct, carelessness with weapons, or hazardous use of heavy equipment including boats and automobiles while intoxicated, whether occurring within a 12-month time frame or a 60-month time frame. It is important to assess for underlying disorders when evaluating alcohol or other substance use. Posttraumatic stress, other anxiety disorders, or depression often underlie alcohol use disorders and need to be assessed in addition to the alcohol use.

Another assessment challenge in aviation is the evaluation of any subtype of attention-deficit hyperactivity disorder (ADHD). Assessment of ADHD in an adult can be challenging when relying on self-report questionnaires or when a history is spotty or nonexistent. Based on professional observation, the phenomenon of prescribing stimulants to children and adults for attentional difficulties without a thorough exploration of all symptoms and differential diagnoses has made subsequent assessment difficult when an individual decides to apply to flight school and current providers state that the previous diagnosis of ADHD was an error. Suddenly, the Adderall that was reported to be helpful in high school, college, or graduate school is now reported to never have been effective. Although the age of onset of symptoms is currently in debate, a retrospective assessment and query of the individual's history is important. If available, collateral information from parents may be obtained in addition to any prior documentation by teachers or medical or behavioral health providers. The importance of accurate diagnostic considerations of ADHD in aviators has been questioned by aviators and some medical professionals with the assumption that ADHD is synonymous with multitasking. Degree of symptomatology is also important, since mild symptomatology that is well managed may be waiverable. Additionally, when applicants with accurate ADHD diagnoses are taking prescribed stimulants and discover that the medication is disqualifying for aviation, they may discontinue the medication, forgetting that the underlying symptoms and diagnosis are the main disqualifying issue.

Finally, a challenge occurs after the assessment when making recommendations. There might be a question about applying results from an evaluation in the structured office to the aviator's performance in the aircraft. The psychologist might be concerned about negatively affecting an aviator's career or alienating the aviation community if results indicate that a return to flying duties would be risky at that time. Consultation with aeromedical psychologists and obtaining specific information about the exact nature of the flying duties and specific airframe from the flight surgeon, the aviator, or the commander can assist with making recommendations in the aeromedical decision-making process. Remembering that safety is paramount and having a frank discussion with the aviator often results in relieving any perceived dilemma in the psychologist's professional growth.

Although this chapter has referred to psychologists in general and not specified aeromedical psychologists, the aeromedically trained psychologist does have a broader perspective of the recommendations to be made based on knowledge and experience in the aviation environment. Most military treatment facilities require the psychologist to be credentialed in aeromedical psychology in order to conduct these evaluations or at least be supervised by an aeromedical psychologist. Aeromedical psychology training for uniformed and civilian Department of Defense psychologists currently occurs at Fort Rucker, Alabama, through the United States Army School of Aviation Medicine (USASAM). Information about this 3-week training can be found in Bowles (1994), on the Internet, or by contacting USASAM directly.

Psychological assessments of aviators present several challenges. First, psychologists must recognize the distinctive nature of aviators, flying duty, and the aviation environment. Then they must understand the assessment question being asked, which may necessitate a phone call to the referring flight surgeon or commander. This helps them select appropriate measures to assess the proper psychological domains for the question at hand. Finally, they can then make effective recommendations about aviators' continued flying or about flight applicants' pursuit of flying. Many psychologists have found that working with aviators and others in the aviation environment is professionally and personally challenging and rewarding.

References

Aeromedical Reference and Waiver Guide of the United States Navy Manual of the Medical Department, NAVMED P-117. (2005). Available at http://www.med.navy.mil/sites/nmotc/nami/arwg.aspx.

Air Force Instruction 48–123. (2011). Medical examinations and standards. Available at http://www.e-publishing.af.mil.

Army Aviation Aeromedical Policy Letters. (2008). Available at https://aamaweb.usaama.rucker.amedd.army.mil.

Bowles, S. (1994). Military aeromedical psychology training. *International Journal of Aviation Psychology, 4,* 167–172.

Bruskiewicz, K. T., Katz, L., Houston, J., Paullin, C., O'Shea, G., & Damos, D. (2007, February). *Predictor development and pilot testing of a prototype selection instrument for Army flight training* (Technical Report 1195). Arlington, VA: Army Research Institute for the Behavioral and Social Sciences.

Kratz, K., Poppen, B., & Burroughs, L. (2007). The estimated full scale intellectual abilities of U.S. Army aviators. *Aviation, Space, and Environmental Medicine, 78*(5), B261–B267.

Orme, D., Brehm, W., & Ree, M. (2001). Armed Forces Qualification Test as a measure of pre-morbid intelligence. *Military Psychology, 13,* 187–197.

Paullin, C., Katz, L., Bruskiewicz, K. T., Houston, J., & Damos, D. (2006, July). *Review of aviator selection* (Technical Report 1183). Arlington, VA: US Army Research Institute for the Behavioral and Social Sciences.

Picano, J. J., Williams, T. J., & Roland, R. R. (2006). Assessment and selection of high-risk operational personnel. In C. H. Kennedy & E. A. Zillmer (Eds.), *Military psychology: Clinical and operational applications* (pp. 353–370). New York, NY: Guilford Press.

Wiener, S. (2005). *Military flight aptitude tests* (6th ed.). Lawrenceville, NJ: Thompson Peterson.

10   MILITARY NEUROPSYCHOLOGY

Mark P. Kelly

ENTRY, CLINICAL TRAINING, AND SCOPE OF PRACTICE

Most military clinical psychologists enter active duty at the internship level. The curriculum across military psychology internships is not uniform, but typically some experience in clinical neuropsychology is offered. For example, Army and Navy interns at Walter Reed National Military Medical Center complete a 3-month rotation that includes didactic training in brain-behavior relationships, instruction in the neuropsychological interview and test administration, experience in performing full and screening neuropsychological assessments, and a subrotation involving assessment of patients with traumatic brain injury (TBI). The rotation is designed to prepare interns for their role as clinical psychologists so that they can conduct neuropsychological screening examinations and make appropriate referrals. Specialty training in clinical neuropsychology is available to active duty psychologists through 2-year postdoctoral fellowships in all three services. Navy and Air Force psychologists receive postdoctoral training in accredited civilian medical centers. In contrast, Army psychology offers clinical neuropsychology fellowship training through American Psychological Association accredited military programs at Brooke Army Medical Center, Tripler Army Medical Center, and Walter Reed National Military Medical Center. Information about military neuropsychology training programs can be found on the websites of the Association of Postdoctoral Programs in Clinical Neuropsychology, Association of Psychology Postdoctoral and Internship Centers, and parent institutions of individual programs.

Military clinical neuropsychology postdoctoral programs must prepare trainees in all aspects of general neuropsychological practice required in civilian settings (including, in some programs, such specialized procedures as Wada testing, cortical language mapping, and cognitive rehabilitation), as upon graduating they may be assigned to settings serving active duty service members, family members, and retired service members with a broad spectrum of neurological and psychiatric disorders. In addition, military neuropsychological training must also provide specific preparation to fulfill diverse military roles including fitness for duty assessments, battlefield neuropsychological evaluations, military aviation neuropsychology consultation, military sanitary board service, and military unique neuropsychological research. Upon completing fellowship training, active duty clinical neuropsychologists may serve in military medical centers or clinics, or apply their neuropsychological skill set to assignments in leadership and program development, telemedicine, special operations, military aviation psychology, and military forensic psychology.

TRAUMATIC BRAIN INJURY

The Department of Defense (DoD) (Department of Defense, 2007) defines TBI as:

a traumatically induced structural injury and/or physiological disruption of brain function as a result of an external force that is indicated by new onset or worsening of at least one of the following clinical signs, immediately following the event:

- Any period of loss of or a decreased level of consciousness;
- Any loss of memory for events immediately before or after the injury;
- Any alteration in mental state at the time of injury (confusion, disorientation, slowed thinking, etc.);
- Neurological deficits (weakness, loss of balance, change in vision, praxis, paresis/plegia, sensory loss, aphasia, etc.) that may or may not be transient;
- Intracranial lesion.

External forces may include any of the following events: the head being struck by an object, the head striking an object, the brain undergoing acceleration/deceleration movement without direct external trauma to the head, a foreign body penetrating the brain, forces generated from events such as blast or explosion, or other force yet to be defined. (p. 1)

DoD classifies TBI as mild, moderate, or severe based on duration of loss of consciousness (LOC), duration of alteration of consciousness (AOC), duration of posttraumatic amnesia (PTA), and findings from structural imaging (if available).

TBI commonly occurs in US military service members. The DoD TBI surveillance system, based on clinician confirmed TBI diagnoses, indicates that from 2000 to 2011 there were 233,425 service members who had sustained a TBI, with approximately 77% categorized as mild TBI or concussion (concussion defined by AOC lasting less than 24 hours; LOC of 30 minutes or less; PTA lasting less than 24 hours; and structural brain imaging [MRI or CT scan] yielding normal results) (Defense and Veterans Brain Injury Center, 2012). TBI has been characterized as the "signature injury" of the conflicts in Iraq and Afghanistan, with postdeployment surveys suggesting a history of TBI in approximately 15–23% of service members deployed to these conflicts, with most categorized as concussion. Recent epidemiological data (for reference see Chapter 5 of Kennedy & Moore, 2010) indicates that of cases of TBI with LOC in Iraq, 79% involved blast injuries. Much has yet to be learned about blast-related TBI sustained in current warfare. Available information indicates that while blast-related TBI may have different biomechanics, pathobiology, and patterns of associated injury to other organs than non-blast TBI, the few neuropsychological studies to date have not detected differences between blast- and non-blast-induced TBI (for references see Kelly, Coldren, Parish, Dretsch, & Russell, 2012).

CONCUSSION ASSESSMENT IN THE COMBAT ENVIRONMENT

Sequelae of concussion include impaired cognition, poor balance, and subjective postconcussive symptoms such as headache and light sensitivity. Studies from sport neuropsychology reveal that athletes experiencing objective cognitive impairment and subjective postconcussive complaints immediately after a single concussion typically recover in 1 to 2 weeks. Prompt identification of concussion in warriors is essential for appropriate medical management and prevention of premature return to combat duty that may put the service member, his unit, and the mission at risk. Evidence indicates there is a period of increased vulnerability for repeat concussion within 10 days of initial injury. Individuals with multiple concussions may have a prolonged recovery and be at risk for second-impact syndrome, a rare disorder resulting in severe neurological disability. Reliance solely on self-report of subjective symptoms in determining return to high risk activity is inappropriate in athletes and warriors, both of whom may be highly motivated to return to their team or unit. Neuropsychological measures have been successfully used in sport medicine to objectively identify initial effects of concussion, track recovery, and assist medical personnel in return-to-play decision making. Recent research has provided initial evidence that a computerized neuropsychological test battery, the Automated Neuropsychological Assessment Metrics (ANAM), could be successfully used to detect the early effects of concussion (including blast-related concussion) (Kelly et al., 2012) and track recovery (Coldren, Russell, Parish, Dretsch, & Kelly, 2012) in the combat zone.

Recognizing the high prevalence of concussion in the Iraq and Afghanistan conflicts, the challenges inherent in diagnosing concussion due to the sometimes subtle signs/symptoms, and the need to assess warriors in the combat zone and close to the time of injury, DoD developed three algorithms for in-theater assessment and treatment (Department of Defense, 2010): a "Combat/Medic Corpsman Concussion Triage" algorithm for use when no medical officer is available, an "Initial Provider Management of Concussion in Deployed Setting" algorithm developed for settings with physicians available, and a "Comprehensive Concussion Evaluation" algorithm for the most comprehensively staffed and equipped facilities available on the battlefield.

All three algorithms make use of the Military Acute Concussion Evaluation (MACE), an instrument adapted from the Standardized Assessment of Concussion, a brief cognitive assessment intended for use in sideline assessment in sports medicine. The MACE includes documentation of acute injury characteristics and symptoms, a focused neurological examination, and brief cognitive examination. Deployed psychologists should be aware that the MACE cognitive examination lacks sensitivity when used more than 12 hours following a concussion, and more comprehensive measures (such as the ANAM) are needed (Coldren, Kelly, Parish, Dretsch, & Russell, 2010). Formal assessment with a neurocognitive assessment tool (NCAT) is stipulated in the "Comprehensive Concussion Evaluation" algorithm, with ANAM as the currently used NCAT. In 2008 DoD mandated that all deployers undergo predeployment baseline ANAM testing, to allow comparison to predeployment testing in concussed warriors and increasing diagnostic accuracy (Kelly et al., 2012). The algorithms provide guidance for management, prescribe rest periods and duty limitations, and recommend patient education as a core treatment component. DoD also provides specialized guidance for care of warriors who have experienced recurrent concussions (i.e., three documented concussions within 12 months), with the mandatory evaluation including a 4-hour neuropsychological assessment including a formal measure of effort and evaluation of attention, memory, processing speed, executive function, and social pragmatics prior to return to duty (Department of Defense, 2010).

FITNESS FOR DUTY: MILITARY MEDICAL FITNESS

DoD policy mandates that a service member will be found unfit for duty if there is a disease or injury preventing performance of duties associated with his or her office, grade, rank, or rating (Department of Defense, 1996). To be found fit, service members must be physically, cognitively, and emotionally able to perform the essential functions of their job effectively in any locale. Medical fitness for duty is a sequential, three-step process:

- Physical Profiles: Completed by providers when an Army or Air Force service member's medical/psychiatric condition impacts job performance. Profiles rate each of six major body systems (including psychiatric) on a four-point scale from 1 (*high level of fitness*) to 4 (*drastic duty limitation*). The Navy and Marine Corps have an equivalent system known as a Limited Duty Board.
- Medical Evaluation Board (MEB): If a service member has a medical or psychiatric condition that, by the explicit service-specific standards, *may* render them unfit, their treating provider or command refers them to an MEB to evaluate the condition and determine if it is severe enough to call into question the ability to continue on active duty according to established retention standards.
- Physical Evaluation Board (PEB): If the MEB determines the service member does not meet retention standards, the case is referred to the PEB (typically composed of a physician officer, a personnel management officer, and a presiding officer) for final determination of medical fitness for duty.

While military neuropsychologists may be involved in writing profiles, their major role in the medical fitness for duty process is likely to be at the MEB level. A core component of the MEB process is the narrative summary (NARSUM) prepared by the provider leading the MEB that details the history of the illness, findings from examinations and laboratory/radiology results, consultant reports, diagnosis, response to treatment, and rationale for conclusions. Depending on the service and the condition, the neuropsychological report may serve as the NARSUM or as an addendum to the NARSUM along with other consultant reports. There are several key points critical to conducting an adequate neuropsychological examination related to fitness of duty:

- All neuropsychological fitness for duty examinations should be comprehensive. Examinations should include a thorough interview, medical record review, and review of Armed Services Vocational Aptitude Battery (ASVAB) scores if available for determination of premorbid cognitive ability;
- Fitness for duty determinations also include interviews with collateral sources (e.g., commanders, family members), and review of officer or enlisted evaluation reports and service records. Depending on the composition of the MEB and referral issue, responsibility for gathering this data may or may not fall to the neuropsychologist;
- Formal assessment of effort and motivation to participate in the examination should be included in all fitness for duty assessments (see Chapter 4 of Kennedy & Moore, 2010);
- Neuropsychological batteries should include assessment of major cognitive domains including intellectual function, verbal and nonverbal learning/memory, language, spatial abilities, attention, executive function, academic skills, sensory/motor skills, and assessment of psychopathology. Regulations do not usually stipulate specific tests;
- The neuropsychological report should include: Chief Complaint/Reason for Referral; History of Present Illness; Past Medical History; Past Psychiatric History; Social History (including educational, occupational, marital, military, and legal history); Family Medical History; Laboratory and Imaging Results; Medications; Mental Status Examination; Tests Administered (and normative system employed); Test Results; Conclusions; Implications of Findings for Day to Day Functioning; Diagnosis; Recommendations;

- Fitness for duty determination is a medicolegal process. The written report normally serves as the sole source of neuropsychological evidence. PEB members adjudicating the case may not have extensive familiarity with neuropsychological assessment, so reports should be written in clear language understandable to a layperson and detail the functional implications of test results;
- Service members performing special duties including aviation, submarine duty, and Special Forces must meet standards beyond the prerequisite general military fitness standards;
- Neuropsychologists should interface with their PEB to determine if specific report components, recommendations, or language are needed for adjudication.

FACTORS TO CONSIDER IN MILITARY NEUROPSYCHOLOGICAL EVALUATION

When conducting examinations in a military setting, neuropsychologists must be keenly aware of several factors that may significantly impact neuropsychological performance. While these issues are not unique to military neuropsychology, they are commonplace in service members with both acute and long-standing battlefield injuries and often adversely affect cognition:

- Sleep deprivation is common during sustained combat operations (see Chapter 11 of Kennedy & Moore, 2010) and in those medically evacuated from distant locations;
- PTSD and other psychiatric disorders are also common in combat veterans and are frequently comorbid with TBI (see Chapter 12 of Kennedy & Moore, 2010);
- Pain—headaches are among the most common symptoms of concussion, and pain is a significant clinical issue in warriors who have sustained serious traumatic injuries including amputation injuries;
- Medications are used to treat a host of problems including sleep difficulties, PTSD and other psychiatric disorders, pain, spasticity, and seizures and may have significant adverse cognitive side effects;
- Because military neuropsychologists frequently evaluate service members with potential for secondary gain (e.g., removal from hazardous duty, monetary compensation for disability), or who may not exert full effort during an evaluation due to factors such as sleep deprivation, pain, or psychiatric illness, all evaluations should include explicit assessment of effort.

Disclaimer

Views expressed in this chapter are those of the author and do not necessarily reflect official policy or position of the Department of the Army, Department of the Navy, Department of Defense, or the United States government.

References

Coldren, R. L., Kelly, M. P., Parish, R. V., Dretsch, M. N., & Russell, M. L. (2010). Evaluation of the Military Acute Concussion Evaluation for use in combat operations more than 12 hours after injury. *Military Medicine, 175*(7), 477–481.

Coldren, R. L., Russell, M. L., Parish, R. V., Dretsch, M. N., & Kelly, M. P. (2012). The ANAM lacks utility as a diagnostic or screening tool for concussion more than 10 days following injury. *Military Medicine, 177*(2), 179–183.

Defense and Veterans Brain Injury Center. (2012). *DoD worldwide numbers for traumatic brain injury.* Retrieved from DVBIC.org: http://www.dvbic.org/TBI-Numbers.aspx

Department of Defense. (2007, October 1). Health affairs memorandum 07–030. In *Traumatic brain injury: Definition and reporting, 2007.* Retrieved from http://www.health.mil/about_mhs/HA_policies_guidelines.aspx?policyyear=2007

Department of Defense. (2010, June 21, incorporating change 4, 2011, November 7). *Policy guidance for management of concussion/mild traumatic brain injury in the deployed setting.* Retrieved from http://www.dtic.mil/whs/directives

Department of Defense. (1996). *Separation or retirement for physical disability* (Department of Defense Directive 1332.18). Washington, DC: Author.

Kelly, M., Coldren, R., Parish, R., Dretsch, M., & Russell, M. (2012). Assessment of acute concussion in the combat environment. *Archives of Clinical Neuropsychology, 27,* 375–388. doi:10.1093/arclin/acs036

Kennedy, C. H., & Moore, J. L. (Eds.). (2010). *Military neuropsychology.* New York, NY: Springer.

11 COMBAT OPERATIONAL STRESS AND BEHAVIORAL HEALTH

Mark C. Russell and Charles R. Figley

MILITARY OPERATIONAL STRESSORS: TYPES AND DEFINITIONS

Occupational hazards of military service routinely involve exposure to a plethora of chronic, inescapable, and uncontrollable stressors as well as potentially traumatic events. The length, intensity, and frequency of exposure to chronic, and/or traumatic stressors (e.g., multiple redeployment to war zones) has led to increasing frequency of behavioral health challenges, despite recent efforts to prevent and treat conditions like PTSD.

Acute and chronic breakdown will inevitably occur when the human resistance threshold is exceeded by duration, intensity, and nature of the cumulative, interrelated effects of (1) deployment-related stressors (e.g., prolonged family separation, chronic boredom, climate exposure, excessive noise, disruption in stress-buffers, sexual harassment, dietary change, sleep deprivation); (2) war-related stressors from exposure to persistent, multiple, visible, and unpredictable threats (e.g., ambush, chemo-bio weapons, mines, IEDS, torpedoes, mortars, long-range missiles, indistinguishable enemy), devastation and injury (i.e., high explosive munitions, armored vehicles, automatic weapons), and comparative lack of safety or controllability (e.g., armor piercing munitions, long-range weapons, real-time surveillance and communications, "bunker busters," night vision, precision-guided weapons,

guerrilla "swarming" tactics); and (3) exposure to combat-related stressors (i.e., killing, being wounded, a buddy killed, "collateral damage," survivor guilt, POW, death of children, handling human remains); as well as (4) other potentially traumatic stressors (e.g., disaster relief, body recovery, witnessing war atrocities, military training accidents, interpersonal violence, and military sexual trauma)—potentially resulting in long-term health problems. The cumulative effects of stressors related to military service, especially during times of war, offers an abundantly toxic environmental context for the full spectrum of war-and-traumatic-related stress injuries (e.g., Figley & Nash, 2007).

COMBAT OCCUPATIONAL STRESS REACTION

Currently, the preferred term applied to any stress reaction in the military environment is "Combat Operational Stress Reaction" (COSR), referring to the adverse reactions military personnel may experience when exposed to combat, deployment-related stress, or other operational stressors. COSR replaces earlier terminology, like "battle fatigue" or "combat exhaustion," used to normalize "acute stress responses" (ASR) related to deployment and war-zone stressors and acute "combat stress reactions" (CSR) associated with exposure to combat. Many reactions look like symptoms of mental illness (i.e., panic, depression, hallucinations), but are only transient reactions to the traumatic stress of combat and the cumulative stresses of military operations. There is a combined and cumulative effect of combat and operational stressors that result in COSR (Veteran's Affairs & Department of Defense [VA/DoD], 2010).

Common COSR Symptoms

Symptoms of COSR may include depression, fatigue, anxiety, decreased concentration/memory, irritability, agitation, and exaggerated startle response. Table 11.1 provides a partial list of signs and symptoms following exposure to COSR including potentially traumatic events:

Spiritual or Moral Symptoms

Service members may experience any of the following acute or chronic spiritual symptoms: (1) feelings of despair, (2) questioning of old religious or spiritual beliefs, (3) withdrawal from spiritual practice and spiritual community, and (4) foreshortened future.

Combat and Operational Stress Behavior

Combat and Operational Stress Behavior is the military terminology used to describe the full spectrum of COSRs ranging from "adaptive" to "maladaptive" behaviors that military personnel may demonstrate when exposed to combat and operational stressors, including potentially traumatic events, throughout their military career. The US military views transient COSR as "universal" responses of human beings adapting to acute, combat stressors, and not signs of psychopathology. Differences in severity, type, and length of COSR is highly individualized and determined by a wide range of risk and protective factors.

BEHAVIORAL HEALTH OUTCOMES: TYPES AND DEFINITIONS

Nearly every written account of war and combat stress, regardless of time, culture, or national origin, describes a wide range of stress-related injuries that can best be divided (albeit artificially) into two major classifications: "neuropsychiatric" (e.g., accepted psychiatric diagnoses of the time) and "medically unexplained conditions," often called "war syndromes," "psychosomatic illness," or "hysteria," that are physical conditions without a known neurological or medical etiology, lumped today into the Veterans' Administration category of "Symptoms, Signs and Ill-defined Conditions (SSID)."

TABLE 11.1. Signs and Symptoms Associated with COSR

Physical	Cognitive/Mental	Emotional	Behavioral
• Chills	• Blaming someone	• Agitation	• Increased alcohol
• Difficulty breathing	• Change in alertness	• Anxiety	consumption
• Dizziness	• Confusion	• Apprehension	• Antisocial acts
• Elevated blood pressure	• Hypervigilance	• Denial	• Change in activity
• Fainting	• Increased or decreased	• Depression	• Change in communication
• Fatigue	awareness of surroundings	• Emotional shock	• Change in sexual
• Grinding teeth	• Intrusive images	• Fear	functioning
• Headaches	• Memory problems	• Feeling overwhelmed	• Change in speech pattern
• Muscle tremors	• Nightmares	• Grief	• Emotional outbursts
• Nausea	• Poor abstract thinking	• Guilt	• Inability to rest
• Pain	• Poor attention	• Inappropriate emotional	• Change in appetite
• Profuse sweating	• Poor concentration	response	• Pacing
• Rapid heart rate	• Poor decision making	• Irritability	• Startle reflex intensified
• Twitches	• Poor problem solving	• Loss of emotional control	• Suspiciousness
• Weakness			• Social withdrawal

Acute Stress Disorder (ASD)

When COSR is associated with traumatic operational or combat stressors, symptoms may involve reexperiencing (i.e., intrusive recollections, nightmares, flashbacks), hyperarousal (i.e., insomnia, exaggerated startle, hypervigilance, irritability), avoidance (i.e., avoiding reminders, restricted range of affect, withdrawal), and dissociation (i.e., emotional numbing, detachment, alexithymia) resulting in clinically significant distress or impairment more than days but less than one month after exposure to a trauma, this may result in a diagnosis of Acute Stress Disorder (ASD; American Psychiatric Association, 2000).

Posttraumatic Stress Disorder (PTSD)

Without effective intervention approximately 70–80% of ASD cases will continue beyond 30 days and develop into acute or chronic PTSD—often related to level of dissociation at time of the event (peritraumatic) and/or cumulative effects of stressors.

Traumatic Grief Reaction

Traumatic grief is generally defined as the abrupt, sudden loss of a significant and close attachment. The intensity of the social bonds that develop between Bands of Brothers and Sisters at war—strongly reinforced through mutual trust, respect, and admiration that have been steeled by the fire of war stress—has been described to rival only that of a mother and child. Conversely, when intimate social ties are abruptly severed, the grief can be as intense as any known for human beings. Symptoms and signs will vary, but may include: (1) reacting with rage, hostility, and/or violence toward the enemy or one's own; (2) risking their lives, "going berserk" or "kill crazy"; (3) avoidance of any new attachments; (4) survivor guilt; (5) suicidal ideation or attempt; (6) social withdrawal; (7) persistent agitation; and (8) numbing against emotions.

Misconduct Stress Behaviors

Misconduct stress behaviors describe a range of maladaptive stress reactions present in any armed conflict from minor to serious violations of military or civilian law and the Law of Land Warfare, most often occurring in poorly trained personnel, but "good and heroic, under extreme stress may also engage in misconduct" (Department of the Army, 2006, pp. 1–6). Examples include: mutilating enemy dead, not taking prisoners, looting, rape, brutality, killing animals, self-inflicted wounds, "fragging," desertion, torture, and intentionally killing noncombatants or other war atrocities.

Medically Unexplained Conditions

Common inexplicable physical symptoms include chronic fatigue, muscle weakness, chronic pain, sleep disturbances, headache, pseudo-seizures, chronic constipation/diarrhea, gait disturbance, pseudo-paralyses, nausea/gastrointestinal distress, shortness of breath, pelvic pain, dysmenorrhea, paraesthesias, fainting, sensory loss, dizziness, rapid or irregular heartbeat, skin rashes, persistent cough and tremors, shaking or trembling.

Spectrum of War and Traumatic Stress Injuries and Comorbidity

The spectrum of war and traumatic stress injuries like combat-related PTSD is evident, with reports of 50–80% of clients diagnosed with a "comorbid" conditions (VA/DoD, 2010). Comorbid medical, medically unexplained, and neuropsychiatric conditions are important to recognize and differentiate because they can modify clinical determinations of prognosis, treatment priorities, selection of interventions, and the setting where care may be provided. Psychotherapists are advised that military personnel will often have one or more coexisting mental health disorders such as phobias, generalized anxiety disorder, depression, substance abuse, insomnia, bereavement, psychosis, seizure disorder, TBI, anger/agitation, guilt, and multiple medically unexplained conditions (i.e., headaches, chronic fatigue, and noncardiac chest pain). Military clients with co-occurring disorders, such as depression and alcohol abuse or depression and PTSD, are at much greater risk for suicide and interpersonal violence than clients with only one type of war stress injury.

The literature on anger and aggression within the military population, particularly among combat veterans diagnosed with war stress injury like PTSD, reveals that they are in a high-risk group for excessive anger, aggression, interpersonal violence, and other misconduct stress behaviors, warranting routine screening and early intervention. Depression is also a very common comorbid condition with a variety of war stress injuries including PTSD, anxiety disorders, substance use disorder, traumatic grief reactions, and so forth (VA/DoD, 2010). All clients with traumatic stress injury including subclinical PTSD, should be assessed for safety and dangerousness, including current risk to self or others, as well as historical patterns of risk.

PRIDE, SATISFACTION, GROWTH, AND TRANSFORMATIVE EXPERIENCES

Resilience in the Military

Discussions of military and war-related stressors are often unfairly slanted toward the negative, aversive, and horrific aspects of going to war or a disaster zone, and hopelessly fail to recognize many positive or adaptive outcomes. The term "adaptive stress reactions," refers to positive responses to COSRs that enhance individual and unit performance whereas "posttraumatic growth" refers to positive changes that occur as a result of exposure to stressful and traumatic experiences such as:

- Forming of close, loyal social ties or camaraderie never likely repeated in life (i.e., "band of brothers" and "band of sisters")
- Improved appreciation of life
- Deep sense of pride (e.g., taking part in history making)
- Enhanced sense of unit cohesion, morale, and *esprit de corps*
- Sense of eliteness
- Existential purpose and altruism from helping others (i.e., liberation)
- Improved tolerance to hardship and pain
- Increased faith or spiritual awakening

- Heroic acts of courage and self-sacrifice
- Profound satisfaction from personal growth, sacrifice, and mastery from accomplishing one's mission under the most arduous circumstances.

Military Resilience and Posttraumatic Growth

In 2010, the US Army reported that 18.9% of deployed soldiers reported "high or very high" individual morale, 14.5% reported "high/very high" unit morale, 74.6% reported marital satisfaction, 71.9% reported high unit cohesion, 79.8% perceived their unit as well trained and combat ready, 45.4% expressed satisfaction with NCO leadership, and 49.2% were satisfied with officer leadership. Of deployed Marines, 24.9% reported they have learned to handle stress better because of the deployment, 63.2% reported greater self-confidence as a result of their deployment, and 50.3% reported feeling proud of their accomplishments during the deployment (Joint Mental Health Advisory Team 7, 2010).

RECOMMENDED BEHAVIORAL HEALTH INTERVENTIONS

Expert consensus highlights the critical importance of early identification and intervention of the spectrum of traumatic stress injuries in order to prevent escalation and long-term suffering and disability. The following are the PTSD treatment recommendations as cited from the October, 2010, VA/DoD *Clinical Practice Guideline for the Management of Posttraumatic Stress:*

1. Offer patients with PTSD one of the evidence-based trauma-focused psychotherapeutic interventions that include components of exposure and/or cognitive restructuring; or stress inoculation training.
2. Select a treatment approach based on the severity of the symptoms, clinician expertise, patient preference, and may include an exposure-based therapy (e.g., prolonged exposure), a cognitive-based therapy (e.g.,

cognitive processing therapy), stress management therapy (e.g., SIT), eye movement desensitization and reprocessing (EMDR), or another of equal or better effectiveness.

3. Select an effective set of relaxation (i.e., self-soothing) techniques that help clients during and between sessions in alleviating symptoms associated with physiological hyperreactivity.

4. Use imagery rehearsal therapy for treating nightmares and sleep disruption.

5. Use a combination of approaches until it works perfectly with the client and avoid trying only one approach.

6. Treatment plans should be comprehensive and individualized for military clients and their families. See Russell, Lipke, and Figley (2011) for a more detailed guide to treating combat stress injuries and associated disorders with military personnel.

References

American Psychiatric Association. (2000). Diagnostic and statistical manual of mental disorders (4th ed., text revision), Fourth Edition, Text Revision. Washington, DC: Author.

Department of the Army. (2006). *Combat and operational stress control: Field manual 4–02.51 (FM 8–51)*. Washington, DC: Headquarters, Department of the Army.

Department of Veteran's Affairs & Department of Defense. (2010). *VA/DoD clinical practice guideline for the management of post-traumatic stress* (Office of Quality and Performance publication 10Q-CPG/PTSD-10). Washington, DC: Author.

Figley, C. R., & Nash, W. P. (Eds.). (2007). *Combat stress injury: Theory, research, and management*. New York, NY: Routledge.

Joint Mental Health Advisory Team 7 (J-MHAT 7) Operation Enduring Freedom 2010 Afghanistan. (February 22, 2011). Office of the Surgeon General United States Army Medical Command; Office of the Command Surgeon HQ, USCENTOM & Office of the Command Surgeon U.S. Forces Afghanistan (USFOR-A).

Russell, M. C., Lipke, H. E., & Figley, C. R. (2011). EMDR Therapy. In B. A. Moore & W. A. Penk (Eds.), *Handbook for the treatment of PTSD in military personnel*. New York, NY: Guilford Press.

12 FORENSIC PSYCHOLOGY IN THE MILITARY SETTING

Paul Montalbano and Michael G. Sweda

DEFINITION OF FORENSIC PSYCHOLOGY

Forensic psychology refers to the application of psychological principles to legal issues. The nature of practice defines whether one is engaged in forensic psychology, not one's training or background. The *Specialty Guidelines for Forensic Psychologists* (American Psychology Law Society [APLS], 2011) state that:

forensic psychology refers to professional practice by any psychologist working within any sub-discipline of psychology (e.g., clinical, developmental, social, cognitive) when applying the scientific, technical, or specialized knowledge of psychology to the law to assist in addressing legal, contractual, and administrative matters. (p. 1)

ETHICAL CONSIDERATIONS IN MILITARY FORENSIC PRACTICE

Forensic practice in the military involves a different set of legal rules and procedural practices from those in state or federal jurisdictions. Without knowledge of the legal parameters operating in the military, a forensic clinician may provide a service that is at best ineffective, and at worst may violate the legal rights of the parties involved. In addition, the US Military embodies distinct cultures and subcultures varying by branch of service and whether one is dealing with officers or enlisted personnel. Accordingly, psychologists considering forensic practice within the military should consider whether they have adequate education, training, or experience before agreeing to take a military forensic case, or whether supervision or consultation is necessary to competently deliver the requested forensic service. In particular, psychologists should be mindful of the American Psychological Association (APA) Ethics Code (2010) Standard 2.01 Boundaries of Competence, concerning provision of services to populations they are able to competently serve; Standard 2.01(f), Competence, recommending knowledge of relevant laws and rules; and Standard 9.06, Interpreting Assessment Results, concerning important situational, personal, linguistic and cultural differences that may affect interpretation of assessment results. Forensic psychologists should be aware of the many situational factors affecting military evaluees, such as: frequent change of residence, 24-hour availability, responsiveness to a hierarchical command structure, deployment to war zones, and exposure to life threatening situations in war.

TRAINING AND PRIVILEGING OF FORENSIC PSYCHOLOGISTS IN THE US MILITARY

Training in forensic psychology has been available through the Walter Reed National Military Medical Center's (WRNMMC) Postdoctoral Fellowship Training Program in Forensic Psychology since 2007. The WRNMMC Forensic Fellowship is the only 2-year training program in forensic psychology in the United States and limits its training to Active Duty psychologists. The WRNMMC Forensic Psychology Fellowship is the first program to gain accreditation through the American Psychological Association as a postdoctoral training program in forensic psychology. The Fellowship provides didactic training and supervision covering a full range of forensic evaluations. WRNMMC Fellows have testified in courts-martial around the globe. Psychologists can obtain forensic psychology privileges at WRNMMC if they meet one of the following criteria: possess the ABFP diplomate; have completed a 1-year postdoctoral forensic training program; or have had 2 years of forensic training and supervised forensic experience. In June 2012, the Army approved an official Forensic Behavioral Science Skill Identifier for officers who have completed 1 year of postdoctoral residency training in forensic psychology or psychiatry, or who can demonstrate knowledge and proficiency in the application of forensic behavioral science to military justice issues.

A COMPARISON OF THE MILITARY AND CIVILIAN LEGAL SYSTEMS

In general, there are far more similarities than there are differences between military and civilian legal systems. For example, the Military Rules of Evidence (MRE) generally mirror the Federal Rules of Evidence (FRE). However, knowledge of differences is crucial in performing competent forensic psychological services in the military environment. Military criminal law is codified through the Uniform Code of Military Justice (UCMJ), MRE, and *Manual for Courts-Martial* (MCM; Joint Service Committee on Military Justice, 2012), and supplemented by the *Military Judges' Benchbook* (2010). The MCM contains Rules for Courts-Martial (Part II of the MCM), the MRE (Part III), Punitive Articles (Part IV), Nonjudicial Punishment Procedure (Part V), and the UCMJ (Appendix 2). The *Military Judges Benchbook* is published as a separate Department of the Army pamphlet (DA 27–9). The Benchbook provides suggested instructions for military juries (known as panels) and procedures for trials by court-martial. The military is governed by its own appellate courts under the authority of the US Supreme Court.

The US Military does not have standing trial courts. A court-martial is therefore assembled, or convened, by a Convening Authority (CA). The CA is a commissioned officer in command. The CA initiates a court-martial by issuing an order and designating the type of court-martial (summary, special, or general) that will try the charges. More serious charges, roughly commensurate with felonies, are tried by general courts-martial, less serious offenses via special courts-martial, and the least serious offenses via summary courts-martial. Before a case goes to a general court-martial, a pretrial investigation under Article 32 of the UCMJ is conducted. This process is generally equivalent to a grand jury. Upon completion of the Article 32 hearing, the investigating officer makes findings and recommendations for referral of charges to the CA, who makes the final decision about what charges will be tried. Similar to civilian courts, the accused may request a trial by a panel (equivalent to a jury) or a military judge alone. A military panel for a general court-martial is composed of 5 to 12 members.

The UCMJ lists crimes that would not be classified as such in the civilian sector. For example, Malingering (Article 115) is a military-specific crime. The forensic evaluator should bear in mind that malingering, as defined by the UCMJ, involves intent to avoid work, duty, or service and is not the same as the DSM-IV-TR (American Psychiatric Association, 2000) diagnosis of malingering.

In contrast to the unanimity required by civilian courts, an accused will be found guilty in any military noncapital case if at least two-thirds of the members vote for a finding of guilt. A vote of less than two-thirds results in acquittal. There are no "hung juries" in a court-martial. The panel members have the responsibility of sentencing the accused. At sentencing a wide range of punishments is available that are unique to the Armed Forces. Hard labor may be imposed as well as reduction in rank, forfeiture of pay and allowances, and a bad conduct discharge (BCD). The latter usually results in the loss of retirement and health benefits. This is often a consideration when a service member is offered a pretrial agreement that includes a BCD. After a guilty finding, certain cases in the US Military (e.g., death penalty, dishonorable discharge, confinement for more than 1 year) are automatically eligible for review by the appropriate military criminal appeals court. In contrast to civilian appellate courts, which only review for legal errors made at trial, the military appellate court reviews for legal error, factual sufficiency of evidence supporting a conviction, and appropriateness of the sentence.

If an accused agrees to enter a guilty plea of any type, they will undergo a detailed "Care inquiry" in court, the purpose of which is to demonstrate that the accused is making a knowing, intelligent, and conscious waiver of their rights (*United States v. Care*, 1969). In some cases, an accused may agree to plead guilty after entering into a pretrial agreement, or plea bargain, with the prosecution that sets a maximum cap on time in confinement.

FREQUENTLY PERFORMED EVALUATIONS

The forensic professional working within the military justice system will address a wide range of issues, including competency to stand trial (CST), criminal responsibility, false confessions, Miranda (Article 31b) waivers, psychological autopsies, evaluation of capacity to form specific intent, violence risk assessments, and evaluations for purposes of sentencing, including death penalty evaluations. As in other settings, competency to stand trial is the most frequent issue addressed by forensic practitioners. This chapter will discuss performing CST, criminal responsibility, and sentencing evaluations.

706 Evaluations

The 706 Inquiry or Sanity Board derives its name from Rule 706 of the Rules for

Courts-Martial (MCM, 2008). In a 706 examination the issues of CST and criminal responsibility at the time of the alleged offense are both addressed through a sanity board inquiry. There is a relatively low bar for ordering pretrial 706 evaluations, which can be summarized as a reasonable concern about the accused's mental state such that it is affecting either CST or responsibility at the time of the offense. The *Military Judges' Benchbook* states that "a good faith non-frivolous request for a sanity board should be granted" (p. 932). The 706 Inquiry is a compelled examination, and failure to comply can result in the exclusion of defense expert evidence. RCM 706 states that the Board consists of "one or more persons" and that "each member of the board shall be either a physician or a clinical psychologist" (p. II-70). One report is generated which is signed by all the participants.

The sanity inquiry typically requires the examiner to "make separate and distinct findings" with regard to four different questions, listed below (p. II-70). As can be seen from the questions, the 706 evaluation demands a complex and wide-ranging inquiry from the examiner that assesses both current mental state and mental state at the time of the alleged offense.

(A) At the time of the alleged criminal conduct, did the accused have a severe mental disease or defect?
(B) What is the clinical psychiatric diagnosis?
(C) Was the accused, at the time of the alleged criminal conduct and as a result of such severe mental disease or defect, unable to appreciate the nature and quality or wrongfulness of his or her conduct?
(D) Is the accused presently suffering from a mental disease or defect rendering the accused unable to understand the nature of the proceedings against the accused or to conduct or cooperate intelligently in the defense?

Forensic evaluations performed in a military setting should adhere to the principles of practice of forensic mental health assessment (FMHA), which guide the preparation, data collection, data interpretation, and communication of the results (Heilbrun et al., 2007). A critical component of FMHA is to use multiple sources of information and to seek convergent validity for the conclusions reached. A clinical interview and psychological testing (when relevant) should be augmented by third-party information from documents and interviews with collateral informants. An assessment of the response style of the accused is often an important component of the evaluation. Detailed reviews of best practices in performing evaluations for CST can be found in Zapf and Roesch (2009) and for criminal responsibility in Packer (2009).

As a safeguard to protect the Fifth Amendment rights against self-incrimination, the examiner is required to generate two reports, often referred to as the "short form" and the "long form" or full report. According to RCM 706 the short form contains "a statement consisting only of the board's ultimate conclusions as to all questions specified in the order" (p. II-70). The short form is submitted to both Trial and Defense Counsel. The full report is sent only to Defense Counsel.

The military standard for CST is analogous to the *Dusky* standard for CST. The *Dusky* standard is "whether he [the defendant] has *sufficient present ability* to consult with his attorney with a *reasonable* degree of rational understanding *and* a rational as well as factual understanding of the proceedings against him" (*Dusky v. United States*, 1960) [italics added]. In *United States v. Proctor* (1993) the Court of Military Appeals upheld the applicability of the *Dusky* standard for military courts. Forensic assessment instruments (FAIs) are specialized instruments designed to assess psycholegal capacities. Several well-developed and widely accepted FAIs are available to assist in assessing CST. Evaluators should keep in mind that these instruments were not specifically developed for use with a military population or for the military legal system.

The current military standard for mental responsibility is codified in Article 50a of UCMJ. The standard is substantively identical to the Federal Statute. According to RCM

916(k) the sanity standard in the military for lack of mental responsibility is as follows:

It is an affirmative defense to any offense that, at the time of the commission of the acts constituting the offense, the accused, as a result of a severe mental disease or defect, was unable to appreciate the nature and quality or the wrongfulness of his or her acts. Mental disease or defect does not otherwise constitute a defense. (p. II-112)

This formulation focuses on the cognitive capacity of the defendant to understand what one is doing at a given point in the past and to grasp that it is wrong.

When notifying the accused of the purpose of the evaluation, the evaluator should bear in mind that the answers to the four questions posed will be sent to Trial Counsel (government counsel) and Defense Counsel. Since diagnoses are listed, this may have import at sentencing. For example, it may be that Defense Counsel utilizes a diagnosis of PTSD to mitigate at sentencing; whereas Trial Counsel may utilize a diagnosis of pedophilia to aggravate at sentencing. In addition, the diagnosis of a mental disorder may have ramifications for a continued military career. If viewed as suffering from a mental disease, the service member may be administratively separated. In our view, the service member should be informed of such potential consequences up front before initiation of the examination.

EXPERT WITNESS TESTIMONY

The purpose of performing a forensic evaluation is to provide input during trial on a specific legal issue. This input may be in the form of consultation, a report and/or testimony. MRE 702 makes clear that psychological expertise must utilize appropriate scientifically based methods and principles. Bear in mind that military courts like federal courts follow *Daubert v. Merrell Dow* (1993) for standards regarding admissibility of expert witness testimony. *Daubert* emphasizes that the expertise must adhere to scientific principles.

When functioning as an expert witness during a court-martial, the expert should be aware of some of the unique aspects of military justice system. Prior to testifying, the expert should be prepared to be questioned by opposing counsel about their methods and conclusions before trial or during trial before testifying. This process is somewhat analogous to depositions in civil cases. There is also no prohibition against offering an opinion on the ultimate issue in military court. The ultimate issue is often conceptualized as the final opinion regarding mental state of a defendant in relation to a specific legal issue. With respect to expert testimony, FRE 704(b) prohibits opining on the ultimate issue, while MRE 704 states that "testimony in the form of an opinion or inference otherwise admissible is not objectionable because it embraces an ultimate issue to be decided by the trier of fact" (p. III-39).

EVALUATIONS FOR SENTENCING

In the US Military, sentencing generally follows directly after a finding of guilt. The forensic psychologist must therefore have arranged to evaluate the accused well in advance of the trial date, and prepare a report that anticipates conviction on one or more of the charges. The MCM indicates that at sentencing Trial Counsel will enter evidence both in aggravation and pertaining to rehabilitative potential, followed by presentation of evidence by Defense Counsel in extenuation and mitigation. With respect to evidence in aggravation, this may include psychological impact on any person who was the victim of an offense committed by the accused. Thus, forensic psychologists may have roles in assessing victim impact and psychological consequences of criminal victimization. With respect to evidence in mitigation, the accused's rehabilitative potential also needs to be addressed. Forensic psychologists have a significant role to play in a case where a violent offense has taken place, and a violence risk assessment and risk management plan may be crucial in addressing rehabilitative potential.

It is the authors' belief that sentencing evaluations represent a greatly underutilized service available from forensic psychologists. The highest military appellate court in

United States v. Stinson (1992) found that, "In a sentencing hearing, an accused's potential for rehabilitation is a proper subject of testimony by qualified experts" (p. 6). Although a discussion of how to perform violence risk assessment is beyond the scope of this chapter, many books are available on the topic, such as the *Handbook of Violence Risk Assessment*, by Otto and Douglas (2010).

References

American Psychiatric Association. (2000). *Diagnostic and statistical manual of mental disorders* (4th ed., text revision). Washington, DC: Author.

American Psychological Association. (2010). *Ethical principles of psychologists and code of conduct with 2010 amendments*. Retrieved from www.apa/org/ethics/codex/index.aspx

American Psychology Law Society. (2011). *Specialty guidelines for forensic psychologists*. Retrieved from www.ap-ls.org/aboutpsychlaw/SpecialtyGuidelines.php

Daubert v. Merrell Dow Pharmaceuticals, Inc., 509 U.S. 579 (1993).

Dusky v. United States, 362 U.S. 402 (1960).

Heilbrun, K. M., DeMatteo, G., & Mack-Allen, J. D. (2007). A principles-based approach to forensic mental health assessment: Utility and update. In: A. M. Goldstein, (Ed.), *Forensic psychology: Emerging topics and expanding roles* (pp. 45–72). Hoboken, NJ: John Wiley & Sons.

Joint Service Committee on Military Justice, United States Department of Defense. (2012). Manual for United States Courts-Martial. (2012 edition).

Manual for United States Courts-Martial, United States (2012 Edition). Joint Service Committee on Military Justice.

Military Judges' Benchbook. (2010, January). Department of the Army, Pamphlet 27–9.

Otto, R. K., & Douglas, K. S. (Eds.). (2010). *Handbook of violence risk assessment*. New York, NY: Routledge.

Packer, I. K. (2009) *Evaluation of criminal responsibility: Best practices in forensic mental health assessment*. New York, NY: Oxford University Press.

United States v. Care, 18 U.S.C.M.A. 535, 40 C.M.R. 247 (C.M.A. 1969).

United States v. Proctor, 37 M.J. 330 (C.M.A. 1993).

United States v. Stinson, 34 M.J. 233, 238 (C.M.A. 1992).

Zapf, P. A., & Roesch, R. (2009) *Evaluation of competence to stand trial: Best practices in forensic mental health assessment*. New York, NY: Oxford University Press.

13   OPERATIONAL PSYCHOLOGY

Thomas J. Williams

Operational psychology fundamentally involves leveraging the expertise of psychologists in support of national security objectives and requirements to protect our nation and our population. The scope and practice of operational psychology are nested within and foundational to the history, growth, and profession of psychology. In the past 10 years, operational psychology has become most associated with the activities by psychologists in support of the operations and/or activities within the military, law enforcement, and intelligence arenas (e.g., Kennedy & Williams, 2010). Operational psychologists need familiarity with, and ability to draw from expertise within several specialty areas within psychology (e.g., social, cross-cultural, personality, perception, police, political, learning, forensics, etc.) and from a vast array of interdisciplinary areas (e.g., anthropology, international law, military science, political science, and sociology). It is the multicultural, multidiscipline scope of practice, along with the importance of the processes, products, and outcomes that combine to offer a fascinating richness and challenge for the success of practitioners of this growing subdiscipline within the profession of psychology (Williams, Picano, Roland, & Bartone, 2012). Just as the early psychologists within the profession were asked to contribute to national security during World War I (c.f., Yerkes, 1918), today's operational psychologists are increasingly being asked to leverage their insights and expertise to promote domestic safety and national security in an era of persistent conflict (Williams, Picano, Roland, & Banks, 2006).

FOUNDATION AND DEFINITION OF OPERATIONAL PSYCHOLOGY

The foundation for operational psychology rests within the profession itself, beginning with Yerkes (1918) first call for psychologists to provide their expertise in support of national security. Other notable psychologists (e.g., Urie Bronfenbrenner, Donald Fiske, John Gardner, David Levy, James G. Miller, O. H. Mowrer, Henry Murray, Theodore Newcomb, Donald MacKinnon, Harvey Robinson, Douglas Spence, Edward Tolman, and Kurt Lewin) provided their expertise in support of the assessment and selection and operational components of clandestine services during World War II. Williams et al. (2006) provided one of the first formal and comprehensive definitions of the various roles of operational psychology:

the actions by military psychologists that support the employment and/or sustainment of military forces…to attain strategic goals in a theater of war or theater of operations by leveraging and applying their psychological expertise in helping to identify enemy capabilities, personalities, and intentions; facilitating and supporting intelligence operations; designing and implementing assessment and selection programs in support of special populations and high-risk missions; and providing an operationally focused level of mental health support. (pp. 194–195)

A recent update to the definition of operational psychology captures the conceptual, functional, and scientific underpinnings that are linked to the applied, the art, and the science of psychology supporting organizational outcomes within the national security, intelligence, or law enforcement areas:

the application of the scientific principles and practices of psychology that involve the operational psychologist's taking actions, performing activities, or providing consultation in support of national security, military intelligence, or law enforcement activities and/or programs. (Williams et al., 2012, p. 38)

A watershed occurred with the 2010 publication by the American Psychological Association of a book by Kennedy and Williams, *Ethical Practice in Operational Psychology: Military and National Intelligence Applications*. That book helped identify and reinforce the need for the ethical practice of operational psychology and addressed several of the areas of practice addressed below.

SUPPORT TO NATIONAL SECURITY: DOMESTIC AND INTERNATIONAL THREATS

Homeland Security and Law Enforcement

Operational psychologists may help secure our homeland by helping law enforcement and intelligence agencies better understand the psychological, personality, and motivational attributes that are linked to actions and behaviors that lead individuals to become terrorists or to act on behalf of terrorists to carry out attacks against our nation or international partners. It might also involve supporting law enforcement and/or intelligence organizations by helping them assess and select those who could serve as informants to warn us of those who are planning to engage in terrorist acts. Operational psychologists have supported law enforcement operations by developing behavioral risk assessments to help law enforcement teams better understand the motivations, personality, and situational factors that provide time for the crisis to defuse or that provide more time for negotiations. They may also help assess, select, and train individuals in the areas of perception, signal detection, and behavior patterns to improve their ability to detect threats posed to our national ports of entry and within our domestic airports.

Counterterrorism and Counterintelligence Operations and Investigations

This practice area often involves assessments of both vulnerabilities and willingness to cooperate with ongoing counterterrorism (CT) or counterintelligence (CI) investigations of threats to national security or operations related thereto (see e.g., Williams et al., 2012). This may involve the identification of terrorists' motivations and intents, and provide support to processes used to educe information and/or determine whether someone is likely to cooperate with ongoing national security operations (e.g., Fein, Lehner, & Vossekuil, 2006).

National Intelligence Operations

Operational psychologists have the expertise to assist national leaders to better understand the developmental trajectories, personalities, motivations, and likely behaviors of domestic threats (e.g., Oklahoma City bombing) and international threats (e.g., Al Qaeda and other terrorist groups) (National Research Council, 2002). Support in this area may also involve indirect assessments of political leaders (Williams et al., 2006, Williams et al., 2012) to help guide our own political leaders in negotiations, to help determine likely actions, or identify the psychological stresses involved in participating in undercover operations as informants to national-level investigations and operations.

Insider Threat Assessments

Threats to our information technology infrastructure are increasingly considered one of our greatest threats to national security. Activities within this threat area can trigger criminal and/or national level intelligence investigations, since they may range from the disgruntled employee to a national effort by another country that seeks to secure an economic advantage (i.e., economic espionage) or to attack the informational resources on which we rely. Because of their expertise in security clearance evaluations and investigations (CI and counterespionage, CE), operational psychologists are increasingly being called on to help identify threats and mitigate risks in this increasingly high-risk area.

SUPPORT TO MILITARY OPERATIONS AND MILITARY INTELLIGENCE

Security-Clearance Evaluations

Operational psychologists often are called on to assess individuals to assist in the determination of whether they should have legal access to classified information. Operational psychologists' expertise in supporting national security provides them a great advantage in assessing vulnerability risk factors of individuals to determine two primary factors: whether any conflicts of interest exist that impact on an individual's commitments to positions of trust and whether the individual is reliable, trustworthy, and capable of protecting classified information. This same expertise is often helpful in discerning whether someone is at risk for espionage.

Counterintelligence (CI) and Counterespionage (CE) Investigations

While very similar to the types of activities described above for National Security Operations, the CI and CE investigation support will focus on operations more specific to military operations. Operational psychologists help increase awareness of adversary intentions and morale of their forces within the context of their cultural and psychological characteristics. Thus, CI operations depend on a good understanding of human nature and needs and motives of adversaries, which, if leveraged properly, could actually avert combat operations. In a similar manner, operational psychologists contribute to effective CE investigations by helping identify and neutralize vulnerabilities, both internally and externally, in our own military forces to guard against adversary efforts to undermine our military operations.

Assessment and Selection

Operational psychologists help develop and are often integral components to assessment and selection programs focused on carefully assessing the identification of attributes, characteristics, and skills of civilians, military, or even citizens of other nations to perform high-risk missions that require the identification of selected attributes deemed critical for mission success and/or to determine baseline personality features established in order to monitor suitability for ongoing operations. Consequently, operational psychologists involved in assessment and selection activities need working knowledge of legal requirements regarding assessment and selection of personnel, psychological testing usage, cross-cultural awareness, and other considerations to ensure an ethical practice.

Support to Interrogations

The involvement of psychologists in providing support to operational activities related to interrogations has been unquestionably one of the most misunderstood, challenging, and "politically" contentious practice areas for the profession of psychology. Psychologists involved in this activity are referred to as behavioral science consultants and support authorized law enforcement or intelligence activities (see Dunivin, Banks, Staal, & Stephenson, 2010). They use psychological insights and science in support of detention and related to intelligence, interrogation, and detainee debriefing operations. The skills and roles required to support these operations most often focus on information-gathering techniques that

maintain fidelity with legal and ethical guidelines while assessing an individual's cooperation within a context informed by their culture and ethnicity.

The practice of operational psychology raises issues regarding how to define actions by psychologists who support societal interests by helping to address fundamental threats to society as a whole. In essence, the issue raised involves how one's duties are defined by law and ethics as well as by the need to address the natural tension between protecting society's interests that are less defined versus an individual's (e.g., a patient or detainee) interests that are easier to determine. Needless to say, the ethical issues raised by psychologists supporting interrogations and consultations are very complex and occur within multiple legal contexts (e.g., US military law, constitutional rights, host-nation laws, US Supreme Court rulings, ethical guidelines, Laws of Land Warfare, International Human Rights, and many others). Dunivin et al. (2010) provide a comprehensive overview of the support provided by operational psychologists involved in interrogation operations, while Benhke and Moorehead-Slaughter (2012, see below), provide a very helpful review of the ethical issues and steps taken by the APA to guide an ethical practice of psychology among those who support these operations.

ETHICAL PRACTICE OF OPERATIONAL PSYCHOLOGY

The ethics of operational psychology achieved a watershed with the publication of *Ethical Practice of Operational Psychology* (Kennedy & Williams, 2010). As psychologists increasingly encountered challenges within their emerging operational roles, as required by the ethics code, they sought guidance and clarification to continue acting within an ethical framework of practice and consultation. Recently, Behnke and Moorehead-Slaughter (2012) provided a very helpful overview of APA's efforts to respond with a policy that assures an ethical practice of psychology and details the debate within the APA membership about whether psychologists have *any* role in support of national security, law enforcement, and intelligence activities.

The scope and practice of operational psychology has appropriately caused the profession to reflect on a healthy, natural tension that results between promoting the welfare of society versus individual protections (e.g., Kennedy & Williams, 2010). Operational psychologists serve both these interests well and in so doing, represent well the initial foundation of psychologists to serve both society and its citizens as originally envisaged by Yerkes (1918).

References

Behnke, S., & Moorehead-Slaughter, O. (2012). Ethics, human rights, and interrogations: The position of the American Psychological Association. In J. H. Laurence & M. D. Matthews (Eds.), *The Oxford handbook of military psychology* (pp. 50–62). New York, NY: Oxford University Press.

Dunivin, D., Banks, L. M., Staal, M. A., & Stephenson, J. A. (2010). Behavioral science consultation to interrogation and debriefing operations: Ethical considerations. In C. H. Kennedy & T. J. Williams (Eds.), *Ethical practice in operational psychology: Military and national intelligence applications* (pp. 85–106). Washington, DC: American Psychological Association.

Fein, R. A., Lehner, P., & Vossekuil, B. (2006). *Educing information-interrogation: Science and art, foundations for the future.* Retrieved from http://www.fas.org/irp/dni/educing.pdf

Kennedy, C. H., & Williams, T. J. (Eds.). (2010). *Ethical practice in operational psychology: Military and national intelligence applications.* Washington, DC: American Psychological Association.

National Research Council. (2002). *Making the nation safer: The role of science and technology in countering terrorism.* Washington, DC: National Academies Press.

Williams, T. J., Picano, J. J., Roland, R. R., & Banks, L. M. (2006). Introduction to operational psychology. In C. H. Kennedy & E. A. Zillmer (Eds.), *Military psychology: Clinical and operational applications* (pp. 193–214). New York, NY: Guilford Press.

Williams, T. J., Picano, J. J., Roland, R. R., & Bartone, P. (2012). Operational psychology: Foundation, applications, and issues. In J. H. Laurence & M. D. Matthews (Eds.), *The Oxford handbook of military psychology* (pp. 37-49). New York, NY: Oxford University Press.

Yerkes, R. M. (1918). Psychology in relation to war. *Psychological Review, 25,* 85–115.

14 WORKING WITH SPECIAL OPERATIONS FORCES

L. Morgan Banks

MISSIONS AND ORGANIZATION

Special Operations Forces (SOF) are "those active and reserve forces of the Army, Navy, Air Force, and Marine Corps that have been designated by the Secretary of Defense and specifically organized, trained, and equipped to conduct and support special operations" (US Department of Defense, 2007, p. 503). What then, are special operations? The Department of Defense defines special operations as those operations

conducted in hostile, denied, or politically sensitive environments to achieve military, diplomatic, informational, and/or economic objectives employing military capabilities for which there is no broad conventional force requirement. These operations often require covert, clandestine, or low visibility capabilities. Special operations are applicable across the range of military operations. They can be conducted independently or in conjunction with operations of conventional forces or other government agencies and may include operations through, with, or by indigenous or surrogate forces. Special operations differ from conventional operations in degree of physical and political risk, operational techniques, mode of employment, independence from friendly support, and dependence on detailed operational intelligence and indigenous assets. (US Department of Defense, 2007, pp. 502–503)

In particular, SOF include a wide variety of service members, including Army Special Forces; Navy SEALS; Army Rangers; very highly trained rotary wing (helicopter), fixed wing, and tilt rotor aircraft aviators; Air Force Combat Controller and Pararescue Jumper (PJ) personnel; Psychological Operations personnel (now referred to as Military Information Support Operations personnel); Civil Affairs personnel; and Marines assigned to the Marine Special Operations Command. Consequently, their missions run the gamut from highly dangerous combat raids into denied enemy territory, to training indigenous fighters, to working to establish clean drinking water in rural areas. Because of this, it would be inaccurate to generalize a type of personality, other than the fact that most are highly motivated to succeed in their jobs. Aviators operate under very rigorous flight conditions, and include the very best rotary wing pilots in the Department of Defense. Special Forces soldiers undergo extensive training in order to operate for extended periods of time with limited conventional support. SEALs are likewise selected to operate under incredibly dangerous and rigorous conditions, and trained to operate in an extremely physically demanding environment. Air Force

Combat Controllers infiltrate denied areas prior to the arrival of US forces in order to properly coordinate air support and delivery.

SOF PSYCHOLOGISTS

In order to successfully complete these diverse missions, SOF are specially assessed and selected for their various organizations. In addition to the physical challenges that are a major portion of most assessment programs, psychological evaluations are essential. In most of these programs, detailed psychological assessments are conducted prior to training and assignment. These assessments may include traditional psychometric instruments and usually include at least a brief intelligence screening. These assessments have historically been based conceptually on those conducted for the Office of Strategic Services during World War II (Banks, 2006; Office of Strategic Services Assessment Staff, 1948).

Because of this assessment process, candidates for SOF will be exposed to psychologists as part of their entry into their respective organizations. Psychologists assisting in this assessment and selection process are ordinarily uniformed active duty psychologists who are assigned to the organization to which the candidate is applying, and because of their duties, are referred to as operational psychologists. Subsequently, the relationships that many in SOF have with psychologists are distinctly different from that of most service members. Although these psychologists may be seen as gatekeepers while the candidate is undergoing selection, once the candidate is accepted into the organization this relationship changes. While the operational psychologist still has a primary duty to the organization, a major role for them from that point on is to support the organization by directly supporting the service member. The operational psychologist, as an embedded member of the organization, will usually be seen as supportive and as having a key role in helping members succeed in the unit.

In addition to providing this support to the individual unit members, operational psychologists will provide direct support to the unit's mission. For example, he or she may assist in the target analysis process, in hostage negotiations, and in helping to assess the indigenous military forces that are being trained. For these reasons, many members of SOF have routine interactions with operational psychologists as part of their day-to-day jobs. Overall these interactions reduce the stigma that often accompanies talking with a psychologist. The psychologist in this role is seen more as simply another staff officer supporting the mission, rather than a dedicated mental health provider.

Like the other members of SOF units, operational psychologists are also screened, selected, and then trained for their jobs. In addition to being licensed clinicians, they will ordinarily be required to attend airborne school, survival school, and other mission related training in order to provide helpful consultative services to the unit. All will have security clearances, and like many others in military service, few will be at liberty to discuss much of their actual work in detail. The ethical issues that arise for these psychologists are similar to those associated with other consultative settings. The largest challenges usually revolve around establishing clear boundaries concerning the multiple relationships that will develop, and ensuring competence in novel areas of practice. Since operational psychologists function as both clinicians and consultants, this requires thoughtful awareness of these ethical issues, and often necessitates frequent consultation when starting out in this field. Many of these psychologists will have worked within SOF organizations their entire career, and will have strong personal bonds with other SOF personnel. This can be essential, as it make take a very long time for an individual operational psychologist to develop the credibility to work with these organizations. Once this credibility is established, however, it may last for a career and beyond. As a Special Staff Officer, the psychologist in this role has very little direct authority, but over time can develop a great deal of influence. This will only happen if the psychologist is seen as an honest broker who can be trusted to speak the truth (tactfully)

while maintaining an open and ethical relationship with all those in the organization. In other words, to be successful, the psychologist must be trusted by all levels in the unit, from the most junior to the commander. His or her integrity must be absolute.

The model for treatment that has developed is one where this assigned operational psychologist is seen as just another staff member of the unit, and is therefore often more easily approached by service members when a behavioral health issue arises. The operational psychologist can then function to initially assess and triage the service member. In some cases, the operational psychologist may be able to provide the treatment directly. In many cases, though, the operational psychologist will assess the service member and then refer him or her to an outside clinical provider. This provider might be one assigned to the local military medical treatment facility, or a local civilian Tricare provider, or in some cases, a clinical provider who is employed directly by the SOF unit. A model that has proven highly successful within SOF is for Brigade level units to have their own dedicated clinical providers, often psychologists. Again, sometimes this is a uniformed active duty psychologist, sometimes it is a federal civilian psychologist, and sometimes it is a contract provider. When the provider is organic to the unit, their sole mission is to provide behavioral health care to the unit, and not the provision of operational psychology support. This allows the operational psychologists to primarily be consultants to the command, and to help ensure the entry of service members into the health care system. This greatly increases the trust afforded to the provider, and increases the likelihood of service members seeking care. If the operational psychologist has been successful at establishing a positive relationship with the unit's members, this can transfer to the clinical provider.

STIGMA

As discussed above, this normalization of working with operational psychologists has produced an increase in the acceptability of talking to psychologists about personal issues. Consequently, actual access to behavioral health services has seen an increase within SOF over the last several years. However, working against this reduction in stigma is the fact that most SOF personnel have security clearances. Because of this, there may be a fear that speaking with a behavioral health provider may damage their ability to keep their clearance. This fear is factually unfounded, as ordinarily only significant mental illness is a mental health disqualifier for maintaining access to classified information. This fear can be overcome by education on the actual regulations that affect clearances, but this takes time and goes against common misperceptions. Regardless of the facts, the fear still exists. The following information may be helpful. In the 4 calendar years of 2008 through 2011, a total of 31 individuals out of 1,192,850 had their security clearances denied or revoked due only to psychological conditions (S. Harvey, Briefing by the US Army Central Personnel Security Clearance Facility: Impact of Counseling on Security Clearances, personal communication, April 26, 1012). As a percentage of clearance determinations, this ran from a low of .002% to a high of .0059%. Most of these were due to the abuse of alcohol or drugs, violence, or personal conduct. In particular, the current guidance specifically states, "Mental health counseling in and of itself *is not a reason* to revoke or deny a clearance" (US Secretary of Defense, 2008, p. 3). A provider may find more detailed information at the following websites: http://www.fas.org/sgp/isoo/guidelines.html http://www.arl.army.mil/www/pages/208/ PolicyImplementation-SF86.pdf http://www.opm.gov/investigate/fins/2008/fin08–01.pdf

TREATMENT IMPLICATIONS

Although the following generalizations are not universal, knowledge of them should help a provider in understanding the unique demands on and characteristics of SOF. In addition to the screening process described above, the training for SOF can be extensive, not uncommonly taking over two years following selection (in addition to the training that is a prerequisite for selection, such as basic training, advanced individual training, airborne school, etc.). This training is often quite challenging, and combined with the initial screening, results in highly stress tolerant individuals. This selection and subsequent training also results in a highly motivated force. This likely occurs because only those who are highly motivated will go through the process, and/or because the process of cognitive dissonance reinforces their motivation.

Once assigned, such personnel often will stay in a particular unit for much of their career, moving on a much less frequent schedule than is common in the rest of the force. For these reasons, SOF are very close-knit, and the unit members may have worked together for many years. This will obviously have an effect on the methods of coping with casualties. Because intelligence is a significant factor in many of the screenings for SOF, they will have higher than average intelligence scores. As discussed earlier, because of the screening and the follow-on training, they usually will have much higher initial resilience than the general forces. Because most of the selection programs look for individuals with a high tolerance of risk, they may be more likely to bend, rather than rigidly follow rules they believe are hurting their ability to do their job. Again, because most selection programs look for individuals with high drive and initiative, they will likely have a very strong streak of independence and internal locus of control.

Over the last 10 years, SOF service members may have deployed for shorter periods of time, but much more often, than the general forces. It is not uncommon for a SOF service member to have over a dozen combat deployments over a span of several years. Although it is hard to predict the future, as this chapter is written it appears that while the deployment tempo may significantly decrease for the general purpose forces, it is unlikely to decrease for SOF.

It should not be surprising that malingering among SOF is rare. One of the common observations among treating clinicians within the military system is the rewarding nature of providing treatment to SOF. It is much more common for SOF to underreport symptoms than to exaggerate them. In general, SOF's rates of PTSD are lower than the general purpose forces, but the current trends appear to be increasing, and a clinician should not overinterpret this in a specific case.

Operational psychologists can be a valuable asset for an outside clinical provider, uniformed or otherwise, in understanding and helping SOF personnel who are seeking treatment outside of their assigned unit. As discussed above, embedded operational psychologists are doctoral level state licensed clinical or counseling psychologists. Although their primary client may be the organization, they are still privileged to provide clinical care to SOF members within the limits of their training and experience. They will usually have access to at least some psychological assessment information, to include personality and intelligence testing, although it may be dated. There will be strict limits to how this information may be shared, but, with an appropriate release of information from the client, it may be possible for the embedded operational psychologist to release useful portions of that information to a properly licensed treating clinician. This will mostly likely be possible when the treating clinician is working in a military facility within the DoD health care system. In addition, the assigned psychologist may be helpful in understanding the culture and background of the unit to which the client is assigned. It is also not uncommon for some SOF personnel to wish to receive treatment without notifying their organization. For this reason, it should be obvious that contact with the embedded operational psychologist must be discussed and approved by the client prior to any initiation of contact by a treating provider. Although the embedded operational

psychologist has a clear duty to the individual client, their primary duty is to their organization. If these obstacles can be overcome, consultation may be helpful. For example, it may be possible to compare preinjury psychological functioning to current functioning. Because of the testing given during assessment, it may be possible to compare preinjury intellectual functioning to current functioning, especially following closed head trauma.

Because of the very high rate of deployments of these organizations, even prior to 9–11, many have very robust family support groups. Because of their history of intensive training, some of these organizations had higher accident rates than the general purpose forces, and have family support group programs for survivors that predate 9–11. These family support groups may be helpful to a treating clinician, especially when dealing with family issues.

References

Banks, L. M. (2006). The history of special operations psychological selection. In A. D. Mangelsdorff (Ed.), *Psychology in the service of national security* (pp. 83–95). Washington, DC: American Psychological Association.

Office of Strategic Services Assessment Staff. (1948). *Assessment of men: Selection of personnel for the Office of Strategic Services*. New York, NY: Rinehart.

US Department of Defense. (2001, as amended though 2007). *Dictionary of military and associated terms, Joint Publication 1–02*. Washington, DC: U.S. Government Printing Office.

US Secretary of Defense. (2008). *Policy Implementation—Mental Health Question, Standard Form (SF) 86, Questionnaire for National Security Positions*. Retrieved from http://www.arl.army.mil/www/pages/208/PolicyImplementation-SF86.pdf

15 COMMAND AND ORGANIZATIONAL CONSULTATION

Paul T. Bartone and Gerald P. Krueger

Whether clinicians or researchers, military psychologists are recognized experts on human behavior. Organizational leaders rely on them for advice on a range of issues related to soldier health and performance including initial selection and classification decisions; determining individual fitness for continued duty; assessing morale, health, and well-being; designing and implementing prevention and treatment programs; and developing policies to protect and enhance individual and group fitness. Military consultants' advice can take the form of informal conversations with leaders, more formal briefings, or published reports and policy recommendations. In the active force, military psychologists can be either uniformed or civilian, while those working with veterans organizations (e.g., the US Veterans Administration) are most often civilians.

Whatever the focus of the consultation, it is important that military psychologists have a good understanding of the unit or organization in which they are consulting. Military units have their own special cultures and subcultures, language, dress, rituals, and norms of behavior. An understanding of the military culture facilitates access to the unit, while also informing the consultant's judgment about what is going on within the unit (Warner, Appenzeller,

Breitbach, Lange, Mobbs, & Ritchie, 2011). The same applies in clinical settings, where the psychologist may provide individual evaluations and/or counseling. Uniformed military psychologists have an advantage, having undergone various military training and often having spent time embedded in military units.

The increasing number of civilian psychologists working in military settings have a greater challenge in this regard (Reger, Etherage, Reger, & Gahm, 2008). Civilian psychologists must work to develop their military "cultural competence" and credibility before consulting with military commanders and organizations. The most useful strategies involve spending time with military units, such as by accompanying them on training exercises. This helps establish relationships and builds trust with unit members. Having a uniformed military psychologist as a mentor and guide can help develop the needed cultural understandings.

HEALTH CARE TEAMS AND ROLES

Some command consultations, such as fitness-for-duty evaluations, are conducted by individual military psychologists. Other types of consultations require close coordination with health care providers throughout the organization. Physicians are more often found in senior leadership roles in the military health care system than are nonphysician specialists. This means a command consultation by a psychologist must sometimes be done under the nominal supervision of a psychiatrist or other physician. Also, the psychologist is often junior in rank to the physician leading the health care team, which creates a power differential on the team. In such situations, the consulting psychologist is aided by his/her demonstrated expertise and consistency, and by carefully coordinating all activities and recommendations with superiors and other members of the health care team.

ETHICAL ISSUES

Military psychologists fill multiple roles, a fact that can pose ethical dilemmas when consulting or advising leaders. The military officer psychologist is sworn to place the interests of the organization first, a priority that may conflict with what appears to be best for the individual service member (see Chapter 23 by Barnett, current volume). The need for confidentiality of information gained during an individual consultation can sometimes present a conflict for military psychologists. While it is usually in the best interests of the individual client that confidentiality be maintained, in many cases the military organization has a legitimate interest and even a legal right to access information relating to the health and performance potential of individuals and groups. In cases, such as unit level surveys, the consultant can minimize this problem by collecting only anonymous data. When individual identifiers must be obtained, the consultant should be candid about any possible lack of confidentiality. While this can reduce the effectiveness of some command and organizational consultations, it is a necessary condition of consulting in military organizations. For a fuller discussion of ethical issues confronting military psychologists, see Johnson (2008).

CONSULTATION SETTINGS FOR MILITARY PSYCHOLOGISTS

Military psychologists work in two primary settings: (1) the community or garrison environment, and (2) the deployed environment. The nature of their activities varies depending on the setting. The community or garrison environment refers broadly to the home base, military posts, and facilities in the home country, to include veterans' hospitals. Most military bases have their own medical facilities, which is often where military *clinical psychologists* are assigned. Military *research psychologists* also may work in the hospital or clinic, but more commonly are found at separate research units on base. In garrison, command consultations tend to focus on (1) individual fitness-for-duty and deployability evaluations (see Budd & Harvey, 2006); (2) education and primary prevention efforts (Warner et al., 2011); (3) testing and assessments for selection and placement (see Rumsey, 2012); and (4) assessing various health, morale, and well-being factors that can affect readiness and performance (see Krueger, 2010).

All of these functions may also occur in the deployed environment, although there is heavier emphasis on maintaining operational effectiveness, sustaining performance, and preventing problems through education efforts and brief interventions. Deployed units commonly experience exposure to a range of stressors not generally encountered in garrison. Stress-related adjustment and performance problems are more prevalent in overseas deployments. There, military psychologists' consulting aims at assessing, preventing, and treating stress-related problems, while also addressing organizational factors that influence how well troops adapt to the stressors of deployment. Psychologists may recommend medical evacuation or repatriation of service members, but the vast majority of those receiving counseling are returned to duty. Current US Department of Defense guidelines for in-theater management of stress reactions emphasize BICEPS factors: brevity and immediacy (brief interventions soon after recognition of symptoms), centrality (in some central location away from wounded), expectancy (with expectation of return to duty), proximity (close to the service members' military unit), and simplicity (simple interventions, e.g., rest, food, and reassurance). More information on managing stress issues in theater is provided by Campise, Geller, and Campise (2006).

EVALUATIONS OF INDIVIDUALS

Military leaders may direct that individuals under their command undergo psychological evaluations to determine their fitness for duty, and occasionally for other reasons. Such command directed evaluations (CDEs) may be requested when individuals are showing adjustment problems, anxiety, depression, cognitive difficulties, or are thought to be a danger to themselves or others. These evaluations are performed by licensed psychologists, social workers, or psychiatrists. The military psychologist's expertise in psychological testing is especially valuable in such assessments. Results are provided to the commander in a written report that can recommend discharge, return to duty, or return to duty with certain restrictions. Budd and Harvey (2006) provide a full description of fitness-for-duty evaluations.

EDUCATION AND OTHER PRIMARY PREVENTION EFFORTS

In both garrison and deployed environments, military psychologists consult with leaders and with other behavioral health experts, social workers, psychiatrists, and military chaplains on the design and delivery of education and training programs aimed at preventing problems and sustaining good health and performance. Consultative efforts address topics important to the health and well-being of troops, including stress management, suicide awareness and prevention, drug and alcohol prevention, smoking cessation, and preventing sexual harassment. Increasingly, military psychologists are involved in providing special consultation and/or training sessions for units preparing to deploy. The focus is on helping military personnel form realistic expectations regarding the deployment and teaching them healthy coping strategies. Military psychologists also consult with leaders at all levels regarding the many challenges faced by personnel returning home from deployment and various approaches to facilitate healthy reintegration and adaptation of troops to the home environment.

SELECTION SCREENINGS

Beginning with the seminal work on standardized intelligence tests for Army recruits during World War I, military psychologists have consulted with leaders on improving methods for selection of military personnel. By the end of World War I, psychological screening tests had been administered to nearly two million men. Since then psychologists have developed, validated, and implemented a wide range of tools and methods to assist leaders and policy makers in the selection and placement of military personnel (see Rumsey, 2012). Military psychologists also provide consulting support to leaders regarding the selection of personnel for high-risk units, such as special operations forces (see Christian et al., 2010).

MILITARY HEALTH SURVEILLANCE AND RESEARCH

Another important role for military psychologists involves conducting research within units to identify conditions that influence the morale, health, well-being, and performance of soldiers. Consultation with leaders happens at every step of the way, beginning with an initial request from unit leaders for research to address particular concerns. For example, during the first Persian Gulf War in 1991, small teams of research psychologists and other specialists deployed into the theater to conduct research on stress, morale, and adaptation in the combat environment. These teams used multiple methods including surveys, observation, and interviews, quickly analyzed their data, and provided rapid feedback on results to commanders in the field, often influencing important personnel policy decisions. Human dimensions research psychology teams deployed to Croatia in 1991, and Somalia in 1993, and because of their successes, were also deployed to work in Kuwait and Saudi Arabia in 1994 and to Bosnia in 1995–1996. After the terrorist attacks of September 11, 2001, and subsequent deployment of US military forces to Iraq, the Army Surgeon General established "Mental Health Advisory Teams" (MHATs) to conduct human dimensions research in-theater with a focus on factors that influence the mental health and operational readiness of military forces. Since then, MHAT teams have deployed to conduct field research with the express purpose of providing rapid analysis and results to guide commanders in their training and policy decisions. For more on the activities of MHATs, see McBride et al. (2010) and Bliese et al. (2011).

With appropriate privacy protections, military psychologists at times are able to access medical records and test scores for research purposes. For example, data from postdeployment health screens (mandatory in the United States since 2003) have been used to identify the impact of various deployment experiences on a range of physical and mental health outcomes. Shen, Arkes, Kwan, Tan, and Williams (2010) accessed military personnel and medical records for their study, which identified length of deployment as a major influence on later diagnosis of PTSD—posttraumatic stress disorder. Such research can lead directly to changes in policy having broad consequences for the health and performance of military forces.

SUSTAINED OPERATIONAL PERFORMANCE

Military psychologists often consult with unit leaders regarding the importance of ameliorating multiple soldier stressors that accompany deployment and combat activities and that threaten to compromise soldier performance and health. In addition to threats of being physically injured or killed, these stressors include combinations of exposures to environmental extremes (high heat, extreme cold, high terrestrial altitude), significant acoustical noise, whole body vibration, rapid acceleration, toxic fumes, carrying heavy loads, sleep deprivation, physical and mental fatigue, occasional cognitive overload, and a press for time-based reactions and responses. Threats of being exposed to chemical, biological, or radiological weapons or to novel agents heighten anticipation and trepidation in soldiers. Women soldiers may experience additional stressors unique to their gender (for a review see Krueger, 2008).

Military psychologists provide consultative assistance on many of the above stressors; but ubiquitously, it is the need for *sufficient sleep*, both in terms of quality and quantity, that pervades so much of what troops must accomplish. Military personnel require 7–8 hours of sleep per 24-hour day to maintain adequate levels of alertness on the job. Since they often do not obtain it, they accumulate a *sleep debt*. Tired soldiers exercise poor judgment, lose situational awareness, make more mistakes, and have more accidents. Likewise, fatigued leaders may find it difficult to continually make sense of an erratic battlefield. Some of the most important guidance a psychologist can give to a commander is to assist in developing a sound *unit sleep discipline policy*, and then verify that the unit is adhering to that policy (see Krueger, 2012).

INTERVENTIONS

Clinical psychologists engage in command consultation when they provide individual assessments and psychotherapy, whether short term in deployed settings, or for the longer term in garrison and at VA centers (see Ball & Peake, 2006). Furthermore, military psychologists are providing *psychological interventions* of sorts when they consult on selection and placement, education and training, and organizational effectiveness. These activities typically aim to (1) preserve individual performance and health (e.g., optimizing soldier cognitive readiness to fight, fostering resilience in individuals and units, developing unit cohesiveness) and (2) prevent problems before they occur (e.g., suicide awareness and prevention, prevention and treatment for alcohol and drug use).

Several recommendations are offered to assist military psychologists serving in consultant roles:

- Get to know the culture, language, and habits of the military organization you consult for; get out of the office/clinic; participate in various military training courses (e.g., airborne, air assault); accompany the unit on training exercises. In addition to developing "cultural competence," this also helps to build relationships, trust, and credibility.
- Take time to find out what commanders need, and understand their questions.
- Be honest and clear with commanders and military personnel regarding ethical issues, any limits on confidentiality, and so forth.
- Include local unit behavioral experts in your consultations; take a team approach. This expands the resources of the military consultant while generating greater cooperation.
- Provide clear advice in a format the commander is familiar with and if asked, offer practical suggestions for how to implement it.

References

Ball, J. D., & Peak, T. H. (2006). Brief psychotherapy in the U.S. military: Principles and application. In C. H. Kennedy & E. A. Zillmer (Eds.), *Military psychology: Clinical and operational applications* (pp. 61–73). New York, NY: Guilford.

Bliese, P. D., Adler, A. B., & Castro, C. A. (2011). Research-based preventive mental health care strategies in the military. In A. B. Adler, P. D. Bliese, & C. A. Castro (Eds.), *Deployment psychology: Evidence-based strategies to promote mental health in the military* (pp. 103–124). Washington, DC: American Psychological Association.

Budd, F. C., & Harvey, S. (2006). Military Fitness-for-Duty Evaluations. In C. H. Kennedy & E. A. Zillmer (Eds.), *Military psychology: Clinical and operational applications* (pp. 35–60). New York, NY: Guilford.

Campise, R. L., Geller, S. K., & Campise, M. E. (2006). Combat stress. In C. H. Kennedy & E.A. Zillmer (Eds.), *Military psychology: Clinical and operational applications* (pp. 215–240). New York, NY: Guilford.

Christian, J. R., Picano, J. J., Roland, R. R., & Williams, T. J. (2010). Guiding principles for assessing and selecting high-risk operational personnel. In P. T. Bartone, B. H. Johnsen, J. Eid, J. Violanti, & J. C. Laberg (Eds.), *Enhancing human performance in security operations: International and law enforcement perspectives* (pp. 121–142). Springfield, IL: Charles C. Thomas.

Johnson, W. B. (2008). Top ethical challenges for military clinical psychologists. *Military Psychology, 20,* 49–62.

Krueger, G. P. (2008). Contemporary and future battlefields: Soldier stresses and performance. In P. A. Hancock & J. L. Szalma (Eds.), *Performance under stress* (pp. 19–44). Aldershot, Hampshire, UK: Ashgate.

Krueger, G. P. (2010). Sustaining human performance during security operations in the new millennium. In P. T. Bartone, B. H. Johnsen, J. Eid, J. Violanti & J. C. Laberg (Eds.), *Enhancing human performance in security operations: International and law enforcement perspectives* (pp. 205–228). Springfield, IL: Charles C. Thomas.

Krueger, G. P. (2012). Soldier fatigue and performance effectiveness: Yesterday, today and tomorrow. In G. Matthews, C. Neubauer, P. A. Desmond, & P. A. Hancock (Eds.), *The handbook of operator fatigue* (pp. 393–412). Aldershot, Hampshire, UK: Ashgate.

McBride, S. A., Thomas, J. L., McGurk, D., Wood, M. D., & Bliese, P. D. (2010). U.S. Army Mental Health Advisory Teams. In P. T. Bartone, R. H. Pastel, & M. A. Vaitkus (Eds.), *The 71F advantage: Applying Army research psychology for health and performance gains* (pp. 209–245). Washington, DC: National Defense University Press.

Reger, M. A., Etherage, J. R., Reger, G. M., & Gahm, G. A. (2008). Civilian psychologists in an army culture: The ethical challenge of cultural competence. *Military Psychology, 20,* 21–35.

Rumsey, M. G. (2012). Military selection and classification in the United States. In J. H. Laurence & M. D. Matthews (Eds.), *Oxford handbook of military psychology* (pp. 129–147). Oxford, UK: Oxford University Press.

Shen, Y., Arkes, J., Kwan, B., Tan, L., & Williams, T. V. (2010). Effects of Iraq/Afghanistan deployments on PTSD diagnoses for still active personnel in all four services. *Military Medicine, 175*(10), 763–769.

Warner, C. H., Appenzeller, G. N., Breitbach, J. E., Lange, J. T., Mobbs, A., & Ritchie, E. C. (2011). Psychiatric consultation to command. In E. C. Ritchie (Ed.), *Combat and operational behavioral health* (pp. 171–188). Washington, DC: Department of the Army, Office of the Surgeon General, Borden Institute.

16 HUMAN FACTORS ENGINEERING AND HUMAN PERFORMANCE

Michael D. Matthews

War and the military have been critical to the growth of almost all areas of psychology. Human factors engineering, a discipline that conducts basic and applied research "on human beings and their interaction with products, equipment, facilities, procedures, and environment used in work" (Sanders & McCormick, 1993, p. 4), is no exception. The birth of human factors engineering as a formal discipline can be attributed to the exponential growth in the speed and complexity of weapons systems (such as fighter aircraft) that occurred in World War II. In order to fully exploit the capability of a given system, developers had to take into account both the capabilities and limitations of the human beings who operated these systems. Reaction time, attention and perceptual processes, memory capacity, and decision making had to be studied in the context of that system in order to maximize performance and minimize risk. In recognition of the overarching importance of these factors to military performance, all branches of the US Military established human factors engineering and performance laboratories shortly following the end of World War II (Krueger, 2012).

In this chapter, several areas of contemporary human factors engineering that are especially relevant to the military are described. For a more comprehensive overview of military human factors engineering, see Matthews and Laurence (2012).

AVIATION PSYCHOLOGY

Modern military aircraft operate near or above the speed of sound. They are capable, with in-flight refueling, of completing intercontinental missions that may last for 24 or more hours. Fighter planes can maneuver so abruptly that they can induce g-forces that exceed the capability of the pilot to withstand. They can fly at high speeds just above the ground—to avoid enemy radar—or several miles above the earth. These aircraft are equipped with state-of-the-art digital command and control systems, and weapons systems that can hit targets with pinpoint precision at great distances.

Military aviation human factors engineers must ensure that all components of the aircraft system are compatible with the ability of the pilot and crew to operate them. Besides the technical complexities of modern military aircraft, the human factors engineer must also consider the psychological component of flying in combat where a lapse of attention or failure to react may result in the death of the pilot and crew. The stress of operating in the *in extremis* conditions of combat can further impair motor, perceptual, and cognitive function and is also a critical component of understanding pilot performance.

An example of contemporary work in this area is experimentation on the effects of sleep deprivation on pilot and crew performance. How long can pilots and crew go without sleep before they experience significant impairment in function? What are ways to minimize the adverse effects of sleep deprivation during long missions? What is the relationship between sleep deprivation and pilot error, and what types of errors are most associated with sleep deprivation? What systems can be designed to mitigate the effects of sleep deprivation on performance?

Aviation human factors engineering addresses many other issues critical to crew performance. The effects of high g-forces on sensory and perceptual processes are critical in understanding pilot performance in fighter aircraft. Designing command and control systems that facilitate situational understanding and decision making under stressful conditions is a major area of research and development.

In summary, the military aviation human factors engineer must consider every aspect of the physical and psychological makeup of the pilot in designing aircraft and their subordinate systems. To the extent that the capabilities of the aircraft and its systems match the capabilities of the pilot and crew, the effectiveness of the total system is maximized.

SITUATIONAL AWARENESS/DECISION MAKING

A major area of research and application in military human factors engineering focuses on decision-making, especially under high stress, high-stakes conditions. Much of this work involves the concept of situational awareness (SA). Situational awareness is a cognitive construct that is viewed as a precursor to fast and accurate decision making. The construct consists of three components. Level I SA is the ability of the person to accurately perceive key elements of the environment. Level II SA is the ability to comprehend the meaning of what is perceived. Finally, Level III SA represents the ability to predict what is about to happen in the near future (Matthews, 2012).

Much of the early research on SA was done in the context of aviation, both military and civilian. In the past decade, the construct has been applied to other settings including

almost every type of military activity. Some of the most interesting work looks at the SA of infantry small unit leaders. It takes considerable skill for a small unit leader to know what elements of the environment to focus attention on in typical infantry operational settings, to understand what it means, and to predict what is likely to occur next. Research shows that experienced platoon leaders establish better SA at all three levels, and focus on different aspects of the battle space than less experienced platoon leaders. By comparing experienced and inexperienced platoon leaders, and breaking down the three levels of SA, it is possible to develop training aids designed to build SA skills in new lieutenants before they deploy into the war as platoon leaders. This should, in turn, lead to better decision making, greater combat effectiveness, and less fratricide.

A good deal of research critically examines the impact of new technologies on SA. For instance, does a newly developed heads-up display for helicopter pilots improve SA, or does it interfere with it? In command and control systems, the organization and method of presenting information to the user (pilot, commander, etc.) may impact SA and therefore performance. Environmental factors including weather and terrain, personal factors such as fatigue and stress, and organizational factors such as doctrine all combine to affect SA.

An emerging area of decision-making research in military contexts involves naturalistic decision making (NDM). As the name implies, NDM focuses on decision making in real situations. This research suggests that in circumstances that require rapid decisions in high-risk settings, leaders do not typically invoke classic decision-making models that involve a systematic analysis of the situation and an assessment of various courses of action. Under these conditions, leaders quickly assess the situation and then pattern-match that assessment to scripts that they have found to be successful in similar situations in the past. If that course of action fails, they repeat the scan and match process and quickly select another tactic. The result may appear intuitive, but in fact is based on extensive experience that allows the leader to rapidly select an appropriate course of action (Kahneman & Klein, 2009).

WORKLOAD/DISPLAYS

Military tasks often require immense physical and mental workloads. The average weight of an infantry soldier's pack is 91 pounds. The impact of carrying such a load on soldier performance is obvious. It limits speed, flexibility, and endurance. A good deal of military human factors engineering research and development looks at ways to reduce the combat load and on designing weight bearing systems (e.g., packs) that distribute the weight in an optimal manner. Training can also be designed to build strength and technique in soldiers to aid them in handling heavy loads.

Less obvious to the casual observer is mental workload. Modern military command and control systems present more information than the user can efficiently or effectively process. Every undergraduate psychology major knows about Miller's magic number of seven, plus or minus two, with respect to the capacity of short-term memory. Command and control systems not only tax memory resources, they also challenge attentional processes. So much information is presented that the user is forced to divide attention among multiple inputs, leading to the possibility of missing vital information and/or failing to respond in a timely manner. This "cognitive" overload can be a major source of operator error in any military context, and is compounded by sleep restriction and high stress.

One solution to minimizing cognitive overload is in the optimal design of displays. Considerable human factors research is aimed at outlining just how to present the right information, to the right user, at the right time. In a cluttered visual environment, it may be better to use nonvisual cues to alert the user to critical information. It may be possible to design command and control systems that automatically adapt to the unique strategies and requirements of individual users. Or some system components may be fully automated, bypassing the human user altogether. These and other related questions are critical in helping the 21st-century soldier fully exploit the capabilities of modern digital systems.

ROBOTICS AND AUTOMATED SYSTEMS

The military is turning increasingly to robots and unmanned systems to complete tasks once assigned strictly to humans or human-operated systems. Current military robotic systems include unmanned aerial vehicles (UAVs) and a host of unmanned ground vehicles (UGVs). Missions include aerial surveillance, use of missiles or other ordinance on high value targets, and detecting and/or removing hazardous materials such as bombs. Robotics are a major part of the Army's Future Combat System, and as such human factors engineers are heavily engaged in understanding the dynamics of all aspects of the human-robot interface.

Currently, humans play a major role in operating robotic systems. These systems are at best semiautonomous given the key role of the human operator in the system. In the future, fully autonomous weapons systems may be fielded. For example, unmanned, fully autonomous aircraft are being designed that have the capability to "loiter" in a battle space for extended periods of time, and to shoot and kill certain targets, for instance, enemy vehicles. Human factors engineers must learn how to design the artificial intelligence (AI) of these systems to allow nearly perfect performance. To that end, the systems must identify and select enemy targets from the myriad of stimuli present on the battlefield, and engage the enemy as necessary. From a signal detection model point of view, they must maximize "hits" (killing the correct target) and "correct rejections" (correctly identifying a target as nonenemy) while minimizing "false alarms" (killing friendly targets) and "misses" (failing to detect enemy targets). This will require a robotic system that can sense, decide, and act—quickly, with no room for error.

The most familiar robotic system at the current time is the UAV. Human factors engineers play a vital role in determining how to design command and control systems that allow the UAV "pilots" to operate the systems with minimal error. Unlike pilots of traditional aircraft, the UAV pilot does not receive physiological feedback such as yaw, pitch, or roll. The absence of such cues necessitates a greater reliance on the displays of the UAV command and control system. This, in turn, raises classic human factors engineering questions pertaining to the optimal design of displays and controls, how to best train the operators, and what are the limits of the operator's ability to control multiple UAVs simultaneously. A large and growing literature exists on this subject.

Although not traditionally a topic for human factors engineers, it is worth noting that UAV pilots experience considerable stress as a result of "flying" the aircraft in combat operations. There are reports of some of these pilots experiencing posttraumatic stress disorder (PTSD) symptoms. This raises the possibility that human factors engineers may need to address ways of selecting and training these personnel, and design systems to lessen the odds of a pathologic response.

NEUROERGONOMICS

A rapidly emerging field relevant to military human factors engineering is neuroergonomics. Neuroergonomics involves engineering an interface between the brain and various psychomotor and behavioral systems. For example, basic research with primates shows that they can be trained to control robotic arms through the use of sensors placed directly into the animal's brain. In general, the objective is to use an understanding of the brain to build interfaces with systems and technologies in the real world.

There are many possible military applications. Past research, for example, has examined the plausibility of using brain waves, as measured by electroencephalographs (EEGs), to control aircraft or weapons systems. A pilot could decide to engage an enemy aircraft, and brain sensors could almost instantly activate a weapon. This might allow the pilot to engage the enemy faster (at Mach 2, every fraction of a second matters), as well as free up the pilot's arms and hands for other tasks. By extension, such a system could be used with UAV pilots to improve performance.

Another—and perhaps more likely—application of neuroergonomics may be in the design of artificial limbs for military (and civilian) amputees. Based on sensors placed into the sensory and motor centers in the brain, it may be possible to design prosthetics that behave like a real arm or leg, and also provide feedback that feels like the missing limb. This would make it easier for amputees to adjust to their injuries, and also enable them to remain on active duty and with fewer restrictions than are supported by current prosthetics.

TRAINING AND SIMULATIONS

Before a US fighter pilot ever engages an enemy aircraft, he or she has flown hundreds of realistic training missions in a flight simulator. In doing so, the pilot has built a large library of scripts—courses of action—to match to almost any tactic than an enemy pilot may employ. It is now possible, with modern simulation technologies, to provide similar training to other military occupational specialties, notably ground troops. This "bloodless" training can produce military members and leaders who can perform at a high level early in their first combat experience.

Human factors engineers identify the emotional, perceptual, and cognitive components of decision-making that is to be simulated. Scenarios that shape and stretch these components must then be integrated into the content of the simulation. There are many currently unresolved questions that impact the design of these simulations. The sights, tastes, and sounds of the battle environment may be vital in training the soldier on how to deal with stress and the emotional component of combat. Realistic and diverse scenarios ranging from traditional firefights to complicated negotiations with tribal leaders will build the mental scripts the soldier needs to prepare for diverse missions. Duplicating the "fog of war," both perceptual and cognitive, will aid the soldier in learning to deal with ambiguity, which is one of the immutable aspects of war.

SOCIAL-CULTURAL FACTORS

Success in 21st-century war hinges as much on understanding and appreciating the social and cultural nature of the enemy as it does the employment of traditional firepower. Human factors engineers may play a significant role in improving military performance in this domain by assisting in the design and use of technologies that facilitate training in these areas. Hand-held language translators must reflect subtle nuances in both the denotation and connotation components of linguistic expression, cultural differences in direct and indirect use of speech, and gender difference in the use of language. Human factors engineers can apply the research methods and analytic skills used in other areas to inform designers how to create immersive simulations that train military personnel about the customs, beliefs, and behaviors of other cultures.

The Human Factors and Ergonomics Society (HFES) has 23 separate technical groups, each of which represents a different area of basic research or application within the field. It is beyond the scope of this chapter to explore all of the possible applications of human factors engineering to the military, and the interested reader is directed to the HFES website (www.hfes.org) to gain a broader appreciation of the general field of human factors engineering. In the end, it is worth remembering that war is a political tool that depends on human beings for success or failure. By systematically considering the role of the human being in military systems, human factors engineering thus plays a pivotal role in modern war.

References

Krueger, G. P. (2012). Military engineering psychology: Setting the pace for exceptional performance. In J. H. Laurence & M. D. Matthews (Eds.), *The Oxford handbook of military psychology* (pp. 232–240). New York, NY: Oxford University Press.

Kahneman, D., & Klein, G. A. (2009). Conditions for intuitive expertise. *American Psychologist, 64,* 515–526. doi:10.1037/a0016755

Matthews, M. D. (2012). Cognitive and non-cognitive factors in soldier performance. In J. H. Laurence & M. D. Matthews (Eds.), *The Oxford handbook of military psychology* (pp. 197–217). New York, NY: Oxford University Press.

Matthews, M. D., & Laurence, J. H. (2012). *Military psychology: Vol. 2. Applied experimental and engineering psychology.* London: Sage.

Sanders, M. S., & McCormick, E. J. (1993). *Human factors in engineering and design* (7th ed.). New York, NY: McGraw-Hill.

17 CLINICAL HEALTH PSYCHOLOGY IN MILITARY SETTINGS

Alan L. Peterson

Clinical health psychology has been one of the fastest-growing specialty areas of psychology over the past three decades (Andrasik, Goodie, & Peterson, in press; Belar & Deardorff, 2009). Clinical health psychology is both a specialty field within clinical psychology as well as a general field applicable to many psychologists working in military settings. Sometimes referred to as behavioral medicine, clinical health psychology involves the assessment and treatment of individuals who have psychological factors that affect their physical condition. Some of the most common conditions seen by clinical health psychologists include nicotine dependence, overweight and obesity, chronic pain, insomnia, cancer, cardiovascular disorders, and gastrointestinal disorders. Clinical health psychologists working in military treatment facilities see both inpatients and outpatients in deployed and nondeployed locations and are often involved in the development and implementation of population health interventions designed to target behavioral health risk factors in military populations as a whole. During times of military conflict, clinical health psychology has increased in importance for military populations in deployed locations and in garrison because of the significant increase in medically injured military patients.

The seminal textbook for clinical health psychologists is *Clinical Health Psychology in Medical Settings: A Practitioner's Guidebook* by Cynthia Belar and William Deardorff

(2009). This book reviews the roles and functions of clinical health psychologists as well as education, training, and personal and professional issues related to practice. The book includes chapters on assessment, treatment, and consultation. The unique legal and ethical issues encountered by clinical health psychologists in evaluating and treating medical patients are also reviewed. A comprehensive review of the practice of clinical health psychology and behavioral medicine in military medical settings has been provided by Peterson, Hryshko-Mullen, and McGeary 2012.

EDUCATION AND TRAINING REQUIREMENTS

The recommended minimum training requirements for individuals identified as clinical health psychologists in military settings is the completion of (1) a doctoral program in clinical or counseling psychology (PhD or PsyD) accredited by the American Psychological Association (APA); (2) an APA-accredited predoctoral internship program; and (3) a one- or two-year postdoctoral fellowship in clinical health psychology or behavioral medicine. It is also highly recommended, but not required, that clinical health psychologists become board certified in clinical health psychology by the American Board of Professional Psychology. The importance of board certification may be more significant for clinical health psychologists because most work in medical settings is done in close collaboration with physicians, for whom specialty board certification is often considered a requirement.

In the past, military psychologists were sometimes sponsored to complete postdoctoral fellowships in clinical health psychology or behavioral medicine at civilian institutions. However, more recently, most military psychologists complete military-sponsored fellowships. There are currently eight APA-accredited specialty practice postdoctoral residency programs in clinical health psychology, and three of these programs are military programs. The US Air Force sponsors a 2-year APA-accredited postdoctoral fellowship at Wilford Hall Ambulatory Surgical Center in San Antonio, Texas. The US Army sponsors 2-year APA-accredited fellowships in clinical health psychology at the San Antonio Military Medical Center (formerly known as Brooke Army Medical Center) and at Tripler Army Medical Center in Hawaii. The US Navy does not currently sponsor psychologists for postdoctoral fellowship training in clinical health psychology.

The provision of clinical health psychology services in military settings is not limited to fellowship-trained clinical health psychologists. Many non-fellowship-trained clinical and counseling psychologists provide tobacco cessation, weight management, chronic pain management, and other services as part of their regular clinical practice. Most clinical and counseling psychologists have some exposure to clinical health psychology coursework and supervised clinical experience during their graduate school training. In addition, many clinical psychology internships at military training sites include clinical health psychology rotations. As a result, many generalist psychologists will do some clinical health psychology work, similar to how some non-fellowship-trained psychologists will perform limited neuropsychological evaluations for traumatic brain injuries.

to: (1) assessment, (2) intervention, (3) consultation, (4) research, (5) supervision and training, and (6) management and administration (France et al., 2008). These competencies are further subdivided into *knowledge-based* and *applied* competencies. For example, a *knowledge-based intervention competency* connotes that an entry-level clinical health psychologist should have knowledge of psychological factors associated with health behavior, illness, and disease, along with their implications for the delivery of biopsychosocial treatments. Indeed, a major emphasis of clinical health psychology fellowship training is extensive didactic instruction in medical and psychophysiological disorders such as headaches, gastrointestinal disorders, cancer, cardiovascular diseases, diabetes, and temporomandibular disorders. This knowledge is particularly valuable when serving in deployed hospital settings, where clinical health psychologists often go on medical/surgical rounds with the attending physicians.

An example of an *applied intervention competency* is that an entry-level clinical health psychologist should be able to implement an evidence-based treatment by integrating the best available research with clinical expertise in the context of patient characteristics, culture, and preferences. For example, thorough understanding of the medical and physiological factors involved in tension-type and migraine headaches can be valuable in the development of cognitive-behavioral interventions for blast-related postconcussive headaches. Many of the medical disorders treated by clinical health psychologists have evidence-based treatment manuals that have been developed and evaluated in randomized clinical trials. A detailed description of all of the knowledge-based and applied competencies is beyond the scope of this chapter, but those interested can review them in the original published manuscript on this topic (France et al., 2008).

CLINICAL HEALTH PSYCHOLOGY COMPETENCIES

Leaders in the field have outlined six competency areas in clinical health psychology related

ASSESSMENT IN CLINICAL HEALTH PSYCHOLOGY

Clinical health psychology is perhaps the psychology specialty with the strongest emphasis on biopsychosocial assessment approaches (Andrasik et al., in press). The biopsychosocial model refers to the influence of biological, psychological, and social factors in psychological and physical health and disease (Engel, 1977). The application of this model within clinical health psychology often includes additional domains such as physical, emotional, cognitive, behavioral, and environmental factors. The unique environmental factors involved in military settings can be particularly important. For example, the biopsychosocial assessment of a Special Operations Forces (SOF) service member with chronic back pain for admission into an interdisciplinary functional restoration program requires an understanding of the unique military culture and cognitive mindset associated with the SOF environment. In addition, the assessment of treatment-outcome goals for SOF service members involved in physical rehabilitation must be set at the high level of fitness standards required for this career field.

A common misperception among health care providers working in medical settings is what is referred to as "mind-body dualism." This misperception is the belief that a particular medical condition is caused by physical *or* psychological factors, rather than both. This can be particularly true in military settings, where significant stigma is often associated with seeking treatment by a psychologist. The true embodiment of the biopsychosocial model within clinical health psychology is that all medical conditions are influenced by a combination of physical, emotional, cognitive, behavioral, and environmental factors. In addition, clinical health psychologists emphasize the bidirectional influences involved in these biopsychosocial factors when conducting a clinical assessment. The primary goal of a clinical health psychology assessment is to determine the degree to which each of these factors contributes to diseases, disorders, and illnesses as well as overall health and fitness for duty.

Clinical health psychologists employ a variety of self-report and diagnostic interview approaches in the assessment of patients in military medical settings. A somewhat unique aspect of assessment approaches for clinical health psychologists is the reliance on biological or physical assessments and measurements such as blood pressure, cholesterol levels, leukocytes, polysomnography reports, and blood glucose levels. Another assessment activity often conducted by clinical health psychologists is presurgical screenings. The most common assessments of this type include screenings for gastric surgery for morbid obesity, spinal cord stimulator implantation for chronic pain, and organ transplant donation. A comprehensive review of assessment approaches for clinical health psychologists is included in the book titled *Biopsychosocial Assessment in Clinical Health Psychology: A Handbook* by Andrasik and colleagues (in press).

The most common *Diagnostic and Statistical Manual for Mental Disorders* (DSM-IV-TR) diagnostic code used by clinical health psychologists is Psychological Factors Affecting Medical Condition. The first criterion for this disorder is that a general medical condition must be present. The second criterion is that psychological or behavioral factors adversely affect the general medical condition in one of a variety of ways. For example, musculoskeletal pain conditions can be initiated or maintained by the use of personal protective equipment (e.g., body armor) during military deployments.

The use of this diagnosis often helps patients who are seen in clinical health psychology clinics to "save face." Patients referred by their physician to a psychologist for the assessment and treatment of a health concern often think this means their physician does not believe they have a *real* medical disorder. Many clinical health psychologists will tell their patients that they only see patients with *real* medical disorders, diseases, or illnesses and that if in the process of their evaluation it is determined that they have a mental disorder, they will be referred from the clinical health psychology clinic to the local mental health clinic for treatment. This discussion and clarification of the biopsychosocial model of disease and illness is often sufficient to allay the apprehension of medical patients seeking assistance from a clinical health psychologist.

TREATMENT INTERVENTIONS IN CLINICAL HEALTH PSYCHOLOGY

As is suggested by the previous discussion, most clinical health psychologists treat patients with primary medical disorders rather than mental disorders. Many military clinical health psychologists will serve as the Chief of Clinical Health Psychology at a military medical center after completion of their fellowship training. Therefore, most military clinical health psychologists receive broad-based training to prepare them to assess and treat any type of medical or dental patient referred to them for inpatient or outpatient care. Some of the diseases, illnesses, injuries, and health-risk behaviors treated by clinical health psychologists include:

- Amputations
- Cancer
- Cardiovascular disorders
- Chronic pain (e.g., back pain, headaches, fibromyalgia)
- Dental anxiety and fear
- Diabetes
- Gastrointestinal disorders (e.g., irritable bowel syndrome, fecal incontinence)
- Physical inactivity
- Sleep disorders (e.g., insomnia, circadian rhythm disorder)
- Spinal cord injury
- Temporomandibular disorders
- Tobacco cessation
- Weight management

The majority of the treatment provided by most clinical health psychologists involves individual treatment of patients using cognitive-behavioral interventions. However, some health conditions are well suited for group treatment programs such as tobacco cessation, relaxation training, weight management, chronic pain management, pulmonary rehabilitation, and cardiac rehabilitation. Clinical biofeedback is another treatment approach that is often conducted by clinical health psychologists in military medical settings. Most military treatment facilities require that psychologists be certified by the Biofeedback Certification Institute of America or meet some other specified educational and supervised training requirements in order to be credentialed to perform biofeedback. Electromyogram and thermal biofeedback are the most commonly used biofeedback approaches in military treatment facilities. Common conditions treated with biofeedback include chronic headaches, irritable bowel syndrome, Raynaud's disease, fecal incontinence, and urinary incontinence.

Military clinical health psychologists are also often involved in the development, implementation, and evaluation of population health interventions with military health care beneficiaries. Population health approaches include clinical applications and interventions targeted at an entire patient population rather than individual patients. Less intensive clinical interventions delivered to entire populations of health care beneficiaries have the potential to have an even greater impact on the overall patient population than more potent treatments delivered to a small percentage of patients. Tobacco cessation is one of the best examples of the potential impact of population health interventions. Primary care providers using a universal brief intervention (e.g., 1–2 minutes) with all tobacco users seen in their clinic can bring about a greater reduction in tobacco use throughout a military installation than the comprehensive, multisession tobacco cessation programs that treat only those individuals who seek help in quitting tobacco (Peterson, Vander Weg, & Jaén, 2011).

CLINICAL HEALTH PSYCHOLOGISTS WORKING IN MILITARY PRIMARY CARE SETTINGS

In the late 1990s there was an emergence of interest in the use of psychologists in military primary care settings. Although the colocation of psychologists and other mental health professionals into primary care settings had occurred for many years, a new model emerged in the field involving psychologists working as *behavioral health consultants* for primary care physicians (Hunter, Goodie, Oordt, & Dobmeyer, 2009). Clinical health psychologists were some of the first military psychologists trained to work as behavioral health consultants in primary care because of their special expertise in working with health risk behaviors such as smoking and excessive weight, as well as medical conditions such as chronic pain and insomnia (Gatchel & Oordt, 2003). This model includes having a psychologist support the primary care managers as behavioral health consultants. Appointment times for behavioral health consultants are modeled after those of primary care providers and usually last no more than 30 minutes with a maximum of about four appointments scheduled several weeks apart. The most recent version of this model to be adopted in military medical treatment facilities is the patient-centered medical home. Similar to the behavioral health consultant model, clinical health psychologists serving in this role do not follow patients for outpatient therapy as they might in a specialty mental health clinic. If more comprehensive psychological assessment or treatment is required, the patient is referred to a specialty mental health or clinical health psychology clinic.

CLINICAL HEALTH PSYCHOLOGY DURING MILITARY DEPLOYMENTS

Clinical health psychologists play an important role during military deployments. In combat surgical hospitals and theater hospitals, for example, clinical health psychologists are well prepared to provide brief behavioral assessments and interventions with severely medically injured patients such as those with amputations, burns, and traumatic orthopedic injuries. Many military mental health professionals are not adequately prepared for the personal exposure to severely injured patients that often occurs during military deployments, such as mass casualty incidents after massive explosions. Military clinical health psychologists with extensive predeployment experience working with severely ill or injured inpatients at military medical centers may be better prepared to withstand the personal health-care-stress exposure that occurs during military deployments. Clinical health psychologists in military settings play a vital role in maintaining military operational readiness in both deployed and nondeployed locations.

References

Andrasik, F., Goodie, J., & Peterson, A. L. (Eds.). (in press). *Biopsychosocial assessment in clinical health psychology: A handbook.* New York, NY: Guilford.

Belar, C. D., & Deardorff, W. W. (2009). *Clinical health psychology in medical settings: A practitioner's guidebook* (2nd ed.). Washington, DC: American Psychological Association.

Engel, G. L. (1977). The need for a new medical model: A challenge for biomedicine. *Science, 196,* 129–136.

France, C. R., Masters, K. S., Belar, C. D., Kerns, R. D., Klonoff, E. A., Larkin, K. T., . . . Thorn, B. E. (2008). Application of the competency model to clinical health psychology. *Professional Psychology: Research and Practice, 39,* 573–580.

Gatchel, R. J., & Oordt, M. S. (2003). *Clinical health psychology and primary care: Practical advice and clinical guidance for successful collaboration.* Washington, DC: American Psychological Association.

Hunter, C. L., Goodie, J. L., Oordt, M. S., & Dobmeyer, A. C. (2009). *Integrated behavioral health in primary care: Step-by-step guidance for assessment and intervention.* Washington, DC: American Psychological Association.

Peterson, A. L., Hryshko-Mullen, A. S., & McGeary, D. M. (2012). Clinical health psychology and behavioral medicine in military healthcare settings. In C. H. Kennedy & E. A. Zillmer (Eds.), *Military psychology: Clinical and operational applications* (2nd ed., pp. 121–155). New York, NY: Guilford.

Peterson, A. L., Vander Weg, M. W., & Jaén, C. R. (2011). *Nicotine and tobacco dependence.* Cambridge, MA: Hogrefe.

18 HOSTAGE NEGOTIATION IN THE MILITARY

Laurence Miller

In the world of emergency mental health, there are few emergencies as critical as a hostage crisis. Lives are at imminent risk of violent death, often at the hands of an unstable and desperate perpetrator, in the midst of a chaotic and uncontrolled environment. To date, however, the Armed Services provide little formal training in hostage and crisis negotiation (Rowe, Gelles, & Palarea, 2006), despite the fact that more and more military service members are being deployed to nontraditional battle sites and in peacekeeping missions. This chapter adapts the principles and practices of hostage and crisis negotiation developed in the field of civilian law enforcement that can be productively applied to the military setting (Greenstone, 2005; McMains & Mullins, 1996; Miller, 2005, 2006, 2008; Slatkin, 2010). All recommendations herein should be reinforced and supplemented by appropriate training.

TYPES OF HOSTAGE CRISES

Although every situation is unique (McMains & Mullins, 1996; Miller, 2005, 2006; Rowe et al., 2006), there appear to be some general categories of hostage crises that military service members may encounter.

Planned operational hostage scenario. In this scenario, the criminal or tactical operational plan includes the deliberate use of hostages, usually when escape is deemed to be virtually impossible otherwise, as with big-score robberies by criminal gangs, guerrilla raids by paramilitary fighters, or planned escapes by prisoners in military or civilian detention facilities.

Planned ideological hostage scenario. The political or religiously motivated hostage taker (HT) has a clear ideological agenda for his actions, which often characterizes terrorist hostage scenarios. This is a particularly dangerous situation, because the HTs may be willing to die for their cause and to kill others with impunity.

Miscalculated robbery. Far more common is the ordinary bank or store robbery gone sour, in which the crooks plan for a quick in-and-out, but law enforcement appears on the scene sooner than expected, and now the robbers are trapped in the building with unwitting employees and customers, who have just become de facto hostages.

Escalating domestic crisis. Here, what may have begun as a fight between a couple escalates to the point where one of the combatants, usually the male, effectively barricades his mate inside a dwelling and refuses to let her leave. In another version of this scenario, an estranged spouse shows up at the home or worksite of his mate, already prepared for a confrontation, and often armed. The hostage crisis then ensnares any family members or coworkers who may be on the scene.

Mentally disordered hostage taker. This may overlap with any of the above categories, where at least part of the HT's motivation is fueled by emotional disturbance and/or delusional ideation. The most common types of mental disorders seen in HTs are psychotic disorders, mood disorders, and personality disorders, especially antisocial and borderline personality disorder. The inherent unpredictability of mentally disordered behavior makes this type of hostage situation one of the most dangerous, often requiring focused and specialized negotiating strategies (McMains & Mullins, 1996; Miller, 2005, 2006).

HOSTAGE CRISIS RESPONSE: BASIC PROTOCOL

While life-and-death crises rarely go by the numbers, there does appear to be a certain uniformity that guides the evolution of most hostage scenarios and that consequently prescribes the measures used to contain it (Greenstone, 2005; McMains & Mullins, 1996; Miller, 2005, 2006; Slatkin, 2010).

Secure the perimeter to isolate and contain the hostage taker(s). As a rule, the perimeter should be large enough to allow freedom of movement of the tactical and negotiating teams, but small enough to be kept under observation and control by the authorities. More than one perimeter, that is, inner and outer, may be necessary.

Control the scene. Often, you will have to work around the realities of the surrounding community, which includes marshaling medical services, controlling local traffic, dealing with the media, and keeping the surrounding community sufficiently informed to protect their safety.

Establish communication with the hostage taker(s). The sooner you begin a dialogue with the HT, the less time he has to stew and consider drastic options.

While face-to-face contact between the negotiator and the HT is categorically discouraged, any safe means of communication—line phone, cell phone, bullhorn, or even text messaging or e-mail—should be established as soon as possible.

GENERAL COMMUNICATION STRATEGIES IN HOSTAGE NEGOTIATION

While customizing your communications approach to the individual HT's motives and personality, there are a number of general recommendations for communicating with HTs (Greenstone, 2005; McMains & Mullins, 1996; Miller, 2005; Slatkin, 2010).

Minimize background distractions. Distractions include more than one person speaking at a time, background radio chatter, road noise, and so forth.

Open your dialogue with an introduction and statement of purpose. "This is USMC Sergeant Bruce McGill of the Fort Pendleton Crisis Response Team. I'm here to listen to you and to try to make sure everybody stays safe."

To build rapport, ask what the HT likes to be called. When in doubt, avoid overfamiliarity and address him respectfully, for example, "Sir," "Corporal," and so forth.

Speak slowly and calmly. People's speech patterns often mirror the tone of the dominant conversation, so provide a model of slow, calm, clear communication from the outset. This implies being able to keep calm yourself.

Adapt your dialogue to HT's vocabulary and cognitive level. Avoid either talking over the head of the HT or patronizing him by talking down to him or trying to mimic his pattern or level of speech too closely. Avoid overfamiliarity or unnecessary profanity.

Encourage venting, but de-escalate ranting. Allow the HT to freely express his frustrations and disappointments; let him "tell his story." But don't let venting become unproductive spewing or ranting, which can lead to further loss of control.

Ask for clarification. Clarity is a central principle of all forms of crisis intervention, and a sign of interest, concern, and respect. Don't respond to, or act on, a HT's statement unless you're reasonably sure you know what he means.

Focus the conversation on the HT, not the hostages. Generally, the less the HT thinks about the hostages, the better, especially where the hostages are family members or coworkers who have been targeted to make a statement. Keep the dialogue focused on the HT's concerns.

Be supportive and encouraging about the outcome. Within the bounds of reality and believability, downplay the HT's actions so far: Remember, the goal is to keep violence from escalating from this point on, and the best way to facilitate this is to encourage the HT to believe that there is still a way out of the worst possible consequence. Compliment the HT for any positive actions he's taken and encourage further constructive efforts.

Avoid unproductive verbal strategies. These include: (1) arguing with the HT; (2) engaging in power plays; (3) moralizing; or (4) diagnosing.

ACTIVE LISTENING SKILLS

Active listening consists of multipurpose communication tools that can be effectively applied to hostage negotiations (Greenstone, 2005; McMains & Mullins, 1996; Miller, 2005, 2006; Slatkin, 2010). These include the following.

Emotion labeling. Help the subject clarify what he's feeling by identifying the emotions your hear him express. This contributes to a state of calmness by reducing internal confusion.

Paraphrasing. Rephrase the subject's statement in your own words. This reinforces empathy and rapport, clarifies what the subject is saying, allows him the opportunity for correction, and encourages him to slow down and listen.

Reflecting/mirroring. Repeat the last word or phrase, or the key word or phrase, of the subject's statement in the form of a question, thereby soliciting more input without actually asking for it.

Minimal encouragers. These are short utterances and questions that let the HT know that the negotiator is listening, but don't interfere with the HT's narrative flow: "Oh?" "I see." "Yeah." "Uh-huh." "When?" "And?" "Really?" "You do?"

Silence and pauses. Periods of silence can be used strategically to buy time and to encourage the subject to fill in the gaps, which keeps him talking. Following your own statement by a silent pause is also a way of emphasizing a point you've just made.

"I"-statements. People under extreme stress often become suspicious and defensive, and any statements that are too directive ("you should... ") may sound like an insult or attack. I-statements clue the subject in on what effect he's having on the negotiator's perception, while at the same time allowing for some subjectivity and personalization of the negotiator: "I have a hard time understanding you when you're going so fast. I want to make sure I get what you're saying."

Open-ended questions. Ask questions that cannot be answered with a simple yes-or-no. This encourages the HT to say more without the negotiator actually directing the conversation. This technique may be used in combination with other active listening techniques, and may be followed or combined with closed-ended queries.

DEMANDS AND DEADLINES

One of the defining characteristics of most hostage crises is the presence of some form of demand, which may range from the concrete and immediately practical (food, transportation) to the more grandiose and expansive (release of political prisoners, access to media) to the abstract and bizarre (freedom from government persecution; emancipation of downtrodden classes). Most demands will be of the first type, and most experts would agree with the following principles regarding such demands in hostage crises (Greenstone, 2005; McMains & Mullins, 1996; Miller, 2005, 2006; Slatkin, 2010).

Quid pro quo. Make the HT work for everything he gets (food, electricity) by extracting a concession in return—for example, keep communication open, better treatment of hostages, release of one hostage—for each demand satisfied.

Don't ask the HT if there are any demands. Let him ask you.

Don't offer anything not explicitly asked for. Exceptions include hostage health and safety issues: "Does anyone need medical attention?"

Avoid saying "no." But this is not equivalent to saying yes. That is, deflect, postpone, and modify: "Okay, you want a car to the airport, right? I'll see what I can do. Meanwhile, tell me..." If a "no" slips out of you, don't sweat it; just continue negotiating.

Prioritize hostages. When negotiating for release of multiple hostages, start with the most vulnerable or the least manageable from the HT's standpoint, such as sick or injured victims, children, or overly hysterical hostages.

Negotiable demands. Negotiable demands include food, drinks, cigarettes, and environmental controls, such as heat, air conditioning, electricity, plumbing, blankets, and so on.

Nonnegotiable demands. Nonnegotiable demands include illegal drugs, weapons, release of friends or relatives in prison, or exchange of hostages.

Gray area demands. "Gray area" demands depend on the special circumstances and judgment of the negotiating team, and include alcohol, money, media access, transportation, or freedom.

Talk through deadlines. If the HT makes a deadline, log it, but don't mention it again to the HT if he doesn't bring it up. Try to ignore the deadline and let it pass by keeping the HT engaged in conversation. If he brings it up, try to deflect the conversation to more here-and-now concerns.

THE SURRENDER RITUAL

Nobody likes to surrender, to give up, to capitulate, to lose. Yet, by definition, the successful resolution of a hostage crisis entails the safe release of the hostages and surrender of the HT to authorities. On the strength of practical experience, a basic protocol, or surrender ritual has evolved to guide negotiators in their efforts to safely resolve a crisis (Greenstone, 2005; McMains & Mullins, 1996; Miller, 2005, 2006; Slatkin, 2010). As with all such guidelines, each negotiator must adapt this system to his or her particular situation and type of HT.

Watch your language. When dealing with the HT, avoid the use of words like "surrender," "give up," or other terms that connote weakness and loss of face. "Coming out" is a preferred term because it implies a proactive decision by the subject himself to resolve the crisis.

Make resolution attractive. To begin the discussion of coming out, emphasize to the HT what he has to gain by this action at the present time. Be realistic but optimistic, and try to minimize any damage done so far.

Make a plan. Discuss various coming-out scenarios and identify a mutually acceptable plan. Let the HT set the pace; if he is agreeing to come out at all, this is not the time to rush things. Make sure the plan is understood and agreed on by everyone: the HT(s), the negotiating team, the tactical team, and the on-scene command staff.

Implement the plan. This is super-high-adrenalin territory; a misunderstanding or misstep could blow the whole deal and cost lives. Basic elements of a surrender scenario include: (1) no weapons, or objects that could be mistaken for weapons, on the person of the HT; (2) hands where they can be seen (usually on the HT's head); (3) no bulky clothing; (4) all movements very slow; (5) speak when spoken to; (6) obey all commands from authorities; and (7) do not resist arrest or restraint by authorities.

Follow up. During and after the arrest, the negotiator should maintain engagement, rapport, and communication with the HT. If possible, a brief informational debriefing with the HT should occur in a secure place close to the scene in order to gather any information that might be forgotten or discarded later on, and to give the negotiator the opportunity to reinforce the subject for his contribution to successfully resolving this crisis. In this way, the credibility of the law enforcement team is maintained throughout the subsequent investigation and trial, and also sends the broader message to the community of the hostage negotiation team as honest brokers, which will serve them well in the next crisis.

TRAINING AND PROFESSIONALISM

As noted in the introduction, the Armed Services have provided little formal training in hostage and crisis negotiation (Rowe et al., 2006). However, military personnel can take a lesson from their law enforcement colleagues (Miller, 2006, 2008) and develop training programs suited to their unique needs. For example, negotiating with armed insurgents in a foreign country may require a specialized skillset with regard to linguistic and cultural factors that differs somewhat from that which has proven useful in negotiating with a stateside bank robber or distraught family member. Nevertheless, the foundational principles of crisis communication outlined in this chapter are universal and can provide the nucleus for training professional negotiators to apply their skills to a wide range of military and civilian settings. In any land, in any language, crisis negotiation is all about saving lives with the power of the human word.

References

Greenstone, J. L. (2005). *The elements of police hostage and crisis negotiations: Critical incidents and how to respond to them.* New York, NY: Haworth Press.

McMains, M. J., & Mullins, W. C. (1996). *Crisis negotiations: Managing critical incidents and situations in law enforcement and corrections.* Cincinnati, OH: Anderson.

Miller, L. (2005). Hostage negotiation: Psychological principles and practices. *International Journal of Emergency Mental Health, 7,* 277–298.

Miller, L. (2006). *Practical police psychology: Stress management and crisis intervention for law enforcement.* Springfield, IL: Charles C. Thomas.

Miller, L. (2007). Negotiating with mentally disordered hostage takers: Guiding principles and practical strategies. *Journal of Police Crisis Negotiations, 7,* 63–83.

Miller, L. (2008). Military psychology and police psychology: Mutual contributions to crisis intervention and stress management. *International Journal of Emergency Mental Health, 10,* 9–26.

Rowe, K. L., Gelles, M. G., & Palarea, R. E. (2006). Crisis and hostage negotiation. In C. H. Kennedy & E. A. Zillmer (Eds.), *Military psychology: Clinical and operational applications* (pp. 310–330). New York, NY: Guilford.

Slatkin, A. A. (2010). *Communication in crisis and hostage negotiations: Practical communication techniques, stratagems, and strategies for law enforcement, corrections, and emergency service personnel in managing critical incidents* (2nd ed.). Springfield, IL: Charles C. Thomas.

19 MENTAL HEALTH ADVISORY TEAMS

A. David Mangelsdorff

Between October 2001 and June 2012, over 1.6 million US military personnel deployed to combat operations in Iraq and Afghanistan in support of the Global War on Terror. Many troops served multiple tours with little time to recover between deployments. To assess the effects of combat operations and the psychological adjustment of troops, Army mental health advisory teams were created and deployed to the combat theaters. The Department of Defense was concerned about numerous factors affecting the combat operations including the operational tempo (pace, intensity, duration of deployment tour), the environment (extreme temperatures, unfamiliar weather, and terrain), stressors (lack of unit cohesion, multiple deployments, uncertainties, stigma, drawdown), and casualties (deaths from accidents, hostile action, illness, self-inflicted). Other considerations were: troop and unit demographics (Active versus Guard/Reserve, age, maturity, family), role (combat versus combat service support), nature of the conflict (urban warfare, unconventional weapons), exposure to appropriate training (Battlemind, readiness, and suicide prevention), and health care support (number, location, patient load, and distribution of behavioral health personnel). Together all of these threats contributed to intensified operational stress reactions. The Department of Defense was consistently concerned about the numbers of personnel deployed, the number and types of casualties,

how military health care was organized and delivered, the stigma of seeking mental health assistance, and the ability of troops to adjust (both in the operational theater, after returning to home stations, and family reunions). The Department of Veteran Affairs intensified tracking efforts looking for potential long-term effects on veterans.

It is necessary to understand the casualty statistics and their potential impacts on policy and training decisions. Inspection of the Defense Casualty Analysis System (DCAS) reports of active duty military deaths from calendar years 1980 to 2010 (see Table 19.1) provides numbers from accidents, hostile activities (combat), illness, and self-inflicted casualties (suicide). Military life is dangerous; historically deaths from accidents and hostile activities generally exceed those from illnesses and self-inflicted causes. The increase in suicide rates among military personnel after 2003 from 11.9 to 15.6/100,000/year generated increased attention within the Department of Defense (Defense Manpower Data Center, 2012). The military casualty data must be considered in the context of experiences, gender, and age adjusted cohorts. Suicide is the tenth leading cause of death in the United States (Satcher, 1999). Work related conditions (such as military service and combat deployments) can contribute to increasing risk factors for suicide.

In 2003 the Department of Defense initiated the Force Health Protection program

TABLE 19.1. Active Duty Military Casualties per 100,000 Serving by Cause

Calendar Year	Mil FTE	Deaths	Accident	Hostile Action	Homicide	Illness	Self-Inflict
1980	2,159,630	2,392	1,556	0	174	419	231
1981	2,206,751	2,380	1,524	0	145	457	241
1982	2,251,067	2,319	1,493	0	108	446	254
1983	2,273,364	2,465	1,413	18	115	419	218
1984	2,297,322	1,999	1,293	1	84	374	225
1985	2,323,185	2,252	1,476	0	111	363	275
1986	2,359.855	1,384	1,199	2	103	384	269
1987	2,352,697	1,983	1,172	37	104	383	260
1988	2,309,495	1,819	1,080	0	90	321	285
1989	2,303,384	1,636	1,000	23	58	294	224
1990	2,258,324	1,507	880	0	74	277	232
1991	2,198,189	1,787	931	147	112	308	256
1992	1,953,337	1,293	676	0	109	252	238
1993	1,849,537	1,213	632	0	86	221	236
1994	1,746,482	1,075	544	0	83	206	232
1995	1,661,928	1,040	538	0	67	174	250
1996	1,613,675	974	527	1	52	173	188
1997	1,578,382	817	433	0	42	170	159
1998	1,538,370	827	445	0	26	174	165
1999	1,525,942	796	439	0	38	154	150
2000	1,530,430	832	429	0	37	180	153
2001	1,552,096	943	461	12	49	197	153
2002	1,627,142	1,051	565	17	54	213	174
2003	1,732,632	1,399	597	312	46	231	190
2004	1,711,916	1,847	605	735	46	256	197
2005	1,664,014	1,929	646	739	54	280	182
2006	1,611,533	1,882	561	769	47	257	213
2007	1,608,226	1,953	561	847	52	237	211
2008	1,683,144	1,440	506	352	47	244	259
2009	1,640,751	1,515	467	346	77	277	302
2010	1,685,178	1,485	424	456	39	238	289

Retrieved from https://www.dmdc.osd.mil/dcas/pages/report_number_serve.xhtml

to maintain and protect military personnel through initiatives supporting health services support, fitness and health promotion, protection, disease surveillance, accident prevention, and medical and rehabilitative care (Winkenwerder, 2003). Individual personnel health and adjustment was monitored in all military operations since Operation Desert Storm in 1991 (Mangelsdorff, 2006).

In July 2003 the US Army Surgeon General established the first Mental Health Advisory Team (MHAT) to assess and provide recommendations related to mental health issues within the Operation Iraqi Freedom (OIF) theater and the evacuation chain. The members of the first team included behavioral health consultants (primarily research psychologists), combat stress control officers, epidemiology support, a chaplain, enlisted, and other behavioral health subject matter experts. The MHAT was deployed to investigate several factors: organizational and resource limitations, increases in OIF suicides, increases in behavioral health patient loads, stress-related issues in the Iraqi theater of operations, and deployment-related health issues at a major deployment installation back in the United States (US Army Surgeon General, 2003). The MHAT mission appears to have followed an "occupational health psychology model," examining work life and organizational stresses and improving soldier well-being (Mangelsdorff, 2006, p. 22).

Beginning in July 2003 the Operation Iraqi Freedom (OIF) Mental Health Advisory Team assessed mental health issues in the Iraqi theater, in Europe, and the United States, and provided recommendations. Soldiers were interviewed in small groups, with over 750 troops in combat surveyed. Notable results included: forward elements of the behavioral health care system in theater successfully assisted soldiers in dealing with operational stressors; and over 75% of soldiers reported none to mild stress and over 95% of troops treated in forward areas of the theater were returned to their units. It was also noted that there was a need for more standardized behavioral health reporting procedures. Furthermore, a number of soldiers in theater reported not receiving the help they felt they needed. There was inconsistent care for some soldiers removed from the Iraqi theater area. Soldiers reported low morale and unit cohesion.

During July 2003, there was a surge in Army OIF evacuations from the Iraqi theater. Because the OIF suicide rate for OIF deployed soldiers from January to October 2003 was higher than expected: there were 15.6 (suicides/100,000 soldiers/year) compared to the 1995–2002 average rate of 11.9 (suicides/100,000 soldiers/year), it was recommended that a theater/Area of Operation behavioral health consultant be appointed to advise the theater surgeon (ASG/HQDA, 2003).

In July 2004 a follow-up Operation Iraqi Freedom (OIF-II) Mental Health Advisory Team (MHAT II) was sent to the Iraqi theater. The MHAT II report noted: the deployment tour length was a concern, unit morale was low, and posttraumatic stress symptoms were higher. Combat service support National Guard and Reserve unit personnel reported higher rates of mental health issues and lower perceptions of combat readiness and training than personnel in other units in theater. The behavioral health care system personnel conducted more outreach and coordination programs. Behavioral health personnel were better distributed throughout the Iraqi theater, and fewer soldiers were evacuated for behavioral health problems. The Army Suicide Prevention Program was showing an impact; fewer suicides were reported during 2004.

Additional mental health advisory teams were deployed in successive years. Assessments of soldier behavioral health, risk and resiliency factors, and behavioral health personnel findings across the time periods generally showed the number of mental health problems declining and combat exposure rates becoming lower. Behavioral health personnel reported fewer symptoms of burnout. Soldiers with multiple deployments reported lower morale and increased mental health problems. Leaders actively promoted suicide education and prevention programs.

The Joint Mental Health Advisory Team 7 (Office of the Surgeon General, 2011) was the first to assess both Army and Marine maneuver unit platoons (war fighters). Individual morale had declined; unit morale remained low. Acute stress rates were higher. Risk factors of increased combat exposure and multiple deployments increased, which was associated with more psychological problems being reported. Additional sleep hygiene and sleep discipline training was recommended. The overall importance of the repeated MHAT assessments between 2003 thru 2010 was that they allowed assessment of changes in the combat environments and policy initiatives across a variety of units and services.

MHAT reports contributed to the creation of behavioral health intervention programs and policy directives. These were created to help facilitate adjustment for combatants and support personnel in the ongoing operations in Iraq and Afghanistan, for military families back home, and for veterans recovering from operational stressors. In addition to the Army MHATs, other organizations (RAND, CDC, PHS, and NIMH) studied soldiers and veterans. Assessments by the RAND Center for Military Health Policy Research examined the psychological and cognitive injuries of troops deployed in the ongoing operations (Tanielian & Jaycox, 2008). The RAND study examined posttraumatic stress disorder, major depressive disorder and symptoms, and traumatic brain injuries through the lenses of prevalence, costs, and the health care system. The RAND study projected that of the 1.6 million armed forces personnel deployed since 2001 there could be approximately 300,000 individuals currently suffering from PTSD or major depression. Recommendations offered from the RAND study confirmed those from MHATs: (1) Increase the number of providers trained to deliver evidence-based care; (2) Change policies to encourage veterans and active duty personnel to seek needed care; (3) Deliver proven care when and where services are offered; (4) Invest in research and planning efforts (Tanielian & Jaycox, 2008). A great need for mental health assessments and interventions among military personnel and veterans was going unmet.

Concurrently the Army senior leadership recognized the need for developing a more resilient force (active duty, Reserves and Guard, family members, and civilian workforce) through preventive and educational enhancement programs. In 2008 the Department of the Army created the Directorate of Comprehensive Soldier Fitness (CSF) to increase the psychological resilience, adjustment, and performance of soldiers and military families by incorporating principles of positive psychology. The CSF principles build on existing individual strengths and emphasize personal growth with the intent of enhancing psychological fitness and adaptive outcomes (Cornum, Matthews, & Seligman, 2011). A special issue of the *American Psychologist* (January, 2011) summarized the background and selected studies of the Comprehensive Soldier Fitness program. The CSF goals were promotion of individual well-being and prevention of adjustment problems. The CSF program components included: assessments, universal resilience training, individual training, and master resilience trainers.

The behavioral health concerns reportedly affecting soldiers (and their families) are no different from those noted during and after earlier conflicts (Korea, Vietnam, and Persian Gulf). The historical evidence suggests 70 to 75% of veterans develop resiliency skills from their military experience; it becomes part of their adult growth and development. Adding adjustment skills from the CSF program has the potential to increase the resiliency percentages higher. Educational, prevention, and fitness-oriented programs are in keeping with the military mission of developing the whole person (physically and emotionally).

The increase in suicides in the United States is recognized as a public health challenge. The United States National Strategy for Suicide Prevention was established in 2001. The Surgeon General of the United States Public Health Service (Satcher, 1999) examined suicide risk and prevention in terms of relative risk factors. In addition, the Centers for Disease Control and Prevention (CDC) and the National Institute of Mental Health (NIMH) investigated the suicides of military personnel in Iraq. A special issue of the *American Journal of Public Health* (AJPH, Supplement 1, 2012) collected some of the lessons learned from mental health enhancement and suicide prevention activities. The increasing number of suicides in the active duty armed forces and veterans parallels the numbers reported in the American public. Studies have asked whether there are higher rates of suicides among active duty members and veterans compared with male adults aged at least 18 years. The findings reported in the AJPH special issue are complicated; veterans of the Global War on Terror campaigns (2001 to present) have higher risks of suicide than those of earlier conflicts. The youngest cohort (ages 17–24) has the highest risk of suicide. The cohorts had different unique mental health stressors and personal experiences. The extent to which unique (high unemployment rates, mortgage defaults, traumatic brain injuries, physical disabilities, multiple operational and combat tours, family stressors, and the downsizing of the armed forces) stressors interact has not been definitely determined.

Projections from the RAND study and reports from the Armed Forces Health Surveillance Center suggest increases in active duty hospitalizations for mental disorders. With the armed services drawing down (decreasing the number of Active Duty and civilian personnel), the Department of Veteran Affairs (VA) must continue adding additional mental health professionals to assess and treat the large number of veterans seeking assistance. In 2011 the VA's overall mental health program provided specialty mental health services to 1.3 million veterans. Since 2009, the VA has increased the mental health care budget by almost 40%. Since 2007, the VA has seen significant increases in the number of veterans receiving mental health services and increases in mental health staff. The independent assessments (RAND, VA, Armed Forces Health Surveillance Center, CDC, NIMH, and PHS) validate many concerns noted by the MHAT reports.

The mental health advisory teams from 2003 to 2010 actively assisted in studying the unique experiences among different military cohorts. Repeated MHAT assessments allowed assessing changes in the combat environments and examining the effects of training and policy initiatives. Stress awareness and suicide prevention strategies include developing and promoting educational programs to increase awareness of stressors, facilitating access to trained behavioral health personnel, and encouraging assessment and treatment programs.

The mental health advisory teams recognized the increased demand for mental health services and groups most vulnerable to not obtaining needed care (support personnel, Reserve and Guard members). Changes evolved in theater policies of how many, what kinds of, and where behavioral health personnel were deployed; recommendations from the MHATs helped reduce some stressors. As the nature and intensity of the operations changed, the number of combat deaths and suicides increased. The MHAT recommendations expressed concern about troops with multiple deployments who reported significant adjustment problems and increased use of medications. The MHAT findings (number of troops at risk, the stigma of troops not seeking care, vulnerable groups, and backlogs of troops not receiving care) are echoed in the RAND study and the special issues of the *American Psychologist* and the *American Journal of Public Health*. The status of the MHAT recommendations since 2003 indicated there are significant increases in behavioral health staff in theater; increases in resilience training for at risk groups occurred; revisions occurred in suicide prevention training to consider theater-specific situations; and more behavioral health personnel supporting

National Guard units were added. The overall question distilled from the MHAT reports concerned whether in the future there will be adequate numbers and appropriately trained mental health support personnel and facilities to access, assess, and provide care for the veterans by the VA, federal, and/or private sectors.

References

American Journal of Public Health. (2012). Supplement 1, Suicide prevention. 102(S1), e1–S159.

Cornum, R., Matthews, M. S. D., & Seligman, M. E. P. (2011). Comprehensive soldier fitness: Building resilience in a challenging institutional context. *American Psychologist, 66*(1), 4–9.

Defense Manpower Data Center. (2012). *Defense casualty analysis system (DCAS)*. Retrieved from https://www.dmdc.osd.mil/dcas/pages/summary_data.xhtml

Mangelsdorff, A. D. (2006). *Psychology in the service of national security.* Washington, DC: American Psychological Association.

Office of the Surgeon General US Army Medical Command, Office of the Command Surgeon HQ, USCENTCOM, & Office of the Command Surgeon US Forces Afghanistan. (2011). *Joint mental health advisory team 7 (J-MHAT 7) Operation Enduring Freedom 2010 Afghanistan.* Retrieved from http://www.armymedicine.army.mil/reports/mhat/mhat_vii/J_MHAT_7.pdf

Satcher, D. (1999). *Bringing the public health approach to the problem of suicide: The Surgeon General's call to action to prevent suicide.* Washington, DC: Department of Health and Human Services.

Tanielian, T. L., & Jaycox, L. (2008). *Invisible wounds of war: Psychological and cognitive injuries, their consequences, and services to assist recovery.* Retrieved from http://www.rand.org/pubs/monographs/MG720.html

US Army Surgeon General & Headquarters, Department of the Army G-1. (2003). *Operation Iraqi Freedom (OIF) mental health advisory team (MHAT) report.* Retrieved from http://www.armymedicine.army.mil/reports/mhat/mhat/mhat_report.pdf

Winkenwerder, W., Jr. (2003). *Force health protection.* Retrieved from http://www.defenselink.mil/transcripts/2003/t03142003_t03131fhp.html

20 COMPREHENSIVE SOLDIER FITNESS

Donna M. Brazil

HISTORY OF CSF

Nine years into the wars in Iraq and Afghanistan the Chief of Staff of the Army, General George Casey, sought to develop a better way to prepare our soldiers, civilians, and family members for the stress and challenges that they encounter as a result of their service as well as the everyday challenges of their complex lives. For years, the Army has had a measure of physical fitness; twice a year soldiers are required to complete the Army Physical Fitness Test (APFT). The APFT gives soldiers a yard stick by which to measure their fitness and a metric by which to know if they are in need of remediation. Daily physical training is encouraged as a way to maintain ones physical fitness.

In 2008 General Casey sought to bring the same level of awareness, assessment, and training to the areas of psychological and psychosocial fitness. To do this he brought together some of the top psychologists and behavioral health experts in the country. Contributors included Martin Seligman, Chris Peterson, Nansook Park, Michael Matthews, Richard Tedeschi, Karen Reivich, Barbra Fredrickson, John Cacioppo, Harry Reis, John Gottman, and Kenneth Pargament. Together with many others, this team of experts developed a program of comprehensive fitness that seeks to prepare soldiers for the challenges that lie ahead by addressing their emotional, social, family, and spiritual fitness. It is important to note that

CSF is not a remedy or a treatment for psychological illness. Instead, it is a program designed to improve performance by better preparing individuals for the challenges that they will face. It is not a single class or a briefing, nor is it a screen for mental illness. Rather, it is a program that teaches skills designed to promote and develop fitness far more broadly defined than ever before (Casey, 2011).

PURPOSE OF CSF

CSF is designed to develop an Army that is as psychologically and psychosocially fit as it is physically fit. The CSF program is designed to assess soldiers, family members, and Army civilians on five dimensions of psychological and psychosocial strength: physical, emotional, social, family, and spiritual fitness; to provide remediation in those areas that might need improvement; and to provide a metric with which individuals can assess their fitness and development (Cornum, Matthews, & Seligman, 2011).

CSF defines these strength dimensions as follows:

Physical
Performing and excelling in physical activities that require aerobic fitness, endurance, strength, healthy body composition, and flexibility derived through exercise, nutrition, and training.

Emotional

Approaching life's challenges in a positive, optimistic way by demonstrating self-control, stamina, and good character with your choices and actions.

Social

Developing and maintaining trusted, valued relationships and friendships that are personally fulfilling and foster good communication including a comfortable exchange of ideas, views, and experiences.

Family

Being part of a family that is safe, supportive and loving, and provides the resources needed for all members to live in a healthy and secure environment.

Spiritual

Strengthening a set of beliefs, principles, or values that sustain a person beyond family, institutional, and societal sources of strength (Comprehensive Soldier Fitness, 2011).

DESCRIPTION OF THE COMPREHENSIVE SOLDIER FITNESS PROGRAM

CSF consists of four components or pillars: assessment, individualized online training, resilience trainers at the unit level, and universal resilience training at every level of military education throughout a soldier's career.

Assessment

Each soldier is tasked to complete an online assessment called the Global Assessment Tool (GAT) annually. After completing the 105-question self-assessment, individuals receive immediate feedback on their levels of emotional, social, family, and spiritual fitness compared with Army-wide norms. Individuals can also compare their results with others of the same rank, gender, occupational specialty, component (active, reserve or national guard), and age. While CSF is an Army-wide program, assessment is intended solely for the individual, and commanders cannot gain access to an individual's scores. Commanders can request a consolidated report on their unit so that they can focus unit training toward those areas that may be of concern, but they do not have access to individual results. The Chief of Staff of the Army is the only one authorized to approve the release of an individual's scores (US Army, 2011).

Online Training Modules

After viewing their GAT results and feedback, soldiers will have access to comprehensive resilience modules (CRM) that are available online. There are currently 36 modules available on the CSF website, and each is tailored toward self-improvement in one of the dimensions (emotional, family, spiritual, or social). Modules are self-contained and take approximately 20 minutes to complete. In some cases the module is a refresher of the Master Resilience Training skills, while others suggest and explain additional activities that have proven helpful in developing that dimension. The modules are designed to be self-paced and may be revisited as often as an individual desires.

Master Resilience Trainers

In November 2009 the Army began training noncommissioned officers (NCOs) as Master Resilience Trainers (MRT) and now requires at least one trained MRT per battalion (US Army, 2010). These MRTs are responsible for the training of all soldiers and Army civilians within the unit. They are trained to provide initial as well as follow-on training. MRT training programs are conducted at the University of Pennsylvania, Ft. Jackson, South Carolina, and at various installations via a Mobile Training Team. In addition to training NCOs for this mission, numerous Army Community Services instructors have also been trained to deliver this instruction to family members in the Army community.

Universal Resilience Training

As the final component of CSF, resilience training has been included in every level of formal education in the Army from Basic Training to Senior Service College. It has also been included at all of the Army's training centers. By infusing resilience training into all educational schools, the Army seeks to ensure that the language of CSF is introduced early and continually reinforced. New recruits take the GAT within weeks of joining the service and receive instruction on resilience as part of their introduction and socialization to the Army during Basic Combat Training (BCT) and Advanced Individual Training (AIT). The training provided early and repeated at each level of military education builds on the last school and reinforces the principles and vocabulary of resilience. The resilience training program itself is discussed in depth below.

US ARMY RESILIENCE TRAINING

The resilience program was developed by Karen Reivich and Martin Seligman from the University of Pennsylvania. The program is based on the Penn Resilience Program that has been effective at training educators and other professionals. The Army course has been modified and tailored for the special needs and concerns of a military audience. The goal of the resilience training program is to strengthen the individual competencies of self-awareness, self-regulation, optimism, mental agility, character strengths, and connection.

The MRT course focuses on enhancing these competencies by building individual resilience skills. The skills are based largely on the work of Dr. Aaron Beck and employ the techniques of cognitive-behavioral therapy. The various resilience skills teach individuals to identify possible counterproductive or faulty thought patterns that lead them to experience negative emotion; to identify deeply held beliefs that might no longer be accurate or useful and hence lead to negative reactions; to identify how their thought patterns might constrain their problem solving; and to minimize catastrophic thinking and to fight counterproductive thoughts. Readers interested in more information on the resilience skills can look in

The Resilience Factor: 7 Keys to Finding Your Inner Strength and Overcoming Life's Hurdles by Karen Reivich and Andrew Shatté (Reivich & Shatté, 2004).

The last two modules of the course focus on strengths and relationships. Students learn to cultivate gratitude by completing an exercise that requires them to list three good things that happened during the day and briefly reflect on what each event means to them. They identify their character strengths by each taking the Values in Action online assessment and then complete a group exercise that requires them to draw on their strengths as well as those of their teammates in order to address an issue. Finally, students learn and practice some basic communication skills to help them in maintaining strong interpersonal relationships.

The 10-day training course consists of 5 days during which the NCOs are taught this basic resilience curriculum followed by 3 days when they are taught to present the material. During the final 2 days NCOs learn how to apply these resilience skills specifically in a military environment and learn how to enhance their performance using techniques that have been developed and validated in the sport psychology field such as controlled breathing and imagery. These final 2 days are presented by instructors from the Walter Reed Army Institute of Research and the Army Center for Enhanced Performance. For more information on the MRT training, see (Reivich, Seligman, & McBride, 2011).

CRITICISM AND FUTURE OF THE CSF PROGRAM

Criticism

Since its inception, several psychologists have argued strongly against the program. The primary criticism revolves around two issues: that the instruments have not been validated on soldiers and that the resilience program is really a massive psychological study being conducted on uninformed participants (Eidelson & Soldz, 2010). While the training and course materials have been modified to meet the needs of soldiers, the base program was validated on a number of populations during the 20-year course of the Penn Resilience Program. As to the training versus psychological study criticism, the alignment of CSF under the training and education directorate of the Army staff rather than the medical command sends a strong message that Army's intent was and remains training focused (Seligman & Fowler, 2011). Despite these criticisms, the Army has continued to move forward with the CSF program.

Future of CSF

A December 2011 program evaluation found that CSF and, in particular, the presence of an active MRT trainer and training program significantly improved the resilience and psychological health of soldiers when compared to a control group that had no MRT. This finding was present regardless of the level of unit cohesion or reported quality of their leadership (Lester, Harms, Herian, Krasikova, & Beal, 2011). To date over 1 million soldiers, family members, and Army civilians have taken the GAT, and over 10,000 MRTs have been trained.

The Army established a resilience training program at Ft. Jackson, South Carolina, that mirrors the University of Pennsylvania program. This program at Ft. Jackson's Victory University runs consecutive 10-day courses and can train approximately 270 MRTs each month. CSF has been designated by the CSA to be the sole resilience training program for the Army. Precommissioning sources such as the United States Military Academy and Reserve Officer Training Corps programs now include resilience as part of their curricula. Each Active Duty unit is required to conduct 2 hours of resilience training per quarter, and new online modules continue to be developed to keep the online presentations fresh (US Army, 2011). The initial results are positive and the feedback from soldiers, Army civilians, and family members is also positive.

AUTHOR NOTE

The views expressed in this chapter are those of the author and do not reflect the official policy or position of the Department of the Army, Department of Defense, or the US Government.

References

Casey, G. W., Jr. (2011). Comprehensive Soldier Fitness: A vision for psychological resilience in the U.S. Army. *American Psychologist, 66,* 1–3.

Cornum, R., Matthews, M. D., & Seligman, M. E. P. (2011). Comprehensive Soldier Fitness: Building resilience in a challenging institutional context. *American Psychologist, 66,* 4–9.

Eidelson, R., Pilisuk, M., & Soldz, S. (2011). The dark side of Comprehensive Soldier Fitness. Retrieved from http://www.counterpunch.org/2011/03/24/the-dark-side-of-comprehensive-soldier-fitness/

Lester, P. B., Harms, P. D., Herian, M. N., Krasikova, D. V., & Beal, S. J. (2011). The Comprehensive Soldier Fitness Program evaluation report #3: Longitudinal analysis of the impact of master resilience training on self-reported resilience and psychological health data December 2011. Retrieved from http://handle.dtic.mil/100.2/ADA553635

Reivich, K. J., Seligman, M. E. P., & McBride, S. (2011). Master resilience training in the U.S. Army. *American Psychologist, 66,* 25–34.

Reivich, K., & Shatté, A. (2004). *The resilience factor: 7 keys to finding your inner strength and overcoming life's hurdles.* New York, NY: Broadway Books.

Seligman, M. E. P., & Fowler, R. D. (2011). Comprehensive soldier fitness and the future of psychology. *American Psychologist, 66,* 82–86.

US Army. (2010). ALARACT 097/2010, Comprehensive Soldier Fitness execution order. Retrieved from http://csf.army.mil/resilience/supportdocs/ALARACT-097-2010-FINAL.pdf

US Army. (2011). ALARACT 086/2011, Mod 02 to ALARACT 097/2010 Comprehensive Soldier Fitness execution. Retrieved from http://csf.army.mil/resilience/supportdocs/ALARACT_086_2011_MOD_02_TO_ALARACT_097–2010_COMPREHENSIVE_SOLDIER_FITNESS_EXECUTION.pdf

US Army. (2012). *Comprehensive Soldier Fitness.* (2012). Retrieved from http://csf.army.mil/

PART III
Ethical and Professional Issues

21 MULTIPLE RELATIONSHIPS IN THE MILITARY SETTING

Jeffrey E. Barnett

Military psychologists are in the unique position of simultaneously being both military officers and practicing psychologists. As commissioned officers in the United States military they have sworn an oath that includes a commitment to comply with all Department of Defense (DOD) and service-specific (Air Force, Army, or Navy) regulations, which is in addition to psychologists' obligation to comply with the American Psychological Association's Ethical Principles of Psychologists and Code of Conduct (APA Ethics Code; American Psychological Association [APA], 2010) and relevant state licensing laws. This adds an extra layer of obligations and responsibilities to the roles of military psychologists that their civilian counterparts typically do not need to address. Additionally, military psychologists serve as members of small, insular, and often isolated communities. Similar to rural practitioners, military psychologists live and work in the same community, providing professional services to many individuals who live in the same small community and with whom they work.

Military psychologists may at times be required to provide professional services to individuals with significant power over the psychologist's career and life, to those with whom they serve and work in the same unit, and to those with whom they and their families interact with socially. While avoiding such conflict of interest situations and multiple relationships might be desirable, military psychologists often live and work without that option. Role, function, and setting factors may at times make participating in challenging and perhaps even undesirable multiple relationships unavoidable. Accordingly, military psychologists need to have an understanding of these issues and challenges, learn to anticipate them and take preventive steps when possible, use relevant decision-making models and processes when faced with such situations, and learn to effectively manage them when they cannot be avoided.

MULTIPLE RELATIONSHIPS

Multiple relationships involve serving in one or more additional relationships with another individual in addition to the primary professional relationship. Examples include being in personal, social, business, religious, or other relationships with clients in addition to the professional psychologist/client relationship. The APA Ethics Code makes it clear that multiple relationships are not always avoidable and that not all multiple relationships are unethical or inappropriate. In fact, the APA Ethics Code makes it clear that those multiple relationships that are not exploitative of the client and that do not impair the psychologist's objectivity and judgment are not unethical.

Some multiple relationships are always unethical and inappropriate. For example, the

APA Ethics Code makes it clear that sexually intimate multiple relationships with current clients, supervisees, and students are always inappropriate, as is providing psychological services to an individual with whom the psychologist has previously been in a sexually intimate relationship. These relationships are seen as an abuse of the power differential in the professional relationship, they take advantage of the client's dependence and trust, they are likely to significantly impair the psychologist's objectivity and judgment, and they hold a great potential for harm to the client. In addition, these relationships are likely to adversely impact the public's trust in psychology and may result in individuals who are in need of assistance forgoing it.

The APA Ethics Code does allow for the possibility of engaging in intimate relationships with former clients under rare circumstances, but for the reasons mentioned above, this type of multiple relationship generally is advised against. The APA Ethics Code requires the passage of a minimum of 2 years since the date of last professional contact with the former client and then provides seven criteria that the psychologist must consider. These include:

1. the former client's previous and current mental health issues and emotional state,
2. the nature of the treatment provided,
3. the nature of the treatment termination,
4. the likelihood of harm to the former client, as well as others.

It is important to consider the best interests and welfare of the former client and not to be motivated by one's own personal interests and needs. Even when the treatment relationship has ended, psychologists maintain a responsibility to those they have served, to the public in general, and to the profession.

Other multiple relationships may not be inappropriate and may even be in a client's best interest. In fact, in some settings, rigidly avoiding all multiple relationships will likely prove ineffective and may actually be inconsistent with psychologists' commitment to serving the best interests and promoting the welfare of clients. When living and working in the same community of which one is a member, certain types of multiple relationships will likely be unavoidable. Examples include providing treatment or other professional services to a psychologist's neighbor, to a coworker, to the parent of one's child's friend or classmate, and others. A psychologist may also find him- or herself serving on a committee or board of a community, civic, or religious organization with a current or former client. As those who live and work in small, isolated, or insular communities will frequently experience, to be a member of one of these communities means to experience these numerous multiple relationships over time.

Efforts to avoid all multiple relationships in a misguided attempt to be ethical will not only be futile, they may have an alienating effect on those most likely to become one's clients. Military psychologists will find that members of the community get to know the psychologist outside of the professional psychology relationship. Members of these communities, to include the military, may be mistrusting of mental health professionals as well as of outsiders. Through other interactions with, and observations of, psychologists in the community and in the work setting, military service members and their families may develop confidence in and trust of the military psychologist and as a result, be more comfortable seeking out psychological services. Through these nonpsychology interactions prospective clients may observe the military psychologist's general competence, professionalism, integrity, personality, and other interpersonal attributes. Interactions in the broader military community and other professional interactions may positively impact service members and their families, resulting in them seeking out the military psychologist for professional services, when otherwise they might not.

Many of these interactions in the community may not constitute multiple relationships but instead may represent incidental contacts; situations in which the military psychologist has contact with a client in the community. Examples include seeing each other in passing at the Post Exchange or Commissary, noticing a client's presence when in the fitness center exercising, sitting near a client at a religious service, or having a client at the same table in a dining facility. While these are not multiple relationships, these situations may range from those that are quite benign, such as walking past a client in a store, to those that may be quite challenging or uncomfortable, such as changing in the same locker room as a client. When serving in an isolated community such as on a Navy ship or at a remote and isolated base, these incidental contacts will likely be a frequent occurrence.

ETHICAL DECISION MAKING

Some multiple relationships are clearly unethical and inappropriate, whereas others may readily be seen as benign or even helpful as is described above. Yet, most often, military psychologists will be faced with situations that are unclear and that present as ethical dilemmas; situations with no readily apparent clearly correct or incorrect course of action. In these frequently occurring situations military psychologists will be well served by the use of an ethical decision-making process. Several useful ethical decision-making models exist that are relevant to sorting through multiple relationship situations (See Cottone & Claus, 2000 for a comprehensive review). One representative model provided by Barnett and Johnson (2008) includes the following steps that may provide a framework for addressing these challenges:

- Define the situation clearly
- Determine who will be impacted
- Refer to the ethical principles and standards
- Refer to relevant laws, regulations, and professional guidelines
- Reflect honestly on personal feelings and competence
- Consult with trusted colleagues
- Formulate alternative courses of action
- Consider possible outcomes for all parties involved
- Consult with colleagues and ethics committees
- Make a decision, monitor the outcome, and modify your plan as needed

Prior to entering into a multiple relationship psychologists should consider factors such as the relative power differential in the primary psychology relationship and the nature of the client's dependency on the psychologist as well as other issues to include considering one's motivations for entering into the anticipated multiple relationship, options and alternatives reasonably available, and the likely impact of each on the client. An additional feature of ethical decision-making models to consider that is of great relevance for military psychologists is the use of consultation with an experienced and trusted colleague prior to making multiple relationship decisions. The nature of the military often places military psychologists in settings and situations in which there may exist few options for making referrals and where total avoidance of multiple relationships is neither prudent nor feasible, thus limiting one's alternatives to entering into a multiple relationship. Often, the question for military psychologists is not "should I enter this multiple relationship?" but rather, "how can I most ethically and appropriately engage in this unavoidable multiple relationship?" As a result, consultation with experienced and knowledgeable colleagues will be especially helpful and important.

THE MILITARY SETTING AND ROLE

In addition to the many characteristics all small and insular work settings have in common, the military setting brings with it additional unique characteristics and challenges relevant to multiple relationships. As commissioned officers, military psychologists are part of an organizational structure in which their "client" in fact can be seen as the Department of Defense, their branch of military service, or the military entity to which they are assigned. The ultimate goal of all military health professionals is to support the mission of the military and to help ensure each service member's combat readiness. This is a very different model than that of civilian psychologists who focus on the goals and needs of their client, the individual to whom they are directly providing the clinical service.

Military psychologists may serve in isolated or remote locations in which they may be the only mental health professional available to provide needed professional services. Options for avoiding multiple relationship situations may be rather restricted in these settings. Further, as part of a command structure, military psychologists may receive a direct order to provide a clinical service to a particular individual. In these situations, both as a result of limited options for referral and the requirement to comply with lawful orders, military psychologists may find themselves required to conduct an evaluation of their immediate supervisor or commanding officer, provide psychotherapy to a subordinate they work with, or be evaluated by their own client (e.g., security clearance evaluation, annual substance abuse screening, etc.). Kennedy and Johnson (2009) address these issues comprehensively and offer multiple specific examples of these situations occurring in the military setting to include being ordered to conduct an alcohol evaluation of one's commanding officer with the only other available mental health officer being over 3,000 miles away.

Some military psychologists are assigned to large medical centers or community hospitals on military installations. In these settings, opportunities for making referrals when multiple relationship situations arise may at times be possible due to the presence of a team of mental health professionals. Yet, many military psychologists are embedded in military units, such as being the sole mental health professional on an aircraft carrier or at a forward operating base for a particular military unit. In these situations, options for making referrals may be quite limited or nonexistent. Further, military psychologists serving in these roles and settings are serving as both military psychologists and military officers. At times, these two roles may come in conflict, such as when one is required to be the boss or direct administrative supervisor of one's clients or perhaps even to live in close quarters with one's clients to include showering, eating, and living in close proximity.

CHALLENGES AND RECOMMENDATIONS

Military psychologists must plan for the regular occurrence of incidental contacts with clients in their small and insular communities and should be prepared for the possibility of regularly occurring multiple relationships. A first step for addressing these challenges is to anticipate them and to discuss them openly with clients in the informed consent process. Some clients may be very comfortable with out-of-office contacts, whereas others may endeavor to keep private their professional relationship with a military psychologist. It is best to discuss these likely events with clients at the outset of the professional relationship, to find out their preferences regarding being acknowledged and greeted in public, and to agree that all such experiences will be discussed at the next scheduled appointment.

Despite the similarities with rural and other small and isolated settings, the military setting brings with it the additional challenges of the command structure and the military psychologist's commitment to fulfilling the military's mission. As a result, there will be occasions when a military psychologist is ordered to enter into a multiple relationship or is presented with a multiple relationship situation in the course of his or her daily activities.

For these situations there are several recommendations that will hopefully help the military psychologist to strike a balance between adherence to the APA Ethics Code and the fulfillment of the military mission.

- From the outset, educate commanders about these situations and sensitize them to the challenges posed by some multiple relationships.
- Develop a flexible approach to multiple relationships and avoid rigidity in your responses to them. Develop resources for making referrals when needed and be creative in your approach (e.g., physicians, nurses, clergy, etc.)
- Don't view all multiple relationships as being the same. Thoughtfully consider the complexities of each situation and apply a decision-making model to reason through the most beneficial and most feasible course of action.
- Openly discuss multiple relationships with clients, clearly articulating the parameters of each relationship. Compartmentalize these relationships so they may coexist. For example, when providing treatment to a coworker,

only discuss treatment issues during treatment sessions and not while in group work areas and only discuss work issues in group work areas and not in treatment sessions. Setting clear boundaries is recommended so that each individual will have appropriate expectations.

- When unsure of how to proceed in any situation and when experiencing confusing or upsetting reactions to participation in multiple relationships, consult with an experienced and trusted colleague. Be sure to utilize technologies such as the telephone, e-mail, and the Internet so as not to be isolated professionally even when isolated geographically.

References

American Psychological Association. (2010). *Ethical principles of psychologists and code of conduct.* Retrieved from www.apa.org/ethics

Barnett, J. E., & Johnson, W. B. (2008). *Ethics desk reference for psychologists.* Washington, DC: American Psychological Association.

Cottone, R. R., & Claus, R. E. (2000). Ethical decision-making models: A review of the literature. *Journal of Counseling and Development, 78,* 275–283.

Kennedy, C. H., & Johnson, W. B. (2009). Mixed agency in military psychology: Applying the American Psychological Association Ethics Code. *Psychological Services, 6,* 22–31.

22 MANAGING CONFLICTS BETWEEN ETHICS AND LAW

W. Brad Johnson

Military psychologists are likely to experience mixed-agency ethical dilemmas when there are conflicts between loyalties or obligations to an individual service member and the larger military organization including the service member's commanding officer. In effect, military psychologists often have at least two clients, the individual sitting before the psychologist and military leaders tasked with achieving a specific mission. The mixed-agency status of military psychologists can make conflicts between ethical obligations and legal requirements especially acute. Military psychologists have obligations to abide by the American Psychological Association's Ethical Principles of Psychologists and Code of Conduct (Ethics Code; American Psychological Association [APA], 2010) as well as the Uniform Code of Military Justice (UCMJ), the US Constitution, and a range of Department of Defense (DoD) statutes and regulations. At times, these ethical and legal obligations may appear to conflict.

On occasion, military psychologists may discover certain incongruities between ethical standards (APA, 2010), and various legal statutes, including DoD regulations. In most instances, these disparities are subtle and easily reconciled. At other times, the differences may be more stark or egregious, leaving the psychologist feeling "stuck" with a choice to follow the law or follow his or her interpretation of an ethical standard. The Ethics Code requires psychologists who encounter such conflicts to make known their commitment to the Code while taking reasonable steps to resolve the conflict in accordance with the Code (APA, 2010). Of course, as the Ethics Code makes clear, no law or statute can ever be used to justify violating human rights. Some of the more common ethical-legal conflicts in military psychology center on the areas of confidentiality and multiple relationships. Surveys of military psychologists reveal that a significant proportion has experienced occasional conflicts between their abiding ethical obligations and their interpretations of federal statutes and DoD regulations (Johnson, Grasso, & Maslowski, 2010).

In some instances, military psychologists have been censured by ethics committees for abiding by military requirements, or conversely, disciplined by the DoD for adhering to ethical standards that seem to conflict with DoD regulations. Jeffrey, Rankin, and Jeffrey (1992) detailed two cases in which military psychologists were sanctioned. In one case, a psychologist was censored by a licensing board for failing to protect a client's confidentiality; another provider released the client's health record long after the psychologist had transferred to a new duty station. In the second case, a psychologist was reprimanded by the DoD for protecting a client's request for confidentiality; the psychologist refused to report the client's improper relationship with a physician in the hospital.

EXAMPLES OF ETHICAL-LEGAL CONFLICT
IN MILITARY PSYCHOLOGY

Confidentiality

"Psychologists have a primary obligation and take reasonable precautions to protect confidential information obtained through or stored in any medium, recognizing that the extent and limits of confidentiality may be regulated by law" (APA, 2010, p. 7). During their training psychologists learn early and often that protecting client confidentiality is a fundamental ethical duty and a genuine hallmark of effective psychological services. Nonetheless, confidentiality is—in many ways—constantly at risk

in military settings. DoD directives related to mental health services have long specified that a legitimate military authority may have access to all records of care provided through military facilities, to include mental health records, when that authority has a legitimate "need to know" for the purpose of determining current fitness for duty or capacity for deployment (Jeffrey et al., 1992). Although the military has worked in earnest to dispel the stigma associated with seeking mental health services, even the latest DoD instruction on confidentiality highlights the potential for conflict between the ethical standard ensuring confidentiality and the DoD regulation that requires psychologists to compromise confidentiality in a number of circumstances not found in civilian contexts (US Department of Defense [DoD], 2011).

DoD instruction 6490.08, clarifies command notification requirements when an Active Duty service member receives mental health care (DoD, 2011). Although psychologists are to follow a general presumption that they should not notify a service member's commanding officer when the service member obtains mental health care, this presumption is overcome, requiring disclosure, when certain conditions are met. Those conditions unique to the military include: (1) when there is serious risk of harm to a specific military operational mission; (2) when services are obtained by "special personnel," or those having mission responsibilities of such potential sensitivity or urgency that preserving confidentiality in the context of evidence of diminished or compromised functioning could place mission accomplishment at risk; (3) when the service member receives inpatient care; (4) when the service member's medical condition could possibly interfere with that person's military duty; and (5) in other special circumstances—determined on a case-by-case basis—in which a psychologist believes that "proper execution of the military mission" outweighs the interests served by protecting confidentiality (DoD, 2011, p. 6). Obviously, there are many exceptions to confidentiality in military settings that a psychologist would not encounter in other settings. Further complicating the ethical-legal tension for uniformed psychologists is the fact that they have taken an oath of office. This commissioned status makes the psychologist bound to promote both a professional code of ethics and the military mission(s) to which he or she contributes. On occasion, the success of a military mission will necessitate sacrifices to the normally sacrosanct confidentiality entitlement. Furthermore, uniformed psychologists must recognize the ambiguous nature of some of these exceptions. In arriving at a decision to violate confidentiality the military psychologist must weigh ethical obligations, legal guidance, and hopefully, good collegial consultation.

Although psychologists in many settings struggle with confidentiality dilemmas, and although military psychologists typically provide detailed informed consent to clients regarding the unique limits to confidentiality in the military (APA, 2010; Jeffrey et al., 1992), many military psychologists will encounter situations in which they feel compelled to choose between protecting client confidentiality and abiding by DoD statutes that grant commanding officers access to client records for a wide range of sometimes ambiguously defined reasons. For instance, Johnson et al. (2010) detailed a case in which a Navy aircraft carrier psychologist entered a psychotherapy relationship with a medical corpsman. After several months of treatment, the client's depression and eating problems had improved dramatically. Then, with no warning, the psychologist was directed by the command to conduct a security clearance evaluation for this client. Because the psychologist had not anticipated being required to engage a client in this additional forensic role, the psychologist had not informed the client about this possibility in advance. In spite of the psychologist's protests about the sudden role shift, no other mental health provider was available and the psychologist was ordered to proceed. When the evaluation was complete—including details about the client's history of sexual abuse and an eating disorder—the psychologist reluctantly submitted the evaluation. Later, the psychologist learned that several people in the client's chain of command read the report, including officers who, in the psychologist's opinion, had no legitimate "need to know" when it came to details about the corpsman's mental health status.

This case serves to highlight the ethical-legal conflicts military psychologists might experience when their interests in adhering to the highest standards of ethical practice—particularly those bearing on confidentiality and privacy—collide with DoD regulations that seem to weaken or undermine these ethical standards. This case further illustrates the nature of mixed agency tension in the military. In many situations the military psychologist has two clients; the service member sitting before him or her and the military command structure. Traditionally, military psychologists have responded to confidentiality dilemmas by engaging in very conservative documentation of the client's history or private concerns—even when these are clinically relevant, providing detailed and exhaustive informed consent regarding the fact that confidentiality can never be guaranteed in the military, and working proactively with commanding officers and other referral sources to resolve confidentiality conflicts by answering key disposition questions (e.g., can the service member deploy to combat, is the service member psychologically fit to perform his or her duties, might the mission be at risk owing to the service member's diagnosis or impairment?) with the minimal level of disclosure necessary.

Multiple Relationships

Another bedrock ethical standard in the APA Ethics Code (APA, 2010) bears on the obligation to avoid potentially harmful multiple relationships with clients:

A multiple relationship occurs when a psychologist is in a professional role with a person and (1) at the same time in another role with that same person…A psychologist refrains from entering into a multiple relationship if the multiple relationship could reasonably be expected to impair the psychologist's objectivity, competence, or effectiveness in performing his or her functions as a psychologist or otherwise risks exploitation or harm to the person with whom the professional relationship exists. (APA, 2010, p. 6)

Military psychologists quickly discover that avoiding uncomfortable and, on rare occasions, multiple relationships that are distressing for clients is nearly impossible. While obligations to engage in multiple relationships may not be codified in DoD statute, the realities of military service—particularly deployment—ensure that psychologists will have multiple roles with many clients. By virtue of their commissioned status, military psychologists may be required—receive a direct order—to suddenly assume administrative, supervisory, or even forensic roles with current or former clients. By virtue of the cramped and isolated conditions that characterize deployment, psychologists will almost certainly find themselves eating with, exercising with, and frequently encountering, clients outside of the professional relationship. At times, military psychologists have had to shower and sleep side-by-side with clients or even allow a client to serve as an official "observer" when providing a urine specimen for mandatory and random substance abuse screening. Because a military psychologist may practice in locations as a solo mental health provider, he or she will inevitably provide services to colleagues, friends, and even direct supervisors. Although ethical guidance bearing on multiple relationships would certainly caution psychologists against mixing so many different kinds of roles with current or former clients (e.g., friend, work supervisor, roommate, forensic evaluator), the military psychologist may be legally required either through direct order or the exigencies of his or her role as an officer in an isolated or deployed unit, to engage in multiple roles.

Multiple relationship dilemmas entail ethical-legal dilemmas for military psychologists when the exigencies of military environment (e.g., deployment or isolated duty) or direct orders from senior military officials place psychologists in unwanted and potentially distressing roles with clients who might reasonably be expected to have a negative response to the multiple role.

NOT ALL ETHICAL-LEGAL DISCREPANCIES ARE CONFLICTS

When military psychologists encounter differences or tensions between laws/regulations

and professional ethics, it is important to remember that differences alone do not constitute conflicts (Johnson et al., 2010). It is important for psychologists to operationally define the term "conflict" as it applies to ethics and law. It will be common for military psychologists to discover routine differences between ethical and legal requirements. In some cases a law may impose requirements that an ethics code does not and vice versa. For an ethical-legal discrepancy to become a conflict, the provider's obligations under the law and the provider's obligations under his or her professional code of ethics must be mutually exclusive (Johnson et al., 2010). In the case of a genuine conflict, a psychologist would perceive that fulfilling legal obligations (e.g., sharing information gleaned during a client's psychotherapy with his or her commanding officer) will necessarily entail violating the Code of Ethics (e.g., protecting confidentiality, minimizing intrusions on privacy).

WHAT DOES THE APA ETHICS CODE SAY ABOUT ETHICAL-LEGAL CONFLICTS?

Standard 1.02 of the American Psychological Association's Ethics Code (APA, 2010) clarifies the ethical responsibilities of military psychologists when they encounter conflicts between ethics and laws, regulations, or other governing legal authority such as DoD statutes or lawful orders issued by a superior military officer:

If psychologists' ethical responsibilities conflict with law, regulations, or other governing legal authority, psychologists clarify the nature of the conflict, make known their commitment to the Ethics Code, and take reasonable steps to resolve the conflict consistent with the General Principles and Ethical Standards of the Ethics Code. Under no circumstances may this standard be used to justify or defend violating human rights. (APA, 2010, p. 4)

There are several key elements of this standard for military psychologists. First, the fact that the standard exists should reassure military psychologists that ethical-legal conflicts occur in many settings and for many civilian psychologists as well; as mentioned earlier, military psychologists are neither unique nor alone when confronting these dilemmas. Second, military psychologists are obligated to try and resolve ethical-legal dilemmas, always keeping in mind the best interests of their clients. Third, military psychologists hold an ethical obligation to speak out and try to create systemic change when conflicts between ethics and law emerge. In this way, they can become agents of change effecting modifications to statutes, regulations, and laws and informing relevant stakeholders. Finally, Standard 1.02 states in unequivocal terms that military psychologists may never use a law, regulation, or military order as justification for harming or otherwise violating the rights of any person.

RECOMMENDATIONS FOR PREVENTING AND MANAGING ETHICAL-LEGAL CONFLICTS

Military psychologists are likely to encounter frequent differences between ethical obligations and legal requirements in their practice of psychology. Occasionally, these differences will rise to the level of conflict in which a psychologist feels obligated to serve either an ethical standard or a legal statute. The following recommendations encompass strategies aimed at preventing or successfully managing ethical-legal conflicts. It is important to keep in mind that successfully resolving these conflicts requires a consistent focus on both the best interests of clients and on the validity, purposes, and morality of the law or regulation in question (Johnson et al., 2010).

1. **Be careful not to elevate ethical-legal differences to conflicts**. In all cases, avoid assuming that a difference between ethics and law means that you cannot effectively serve both the ethical and legal requirement. Rarely are ethical standards entirely incongruent or mutually exclusive with legal requirements. For instance, a psychologist might account for the very liberal DoD regulation bearing on disclosure of client information (DoD, 2011), a much less rigorous standard than the one found in the APA Ethics Code (APA, 2010), by working informally with commanding officers to minimize the volume of information disclosed while simultaneously providing rigorous informed consent to clients so that they fully understand the implications of the DoD regulation.

2. **Be conversant with both the ethics code and relevant federal laws**. Military psychologists who are unfamiliar with specific ethical principles and standards or who lack a clear understanding of valid federal statutes and regulations governing their work are at greater risk in this area. It is imperative that military psychologists frequently review the APA Ethics Code and relevant laws, attend continuing education workshops bearing on ethics in military practice, and consult with military psychology colleagues or military lawyers when apparent ethical-legal conflicts arise.

3. **Remember that one's military service does not override one's obligations as a professional psychologist**. Military psychologists can help avoid harm to clients and themselves by remaining attuned to their unequivocal obligation to abide by the Ethics Code (APA, 2010). There are several risk factors in this regard. Military psychologists serving in embedded billets, those serving in operational (war-fighting) roles, and those serving for extended periods of time in military environments, must guard against "drift" in the direction of primary allegiance to military tradition and regulations at the expense of adherence to professional ethics (Johnson et al., 2010).

4. **Always attempt to balance client best interests with DoD regulations**. When an ethics committee considers a complaint against a psychologist, committee members are often favorably impressed when that psychologist can show clearly how he or she considered how best to promote the client's best interests and minimize harm to the client while working to resolve the conflict. Principle A, Beneficence and Nonmaleficence, of the Ethics Code (APA, 2010) should be a paramount concern as the military psychologist looks for creative, informal, and defusing strategies for

protecting client interests while also assisting military commanders in their efforts to successfully carry out military missions.

References

American Psychological Association. (2010). *Ethical principles of psychologists and code of conduct.* Retrieved from http://www.apa.org/ethics

Jeffrey, T. B., Rankin, R. J., & Jeffrey, L. K. (1992). In service of two masters: The ethical-legal dilemma faced by military psychologists. *Professional Psychology: Research and Practice, 23,* 91–95.

Johnson, W. B., Grasso, I., & Maslowski, K. (2010). Conflicts between ethics and law for military mental health providers. *Military Medicine, 175,* 548–553.

US Department of Defense. (2011). *Department of Defense Instruction 6490.08: Command notification requirements to dispel stigma in providing mental health care to service members.* Washington, DC: Author.

23 MIXED-AGENCY DILEMMAS IN MILITARY PSYCHOLOGY

W. Brad Johnson

Psychologists working in a variety of settings may occasionally find themselves struggling with a *mixed-agency dilemma* or a dilemma involving the psychologist's simultaneous commitment to two or more entities. Most often, mixed-agency dilemmas present as conflicts between loyalties to individual clients and loyalties to an organization or even to the larger society (Kennedy & Johnson, 2009). For instance, mixed agency conflicts may emerge when a psychologist feels tension among obligations to a minor child and both parents in the context of a child custody dispute; when a psychologist evaluating a commercial pilot's fitness must consider both the pilot's personal interests and those of the flying public; and when a school district's policies seem to inhibit a psychologist's capacity to render a fair and accurate diagnosis when assessing children.

In military settings, psychologists are most likely to experience mixed-agency dilemmas when their unique obligations as military officers collide with their more traditional professional and ethical obligations as psychologists. Whether a military psychologist serves in traditional clinical and hospital roles, combat clinical roles, or in operational (war-fighting) jobs, they often wrestle with occasional incongruity and conflict resulting from their dual identities as psychologist and military officer (Jeffrey, Rankin, & Jeffrey, 1992). As in many other settings, military psychologists are most often inclined to experience mixed-agency conflict centering on the best interests of individual clients (e.g., soldiers, sailors, Marines, airmen) and the immediate operational needs of the individual's military unit or a military mission.

Certain elements of military service may exacerbate mixed-agency dilemmas for uniformed psychologists. Some of these bear on the psychologist's dual identity as officer and mental health professional and others bear on the broader military culture. Military psychologists should consider how each element might intensify mixed-agency conflicts.

IDENTITY CONFUSION: MILITARY PSYCHOLOGISTS WEAR TWO HATS

Military psychologists literally wear hats—and accompanying uniforms—that identify them with a branch of military service, clarify their rank, and reveal their status as a commissioned officer. Unlike psychologists in other settings, military psychologists take on a legally binding identity with an oath of office and subsequent obligation to promote the fighting strength and combat readiness of military personnel. Commissioned officers are obligated to hold subordinates accountable to behavioral standards while promoting good order and discipline. At times, this military identity may exacerbate conflicts with the psychologist's professional identity. Because military psychologists must achieve state licensure in order to continue beyond the internship and residency stages of service, they are accountable to a code of ethics, most often the American Psychological Association's Ethical Principles of Psychologists and Code of Conduct (American Psychological Association [APA], 2010). This simultaneous allegiance to both professional and military obligations may generate mixed-agency problems. For example, when a senior officer demands to see client records, or when a military regulation appears to be incongruent with ethical standards, the psychologist's dual identities may intensify these mixed-agency dilemmas.

THE MILITARY MISSION IS THE TOP PRIORITY

Within military culture the "mission" is considered superordinate. All military personnel are trained in a milieu that respects and honors a tradition of placing personal comfort and individual interests secondary to the immediate operational objective; most often this means winning a war. In a combat theater, officer priorities and personnel matters that do not directly contribute to the essential military mission—at times this may include routine clinical care or even the best interests of a single service member—may be seen as superfluous (Johnson, 2008). For instance, a military psychologist providing clinical care for a mission-critical service member (e.g., an extremely effective sniper) who is currently on his 5th deployment in 8 years and showing significant symptoms of PTSD, may struggle with ethical obligations to the individual service member (e.g., what might be most "therapeutic" for the client?) versus the overarching military unit (e.g., an upcoming mission hinges on the effectiveness of this particular sniper).

IT IS NOT ALWAYS EASY TO IDENTIFY THE PRIMARY CLIENT

Although professional psychologists are trained to identify who the client is at the outset so that they can proceed to determine their obligations to clients and clarify the nature of the professional relationship through a process of informed consent (APA, 2010), it may not always be easy to identify the primary "client" in military contexts. At times, military psychologists do not enjoy the luxury of deciding to serve exclusively—or even primarily—as an agent for the system or the individual. Individual service members are typically referred by a chain-of-command populated by officers that are senior in rank to the psychologist. The individual's command may have specific consultation questions with heavy bearing on an upcoming military mission (e.g., is this soldier fit for deployment to a combat zone? Can we trust this sailor with top secret information? Is this airman a danger to others on this highly sensitive mission?). In attempting to address these questions, the psychologist will naturally feel a sense of obligation to the individual client, the referring command, and the assorted persons likely to be directly affected by the psychologist's recommendations. In military psychology, more than many other contexts, the psychologist may struggle with perceived ethical obligations to multiple parties in nearly every case. It is important to keep in mind that having more than one client is not the preeminent challenge here. Rather, the effective military psychologist will ask: "To whom do I owe what obligations and in what measure?" In other words, once it is clear that the psychologist has obligations to more than one party, he or she must quickly focus on navigating and negotiating these obligations so as to minimize risk of harm and maximize benefits to all parties concerned.

ROLES WITH CLIENTS MAY SHIFT WITH LITTLE OR NO NOTICE

Military psychologists may find themselves in situations in which roles with clients shift unpredictably. Few experiences may highlight the mixed or dual agency nature of military psychology more acutely than suddenly having to accept an unexpected new role with an existing client (Johnson, 2008). For instance, uniformed psychologists may be ordered to conduct a formal evaluation for fitness for duty, a security clearance, ability to deploy, selection for special assignment, or even capacity to stand trial, with a current or former client. Alternatively, a military psychologist may be required to assume administrative or supervisory duties with current clients—especially those in his or her chain of command—leaving both in awkward yet unavoidable new roles. Ethical risks are exacerbated when these role shifts are sudden, unanticipated, and entirely beyond the control of the psychologist, who may not be able to fully anticipate the various ways such forensic or administrative roles may intrude on or damage the clinical relationship.

Although psychologists in many contexts, upon receiving such a request, might simply refer the case to another provider in order to avoid uncomfortable or even harmful dual roles with a client (APA, 2010), this may not be possible in the military. In many solo or deployed psychologist jobs in the military, other providers may not exist in theater. The military psychologist will need to strike a balance between caring for the client, including his or her best interests, and honoring the needs of the military to address mission-relevant evaluation questions or have the uniformed psychologist serve in an important unit leadership role, even if it means uncomfortable new roles with current and former clients.

MILITARY PSYCHOLOGISTS ARE INCREASINGLY EMBEDDED WITHIN MILITARY UNITS

Embedded practice in military psychology occurs when a psychologist is intentionally deployed as part of a unit or force when the psychologist is simultaneously a member of the unit and legally or otherwise bound to place the unit mission foremost (Johnson, 2008). Military psychologists are increasingly deployed to war zones as members of an Army brigade or as members of a ship's crew as in the case of aircraft carrier psychology. On the upside, embedded positioning of psychologists allows the practitioner to apply his or her tools to the immediate prevention, assessment, and treatment of combat operational stress and psychological disorders. A psychologist's embedded status may also afford him or her greater credibility with warfighters and better perceived approachability. On the downside, it may be increasingly difficult for a psychologist to remain focused on professional ethical obligations to individual clients. For instance, an aircraft carrier psychologist cannot easily honor standards proscribing problematic multiple relationships—simultaneous clinical and personal relationships—with clients when he or she must eat, sleep, and otherwise live with patients around the clock. Of primary concern is the danger of *identity drift* the longer a psychologist remains embedded with military units. Identity drift involves increasing identification with one's military officer identity and simultaneous weakening of one's professional psychologist identity simply as a result of the psychologist's thorough immersion in the military mission and accompanying isolation from other mental health professionals.

RECOMMENDATIONS FOR MANAGING MIXED-AGENCY ETHICAL DILEMMAS

Although military psychologists will not be likely to avoid ethical dilemmas that are caused or exacerbated by their dual identities as officers and psychologists, there are a number of steps they can take to decrease the risk that these dilemmas will result in harm to either individual service members or the military at large. Each of the following recommendations is designed to help military psychologists prevent and address mixed-agency ethical dilemmas (Kennedy & Johnson, 2009).

1. **First, Understand Your Ethical Obligations**. Military psychologists should be thoroughly familiar with the Ethical Principles of Psychologists and Code of Conduct (APA, 2010). Because military deployment can be professionally isolating for the psychologist, and because of the real danger of identity drift involving the waning of one's sense of self as a psychologist versus an officer, it is imperative that the uniformed psychologist begin with a strong foundation in both the Ethics Code and the literature bearing on ethical decision making. A well-formulated approach to ethical decision making will include steps such as carefully defining the ethical question or dilemma, discerning who will be impacted by your actions, considering both ethical and legal obligations, consulting with trusted colleagues, and then formulating alternative courses of action. It is important to keep in mind that legal obligations may include both federal and state laws, DoD statutes, and even service-specific regulations and policies.

   One question the military psychologist should ask early and often when confronting mixed-agency dilemmas is this: *How can I serve the best-interests of my client(s)?* For instance, consider the case of a commanding officer who is adamant about keeping a service member on deployment status in spite of the fact that the psychologist has discovered clear evidence of severe posttraumatic stress disorder (PTSD) that is interfering with the service member's ability to function effectively. Rather than exacerbate this mixed-agency dilemma by polarizing the situation (e.g., ignoring the commanding officer's concerns, overstating the severity of the client's syndrome in formal documentation, minimizing the client's distress in order to appease the commanding officer), it will be important to ask, *to whom am I obligated and in what measure? How can I best serve the needs of the individual service member and the larger military mission? Is there a way to achieve some middle ground without causing harm to the individual client?*

2. **Be Proactive in Seeking Consultation**. As a military psychologist, it is important to keep in mind that other health care professionals face mixed-agency dilemmas in their daily work. Military psychologists should be very active in seeking consultation and supervision from assorted colleagues and subject matter experts. These might include senior psychologists—both military and civilian—who might be available locally or through telecommunication, lawyers, and other mental health service colleagues within the military community. It is especially wise for psychologists who are preparing for deployment to arrange peer consultation relationships with other deployed psychologists or others with deployment experience. When questions or quandaries bearing on the unique obligations of the uniformed psychologist arise, these consultation relationships should be activated.

3. **Seek Prevention through Strong Collaboration**. Quite often, military psychologists can help to diminish or even prevent mixed-agency conflicts. Perhaps the clearest way to achieve this is through forming strong working relationships with the military commanders one serves. For instance, the more a psychologist can provide psychoeducation and prevention services, the less time he or she will need to spend declaring service members unfit for duty or cutting short their deployments. Further, strong interpersonal and collaborative connections between psychologists and commanding officers will lead to greater mutual understanding and smoother dispositional outcomes for impaired service members.

4. **Engage in Self-Care to Promote Good Decision Making**. A final recommendation involves the need to engage in a program of self-care so that military psychologists are consistently able to execute effective ethical

decision making (Kennedy & Johnson, 2009). Because life during deployment can be profoundly stressful, unlike anything civilian practitioners are likely to encounter, and because effective decision making is most likely to occur when the psychologist is reasonably rested and connected to colleagues, it is recommended that military psychologists place a premium on maintaining their own psychological health. Even during deployment, psychologists should pursue opportunities for physical fitness, sleep, moments of pleasure in contact with home or personal hobbies, close relationships with a small network of other medical providers, chaplains, and others. Thinking about mixed-agency conflicts should be enhanced by quality self-care.

References

American Psychological Association. (2010). *Ethical principles of psychologists and code of conduct.* Retrieved from http://www.apa.org/ethics

Jeffrey, T. B., Rankin, R. J., & Jeffrey, L. K. (1992). In service of two masters: The ethical-legal dilemma faced by military psychologists. *Professional Psychology: Research and Practice, 23,* 91–95.

Johnson, W. B. (2008). Top ethical challenges for military clinical psychologists. *Military Psychology, 20,* 49–62.

Kennedy, C. H., & Johnson, W. B. (2009). Mixed agency in military psychology: Applying the American Psychological Association ethics code. *Psychological Services, 6,* 22–31.

24   PROFESSIONAL EDUCATION AND TRAINING FOR PSYCHOLOGISTS IN THE MILITARY

Don McGeary and Cindy McGeary

Training and education are vital components of military psychology not only as a way of developing future military psychologists but also as a way of recruiting mental health professionals into military jobs. Military psychology positions can be highly desirable based on the high quality of training and fellowship opportunities as well as the diversity of military psychology activities. Military psychologists, unlike civilian psychologists, play a unique role including not only patient care and research roles typical of the psychology profession, but also as valuable resources for consultation to command and policy advisers on both small and large scales. Now, more than ever, psychology training is an important way to maintain and strengthen mental health assets throughout the military. A 2009 report in the American Psychological Association's Monitor on Psychology revealed a significant gap between the number of psychologists employed in military jobs and the number that are needed (Munsey, 2009). According to the report, in 2009 the US Army filled only 70% of the available psychologist positions, a number similar to that of the Air Force (83%) and the Navy (81%).

To improve recruitment and participation in military mental health, most branches offer incentives for military psychologists including loan repayment, enlistment/accession bonuses, and relatively high pay for internship training (e.g., most military interns make $56,000/yr versus approximately $20,000/yr for civilian interns). It is important for both current and future military psychologists to understand the breadth of psychology training offered through the military, recognize potential differences in training and education policy and opportunities across branches, and to better understand the unique benefits of psychological training in the military. The next three sections describe training and education opportunities and requirements throughout the psychology career cycle.

GUIDANCE AND POLICIES

All military psychology training and education programs are structured based on guidance and instructions applicable to specific military services as well as the Department of Defense as a whole. Relevant policies and guidance are included below, though this list is likely to be incomplete due to the significant breadth and complexity of psychology education and training in the military (requiring policy and guidance across numerous domains). Consulting these policies can offer insight into the general requirements for military psychology training as well as the administrative supports in place to ensure quality and longevity of training programs. Military psychologists involved in training should familiarize themselves with applicable policies to ensure that their training activities meet the needs and requirements of the military. These policies can be accessed through military publishing websites. The most immediately relevant policies and instructions are summarized in Table 24.1.

ORGANIZATION AND OVERSIGHT

Service Oversight

Psychology training is overseen on a national level through service-branch-specific organizations. For example, Air Force psychology is governed by the Air Education Training Command (AETC), headquartered at Randolph Air Force Base (San Antonio Texas). Army psychology is overseen by the Army Medical Command (which is broken down into several regional medical commands) and the AMEDD Center and Schools (responsible for tactics, doctrine, and organization of all Army medical programs). Psychology in the Navy is governed by the Navy Medicine Professional Development Center, located in Bethesda, Maryland (with the mission of educating, training, and supporting Navy medicine personnel).

Each of these organizations is responsible for oversight and support of psychology training programs, and often plays a role in determining the number of training slots available at each level (i.e., internship, fellowship). Interestingly, the United States Marine Corps does not maintain a training program for psychology, so there is no organizational oversight for Marine Corps psychology training. Naval clinical psychologists treat both sailors and Marines, because the Marine Corps does not maintain its own medical care system (relying instead on the Navy system, including psychology trainees). Some Marines are being trained in mental health skills through the Operational Stress Control and Readiness Program (OSCAR), which is designed to train Marines who work at battalion and squadron levels to intervene with other Marines who are experiencing early symptoms of stress that could develop into posttraumatic stress disorder and suicide risk. Most military psychology training programs are accredited by the American Psychological Association's Commission on Accreditation (APA COA), which provides guidance and oversight on the content of training activities and the methods of competency assessment used to graduate trainees. Although the vast majority of internships maintain APA accreditation, unaccredited postdoctoral fellowships are not uncommon. Trainees

TABLE 24.1. Military Policies and Procedures Regarding Education

Department of Defense (DoD) Instructions	
DoD Instruction 1322.24	SUBJECT: Medical Readiness Training
	Prescribes procedures for medical readiness training and medical skills training.
DoD Instruction 6000.13	SUBJECT: Medical Manpower and Personnel
	Prescribes procedures to carry out medical manpower and personnel programs.
Air Force Instructions	
AFI 36-2301	SUBJECT: Developmental Education
	Refers to all military education, to include internships and fellowships.
AFI 44-119	SUBJECT: Medical Quality Operations
	Outlines roles and responsibilities in clinical performance improvement, credentialing, privileging, and scope of practice in health care delivery.
AFI 41-110 (section 7.9)	SUBJECT: Medical Health Care Professions Scholarship Programs
	Provides guidance for scholarship programs to obtain qualified medical commissioned officers on Active Duty.
Army Regulations	
AR 40-68 (section 7-9)	SUBJECT: Clinical Quality Management
	Establishes peer review process, credentialing, and clinical privileges in health care delivery.
AR 351-3	SUBJECT: Professional Education and Training Programs of the Army Medical Department
	Outlines military professional training, graduate professional education, and health care incentive programs.
Navy Instructions	
BUMEDINST 1524.1B	SUBJECT: Policies and Procedures for the Administration of Graduate Medical Education (GME) Programs
	Directs Navy GME programs and responsibilities.

seeking a fellowship are required to complete an accredited internship to qualify for most military and VA positions. Fellowship accreditation is not required, though trainees seeking fellowships should ensure that the program tracks clinical supervision hours to meet licensure requirements (as recommended by: http://www.apa.org/gradpsych/2004/01/postdoc-skinny.aspx).

Institutional Oversight

As with any military program, it is important to know the chain of command and institutional oversights that exist. The majority of training and education opportunities in military psychology are organized under medical education programs like Graduate Medical Education (GME) and Allied Health Education (AHE). Although AHE is subsumed under GME in most institutions, there are some cases in which these two organizations are separate. Typically, the best organizational representative

for specific training information is the program training director. Training directors for military psychology programs generally report to the GME or AHE Committee, the installation commander, and the APA COA (if their training program is accredited).

PREDOCTORAL TRAINING EDUCATION

Health Professions Scholarship Program

The Health Professions Scholarship Program (HPSP) is offered by the Army, Air Force, and Navy as a way to defray costs for doctoral psychology education in return for an active duty service obligation. Although the details of HPSP scholarships vary across service branches, most offer 100% tuition coverage (including officer's pay during 45 days of annual training when the student is considered Active Duty) and a monthly stipend of approximately $2,000. To qualify, HPSP applicants need to be US citizens enrolled in an APA-accredited psychology doctoral program who maintain full-time student status and meet qualifications for commission as an officer in the United States military. Some military internship programs show preference for HPSP applicants during the internship match, though there is no formal guarantee for HPSP applicants to be selected for a military internship.

Uniformed Services University of the Health Sciences

The Uniformed Services University of the Health Sciences (USUHS) offers doctoral psychology degree programs for both military and civilian applicants. Eligibility criteria for USUHS positions vary by service branch, and most military applicants apply for military commission before starting their doctoral training. There are no tuition costs at USUHS, though students are responsible for the cost of books. If selected for a military doctoral slot, the service member will incur a 7-year service obligation after completion of their internship year. USUHS offers two degree options including clinical psychology and medical psychology. The medical psychology degree emphasizes health psychology and behavioral medicine. There are civilian slots available for USUHS doctoral programs as well. Civilian USUHS students do not pay tuition costs and they do not incur a service obligation. Most civilian students are receiving some form of financial support (e.g., scholarships, stipends, grant funding).

Internships

Psychology internship training programs are available through the US Air Force, the US Army, and the US Navy. All three service branches require that an applicant for internship apply from a doctoral program in clinical or counseling psychology that is accredited by the American Psychological Association. Although military internships typically require an active-duty service commitment after completion (3 to 4 years of active duty service), there are occasional opportunities for civilian internships that do not require a postinternship commitment. Individuals interested in becoming an Active Duty military psychologist generally begin by contacting a military recruiter to explore their eligibility for Active Duty service. Psychology interns participating in a military internship with the service commitment will typically begin their internship with the rank of O-3 (Captain in the Army and Air Force; Lieutenant in the Navy).

Most military psychology predoctoral internship programs adhere to a scientist-practitioner or practitioner-scholar model of training and emphasize a generalist curriculum. Though there is some debate about the best theoretical fit for clinical and counseling psychology training (see Stoltenberg et al., 2000 for an example), there is reason to believe that both models offer similar benefits (especially regarding clinical competence) (Cherry, Messenger, & Jacoby, 2000). Military internship training experiences vary by site and can include general mental health assessment and treatment skills, consultation, inpatient intervention, primary care consultation, health psychology, neuropsychological screening, drug and alcohol abuse counseling, research, and military-specific mental health practice (command-directed evaluations, medical evaluation board assessment). All military internships are accredited by the American Psychological Association as follows:

- Air Force-Wilford Hall Ambulatory Surgical Center (Lackland Air Force Base-San Antonio, Texas)
- Air Force-Wright-Patterson Medical Center (Wright-Patterson Air Force Base-Dayton, Ohio)
- Air Force-Malcolm Grow Medical Center (Andrews Air Force Base-Washington, DC)
- Army-Tripler Army Medical Center (near Fort Shafter-Honolulu, Hawaii)
- Army-Brooke Army Medical Center (Fort Sam Houston-San Antonio, Texas)
- Army-Madigan Army Medical Center (Fort Lewis-Tacoma, Washington)
- Army-Dwight D Eisenhower Army Medical Center (Fort Gordon-Atlanta, Georgia)
- Navy and Army-Walter Reed National Military Medical Center (Bethesda, Maryland) Though both are located at Walter Reed, the

Navy and Army psychology internships are not a consortium. They are separate programs, each with their own accreditation from the APA that share curriculum and faculty
- Navy-Naval Medical Center (near Naval Base Coronado-San Diego, California)

POSTDOCTORAL TRAINING AND EDUCATION

Many military psychologists, upon completing their internship, will go directly into practice at a military installation. Some, however, will eventually choose to continue their training through postdoctoral education. The US Military offers multiple opportunities for postdoctoral education and psychology.

Residency

Postdoctoral residency training is typically designed to bridge the gap between predoctoral internship and postdoctoral fellowships (Kaslow & Webb, 2011). Military psychologists who complete their internship may contemplate a residency to acquire an additional year of experience toward licensure before assuming independent practice. There is some confusion about the differentiation between psychology interns and residents, mostly because the terms are used interchangeably in many medical center-based internships. It is not uncommon for psychology Internship Training Directors to insist that interns refer to themselves as "Psychology Residents" as a way to communicate their experience and competence to medical providers. In formal training terminology, however, a psychology resident is an individual who has completed internship and attained a doctoral degree and is now seeking additional training without the specialization of a formal postdoctoral fellowship. Brooke Army Medical Center currently maintains a Clinical Psychology Residency Program meeting this description.

Fellowship

Several sites throughout the United States offer embedded military experiences for psychology postdoctoral fellowships. Active Duty psychologists are not limited to embedded Active Duty postdoctoral fellowships. They can access nonmilitary fellowships through programs like the Air Force Institute of Technology (AFIT) or the Naval Postgraduate School (NPS), which provide scholarships for postdoctoral training through civilian institutions. This allows for military psychologists to obtain a greater variety of postdoctoral fellowship opportunities than is offered by the military.

The US Army offers over 11 postdoctoral training positions across four sites. Most of the Army fellowships are APA accredited including:

- Tripler Army Medical Center—8 positions (pediatric psychology, clinical health psychology, neuropsychology)
- Madigan Army Medical Center—two positions (pediatric psychology)
- Walter Reed Military Medical Center—one position (neuropsychology)
- Brooke Army Medical Center—variable positions (clinical health psychology, neuropsychology, pediatric psychology)
- Southern Regional Medical Command/ Warrior Resiliency Program—three positions (trauma, risk, and resiliency)

The US Air Force offers only one embedded military postdoctoral fellowship position, which is APA accredited:

- Wilford Hall Ambulatory Surgical Center— one position (clinical health psychology)

The United States Navy offers two positions in clinical psychology for their APA accredited postdoctoral fellowship:

- Naval Medical Center Portsmouth—two positions (clinical psychology)

SPECIALTY TRAINING, CERTIFICATION, AND CONTINUING EDUCATION

As psychology practice continues to diversify, it is becoming increasingly important to establish competence in both general psychological practice (e.g., licensure and clinical credentialing) and specialty practice (Kaslow et al., 2004). The US military has recognized board certification through the American Board of Professional Psychology (ABPP) as the preferred way of establishing specialty competence. Most service branches offer incentives for Active Duty psychologists to achieve board certification, often in the form of a financial stipend (typically $6,000 per year). Though the military does not independently certify psychological specialists, in many cases it does reimburse service members for applying and completing ABPP certification. Currently, ABPP credentials 14 psychology specialties including (but not limited to) clinical health psychology, clinical neuropsychology, clinical psychology, cognitive and behavioral psychology, and forensic psychology. Certification through ABPP typically requires some postdoctoral practice, supportive references from other board certified psychologists, submission of work samples, and a comprehensive oral examination (see www.abpp.org for additional information).

Because military psychologists treat patient populations with needs that are unique to military service (e.g., posttraumatic stress disorder, traumatic brain injury), there are additional training opportunities throughout the career cycle to help improve competence in treating these issues. One example of such a program is offered through the Center for Deployment Psychology (CDP), which provides both online and in-person trainings for numerous military topics including: military culture, trauma and resilience, deployment and families, serious medical injury, and intensive workshops for gold-standard treatments for PTSD and insomnia. Most military psychologists can attend these trainings free through stipends to support their travel to CDP (Bethesda, MD) and through local trainings at military installations throughout the continental United States.

Those who work as military psychologists have access to numerous opportunities for continuing education (CE), which is a major benefit of working in the military system. Most sites offer continuing education through multiple sources, with many treatment facilities independently authorized as official APA CE sponsors. Distinguished visiting professors (DVPs), who are nationally recognized experts in their field of study, are often invited to provide seminars, trainings, workshops, and lectures. For example, the military will frequently have workshops on the treatment of PTSD (prolonged exposure and cognitive processing therapy).

References

Cherry, D. K., Messenger, L. C., & Jacoby, A. M. (2000). An examination of training model outcomes in clinical psychology programs. *Professional Psychology: Research and Practice, 31*, 562–568.

Kaslow, N. J., Borden, K. A., Collins, F. L., Forrest, L., Illfelder-Kaye, J., Nelson, P. D.,…Willmuth, M. E. (2004). Competencies Conference: Future directions in education and credentialing in professional psychology. *Journal of Clinical Psychology, 60*, 699–712.

Kaslow, N. J., & Webb, C. (2011). Internship and postdoctoral residency. In J. C. Norcross, G. R. Vandenbos, & D. K. Freedheim (Eds.), *History of psychotherapy: Continuity and change* (2nd ed., pp. 640–650). Washington, DC: American Psychological Association.

Munsey, C. (2009). Needed: More military psychologists. *Monitor on Psychology, 40*, 12.

Stoltenberg, C. D., Pace, T. M., Kashubeck-West, S., Biever, J. L., Patterson, T., & Welch, I. D. (2000). Training models in counseling psychology: Scientist-practitioner versus practitioner-scholar. *The Counseling Psychologist, 28*, 622–640.

THE DEPARTMENT OF DEFENSE
25 PSYCHOPHARMACOLOGY DEMONSTRATION PROJECT

Morgan T. Sammons

The acquisition of the legislative right to prescribe psychotropic medications has been a goal, albeit a somewhat controversial one, for the psychological profession in the United States since 1989. Previous to that year, psychologists had been extensively involved in the field of psychopharmacology as both clinicians and researchers, but had in general not sought to expand their clinical scope of practice to include the direct provision of pharmacological agents. A few exceptions existed. Dr. Floyd Jennings, while working for the Indian Health Service in the Southwestern United States, was given by informal agreement the ability to prescribe psychotropics in response to a shortage of psychiatrists or others qualified to prescribe such medications (DeLeon, Folen, Jennings, Willis, & Wright, 1991). Such shortages continue to exist and form the basis of the argument that psychologists should seek this expansion of privileges. A few other clinical psychologists also prescribed in similar settings, but until the 1990s none had a formal mechanism to allow them to do so, and most of psychology's involvement with psychopharmacology was limited to the research laboratory or lecture hall.

BACKGROUND

In the late 1980s and early 1990s, the circumstances were ripe for the profession of psychology to seriously consider the acquisition of prescriptive authority. I have argued elsewhere (Fox et al., 2009) that several phenomena created a situation amenable to this endeavor. The growing recognition that combined treatments for mental disorders might have unique efficacy and a new understanding of the limitations of the placebo effect in drug treatment for mental disorders were colliding with the increasing medicalization of mental health treatment in the United States and around the world.

Although the profession of psychiatry had long been a biologically oriented field, and the use of psychotherapy as a psychiatric intervention had been eroding for some time, the introduction of the selective serotonin reuptake inhibitor (SSRI) class of antidepressants in the late 1980s saw an explosion in the use of pharmacological treatments for mental disorders. That the SSRIs lacked the noxious and potentially lethal side effects of earlier antidepressants meant that they were increasingly used, and increasingly prescribed by nonpsychiatrists in nonspecialty mental health settings. Additionally, the 1980s saw a very rapid expansion of the scope of practice of nonphysician health care providers other than psychologists. Registered nurses began to be trained as advanced practice nurses with increasing specialization in their training and clinical practice, and these nurses began to use psychopharmacological agents, generally on the basis of very little formal training. By the early 1990s several jurisdictions had granted advanced practice nurses independent prescriptive authority. Physician assistants were also increasing in number and representation in specialty areas such as mental health, and while, then as now, physician assistants did not seek independent prescriptive authority, their training and placement in medical specialties, including primary care and mental health, dramatically changed the health care landscape.

The 1980s was also the decade in which clinical psychology came into its own as a health care delivery profession. The success of landmark cases, such as the celebrated Virginia Blues case (Resnick, 1985) allowing psychologists to directly bill third-party payers for clinical services, marked a significant expansion in the role of the profession in the health care marketplace. Finally, but by no means unimportantly, psychology had a strategically placed advocate who worked closely with the American Psychological Association and various state psychological associations on scope of practice issues. Dr. Patrick DeLeon, a psychologist who was also trained as an attorney and had a master's in public health, was at the time chief of staff to US Senator Daniel K. Inouye (D-Hawai'i). Senator Inouye was a WWII veteran who was severely wounded in WWII (he later received the Congressional Medal of Honor for his valor in combat) and maintained an abiding interest in improving health care delivery in the US military. He was instrumental in establishing the Uniformed Services University of the Health Sciences and in ensuring that its role focused on training not only physicians but also nurses and other health care providers. Dr. DeLeon, in his role as chief of staff, worked closely with the senator in drafting legislation and ensuring that professional associations were involved in advocacy on behalf of their clinicians. Dr. DeLeon's initiative, combined with changes in practice and professional responsibilities outlined above, created an atmosphere of receptivity for the expansion of psychology's scope of practice.

While organized initiatives to train psychologists to prescribe occurred in the late 1980s, it is important to recognize that the concept had deeper historical roots. Almost 40 years earlier, the psychoanalyst Lawrence Kubie had called for the creation of a "new profession" that bridged the overly biological orientation of psychiatry with the almost exclusively nonmedical training of psychologists (Kubie, 1954). Kubie's rational assessment was that a new profession, trained in elements of both medicine and psychology, would provide the best avenue for comprehensive treatment of patients with mental disorders. It should be pointed out that when this report was written, clinical psychopharmacology was in its infancy, and nowhere did Dr. Kubie directly address the provision of pharmacological services by psychologists. But it was his clear vision that this new profession, which he called the "Doctorate in Mental Health" could accommodate the shortfalls in clinical training in both professions.

Dr. Kubie's scheme was published in a very obscure journal and did not elicit much comment at the time. In the 1960s however, his notion was revitalized by psychoanalytic colleagues at the University of California in San Francisco. This led to the establishment of a program that set out to actualize Kubie's vision: The Doctorate of Mental Health program at Langley Porter Psychiatric Institute of the Mt. Zion Hospital in San Francisco (see Wallerstein, 1991, for a comprehensive analysis of this program). As Kubie had envisioned, this 5-year curriculum integrated medical and psychological training. It operated between 1976 and 1986, but after graduating 9 classes and a total of around 80 trainees who possessed the new degree, intense opposition from psychiatry of the same type that later led to the demise of the PDP (see below), led to its closure. Afterward, graduates of this program had difficulty finding employment, as their degree was not recognized by either boards of psychology or medicine, and most graduates sought retraining as either psychiatrists or psychologists.

THE PSYCHOPHARMACOLOGY
DEMONSTRATION PROJECT

In 1989 congressional language mandated the start of a program to train military psychologists to prescribe. Initially envisioned as a program that would give psychologists the necessary skills to treat combat stress disorders, it was soon recognized that a program set up to treat a single disorder was not well thought out. After considerable renegotiation involving consultation with military medicine, the American Psychological Association, the American Psychiatric Association, and other external agencies, initial plans to train these psychologists by utilizing the curriculum for physician assistants was abandoned. Variations on a unique curriculum were discussed, but by 1991, when no appreciable progress had been made toward finalizing a curriculum, it was decided to enroll the initial class of Fellows into the 1st-year medical curriculum at the Uniformed Services University of the Health Sciences. These participants (including the author) were expected to take the majority of the first 2 years of medical school and then complete the 2nd year (predominantly inpatient) of a psychiatric residency (this curriculum is detailed in Sammons, 2003, and Newman, Phelps, Sammons, Dunivin, & Cullen (2000). A comprehensive list of documents pertaining to the project may be found in Sammons (2010). Principally as a result of the opposition of organized medicine to the program, an external evaluative component was added to the project. The American College of Neuropsychopharmacology (ACNP) was the successful bidder, and they provided programmatic evaluation, wrote and administered examinations for the candidates, and provided periodic reports to the Congress and military medicine. The first cohort of students took 3 years to graduate, completing the bulk of the first 2 years of the medical school curriculum along with specialized practica and training experiences. Because no structured clinical experience existed in the medical school curriculum, the first class completed a year of psychiatric residency on the inpatient services at Walter Reed Army Medical Center. By the end of the training, the initial class of four fellows had shrunk by 50%, one fellow deciding to go to medical school and the other leaving the program and the military.

The subsequent iterations of the program had a more rationally designed curriculum, and the fellows were presented with more tailored didactic and clinical experiences. While the program was heavily oriented toward the medical school experience, these fellows were able to complete a program in 2 years that, unlike medical school, was designed for students who had already completed a doctoral degree in a health care professional field.

A total of 10 fellows graduated from the program before it succumbed to political pressure in 1998. All of these fellows had highly scrutinized clinical experiences and were subject to rigorous examination by the external evaluator. It was the unequivocal opinion of the ACNP that the program had succeeded in training these psychologists well, however, the cost of the program was resoundingly criticized by other evaluators (see Sammons, 2010, for a further discussion of external oversight of the PDP). Military psychiatry remained unrelentingly opposed to the program, arguing that there was no shortage of military psychiatrists (astonishingly, even while making such claims military psychiatry was requesting that Congress increase their numbers on the basis of provider shortages, Sammons, 2010).

Although the program ended in 1998, the 10 graduates continued to provide psychopharmacological services for the remainders of their military careers. All such graduates did well professionally, most retiring at senior officer grades (of Colonel/Navy Captain or Lieutenant Colonel/Navy Commander). After the demise of the program, the military continued to train psychologists to prescribe via a 2-year postdoctoral fellowship at Tripler Army Medical Center in Honolulu, and also sent Active Duty psychologists to several of the civilian training programs that by then existed. By 2013 there were approximately 25 uniformed prescribing psychologists on active duty, and all three military branches credentialed appropriately trained psychologists to prescribe (references to service-specific credentialing instructions may be found in Sammons, 2010).

AFTERMATH

Since the PDP, there has been increasing activity at the state level to obtain prescriptive authority (Fox et al., 2009). Between 2004 and

2010, a total of 57 bills were introduced into the various branches of state legislatures (Deborah Baker, 2010, personal communication). In two states, New Mexico and Louisiana, psychologists battled considerable resistance by psychiatry and succeeded in passing legislation. In two other states, Oregon and Hawaii, legislatures passed such bills, only to have them vetoed by their respective governors.

More recently, the pursuit of prescriptive authority has been influenced by the recognition that the provision of mental health services in nonspecialty settings (e.g., primary care settings) is required in order to accommodate growing demand for such services. As previously noted, the vast preponderance of psychotropic agents are prescribed in the primary care environment by nonspecialists. Accurate and complete differential diagnosis for many mental disorders is also lacking in the primary care environment, and, as we've seen, psychopharmacology is in general the only treatment offered. Regardless of whether psychologists or another health care provider specialty provides such services, integrated services in the primary care environment is essential for optimum mental health care delivery (Carey et al., 2010).

It is a necessary but unfortunate observation that many of the issues that led to the initiation of the PDP over 20 years ago continue to be hallmarks of the American health care system. We remain plagued by an overreliance on psychotropic medication, often prescribed for nonindicated conditions. In 1991 as today, a shortage of specialty trained prescribers of psychotropic agents (e.g., psychiatrists, prescribing psychologists, and some advance practice nurses) exists, a shortage that reaches critical proportions in traditionally underserved areas. The profession of psychiatry continues to enroll fewer and fewer residents into training, all the while maintaining opposition to extending prescriptive authority to psychologists and other nonmedically trained prescribers. Organized medicine remains fervent in their opposition, in spite of an impeccable safety record accumulated over two decades; and well-funded lobbying efforts on behalf of organized medicine have stopped numerous legislative initiatives. Academic psychology continues to be lukewarm in its support of prescriptive authority; it is a feature peculiar to the profession of psychology that opponents of prescriptive authority from within actively attempt, via testimony and other mechanisms, to thwart legislative initiatives. Thus, the vast majority of psychotropics are prescribed by primary care providers with no special training in psychopharmacology, and most recipients of mental health care get a prescription for a medication and no other form of intervention (Olfson & Marcus, 2009). This has led to what I have referred to as the central paradox in modern psychopharmacology: Although it is clear that combined pharmacological and nonpharmacological interventions yield optimum outcomes for most mental disorders, very few patients are afforded such treatments.

Despite the fact that the PDP trained only 10 psychologists, it has become the de facto reference point for psychopharmacological training. Why this is so remains rather mysterious, as approximately 1,000 psychologists have now been so trained in other venues (including the three programs that have, as of 2012, received APA designation—Fairleigh Dickinson University, New Mexico State University, and the California School of Professional Psychology). In part, the emphasis on the PDP is due to its groundbreaking nature, the widespread professional and political debate it engendered, and the fact that its curriculum was so closely evaluated by a recognized external body. In spite of this scrutiny, what has been lost is the recognition that the PDP presented a less desirable training curriculum for psychologists than do many of today's programs. The PDP was a creature of medicine and politics, not of rational design. Its curriculum was formulated almost exclusively from a medical perspective, and was intentionally designed to resemble psychiatric training as closely as possible. Particularly in the early versions of the PDP, participants did little more than complete a truncated medical school and psychiatric residency experience.

Current training programs are instead deliberately designed from a psychological perspective of pharmacotherapy. Thus, rather than focusing on biological etiologies and biological interventions, these programs present a more nuanced view of the etiology of mental disorders and their treatment. By understanding the nonspecific nature of most psychopharmacological intervention, firmly incorporating the psychosocial etiology of mental disorders into diagnostic formulation and treatment conceptualization, and by focusing on the collaborative, dynamic, and interactional nature of the treatment process, a psychological model represents a different heuristic and one that is definably separate from medicobiological models that undergird psychiatric training. Evidence that combined treatments for most common mental disorders (even severe ones) yield improved outcomes over either psychological or biological interventions buttresses the validity of the psychopharmacological model. It is on these models and training programs that the profession now needs to focus, for therein lie the keys to more effective treatments for our deserving patients.

References

Carey, T. S., Crotty, K. A., Morrissey, J. P., Jonas D. E., Viswanathan, M., Thaker, S., ... Wines, C. (2010). *Future research needs for the integration of mental health/substance abuse and primary care* (Future Research Needs Paper No. 3, AHRQ Publication No. 10-EHC069-EF). Rockville, MD: Agency for Healthcare Research and Quality. Available at www.effectivehealthcare. ahrq.gov/reports/final.cfm

DeLeon, P. H., Folen, R. A., Jennings, F. L., Willis, D. J., & Wright, R. H. (1991). The case for prescription privileges: A logical evolution of professional practice. *Journal of Clinical Child Psychology, 20,* 254–267.

Fox, R. E., DeLeon, P. H., Newman, R., Sammons, M. T., Dunivin, D. L., & Baker, D. C. (2009). Prescriptive authority and psychology: A status report. *American Psychologist, 64,* 257–268.

Kubie, L. S. (1954). The pros and cons of a new profession: A doctorate in medical psychology. *Texas Reports on Biology and Medicine, 12,* 692–737.

Newman, R., Phelps, R., Sammons, M. T., Dunivin, D. L., & Cullen, E. A. (2000). Evaluation of the Psychopharmacology Demonstration Project: A retrospective analysis. *Professional Psychology: Research and Practice, 31,* 598–603.

Olfson, M., & Marcus, S. (2009). National patterns in antidepressant medication treatment. *Archives of General Psychiatry, 66,* 848–856.

Resnick, R.J. (1985). The case against the Blues: The Virginia challenge. *American Psychologist, 40,* 975–983.

Sammons, M. T. (2003). Introduction: The politics and pragmatics of prescriptive authority. In M. Sammons, R. Levant, & R. Paige (Eds.), *Prescriptive authority for psychologists: A history and guide* (pp. 3–32). Washington, DC: American Psychological Association.

Sammons, M. T. (2010). The Psychopharmacology Demonstration Project: What did it teach us and where are we now? In R. E. McGrath & B. A. Moore (Eds.), *Pharmacotherapy for psychologists: Prescribing and collaborative roles* (pp. 49–68). Washington, DC: American Psychological Association.

Wallerstein, R. S. (1991). *The doctorate in mental health: An experiment in mental health professional education.* Lanham, MD: University Press of America.

26 PSYCHOLOGISTS ON THE FRONTLINES

Craig J. Bryan

The sustained combat operations in both Iraq and Afghanistan over the past decade have dramatically changed the role of military psychologists. Ever-expanding empirical evidence demonstrating the significant emotional and psychological cost of combat and other military operations on service members has highlighted the critical nature of psychologists' skills and knowledge for the military's health and success. Given the clear link between combat exposure and the full spectrum of psychiatric morbidity, military psychologists have not only found themselves placed closer and closer to the "point of injury" (i.e., within combat zones), but they have also found themselves asked to engage in a much broader range of professional activities that extend beyond the traditional clinical services of assessment and treatment. Military psychologists must therefore have a basic understanding of these common roles and how the realities of a deployed context can influence or shape their ability to succeed in these distinct but overlapping roles.

THE MILITARY PSYCHOLOGIST AS PREVENTIONIST

It is now well established that deployments with more intense exposure to combat are a risk factor for a number of subsequent psychological health problems, the most prominent of which include posttraumatic stress disorder (PTSD), traumatic brain injury (TBI), and substance abuse. The steady rise in military suicides since the initiation of OEF/OIF has similarly raised concerns about the deleterious effects of deployment and combat exposure. Although combat exposure has generally not been found to be *directly* related to suicide risk, recent evidence has supported an *indirect* association through psychological symptoms including PTSD, depression, and social isolation (Bryan et al., 2012). In light of the clear association between combat exposure and psychological distress, considerable interest in the prevention of psychiatric morbidity has emerged and sparked a great deal of professional discussion and program development.

"Resiliency" is a term that has been used within the military with increased frequency during the past several years, many times as a catchall and arguably an ill-defined term that is often synonymous with the notion of preventing psychiatric morbidity. Because the concept of "prevention" is not as thoroughly fleshed out within the mental health professions, military psychologists often have considerably less guidance regarding how to effectively "prevent" the onset of psychiatric conditions among service members. Nonetheless, military psychologists are frequently called on during deployments to develop and provide programs designed to prevent mental health problems and/or enhance resiliency. In the absence of clearly developed, empirically supported prevention programs, it is easy for military psychologists to fall into the trap of hastily creating (and re-creating) generic mental health briefings that focus on stereotypical mental health issues (the most common of which tend to be stress, depression, PTSD, and suicide) presented from a clinical perspective that often uses language consistent with deficiency, illness, and injury (e.g., signs and symptoms of disorders). Such resiliency programs often take the form of PowerPoint briefings using standardized slide formats that are also used for administrative purposes (i.e., white background, black text, unit logo in upper left-hand corner, unit motto along bottom, etc.), and are often delivered in very large groups with minimal audience participation and little, if any, incorporation of skills training. Such "death by PowerPoint" prevention efforts are unlikely to be successful, however.

When developing prevention programs or efforts, military psychologists should therefore present material that aligns with the background and training of the target audience. For example, because the military culture in general stresses mental toughness, autonomy, strength, and elitism, prevention programs should similarly take a strengths-based approach that assumes that the target audience is inherently resilient and already possesses many of the basic skills necessary for maintaining psychological health (Bryan & Morrow, 2011). Similarly, given the absence of empirically supported "resiliency" or prevention programs, military psychologists should base the development of any such program on well-established and scientifically supported clinical interventions. Furthermore, prevention programs developed and implemented downrange (i.e., in deployed settings) must be flexible and practical to the deployed context. For example, a prevention program might aim to enhance resiliency by seeking to improve sleep quality through the explicit instruction of sleep restriction and sleep hygiene principles. However, the realities of the deployed context might interfere with or limit implementation of these strategies (e.g., large numbers of people sharing sleeping and living quarters, very restricted personal space, unpredictability of mission demands and daily schedules). Military psychologists must therefore be prepared to problem-solve or adapt to these natural barriers to increase the likelihood of their program's effectiveness.

THE PSYCHOLOGIST AS ORGANIZATIONAL CONSULTANT

Consultation with military commanders and leaders is a critical role for military psychologists regardless of setting, and in many cases can be seen as a natural extension of the psychologist's role as preventionist. From a public health and prevention perspective, the military is unique in that the capacity to impact the lives of many is often centralized within the authority or purview of so few. Within the deployed context, the military psychologist can potentially have an indirect impact on the health and well-being of hundreds, if not thousands, of deployed service members via effective consultation with military commanders and leaders. For example, educating commanders on the detrimental effects of frequent shift changes and sleep deprivation can result in the scheduling of missions in ways that are less disruptive to service members' sleep cycles. Psychologists might also assist commanders in effectively administering discipline and reinforcement in order to better shape service members' behaviors and conduct over time without inadvertently contributing to boredom, frustration, and degradation of morale.

When consulting with commanders and military leaders, the military psychologist should ensure that recommendations are empirically supported, to the point, and actionable. To help shape recommendations consistent with these criteria, the military psychologist can ask himself or herself the following questions:

1. **What evidence do I have to back up this recommendation?** Recommendations based on empirically supported principles as opposed to subjective perspectives or personal opinions have been subjected to systematic evaluation and testing, and are therefore more reliable. In situations where clear scientific evidence is lacking

or completely absent, military psychologists should turn to the closest available line of empirical evidence to guide their recommendations.

2. **How can I say this in one simple sentence?** Recommendations should be to-the-point so they are more easily digested and therefore more likely to be implemented. Although most psychological principles for which consultation is requested are complex and multifaceted, military psychologists who rapidly provide straightforward suggestions or recommendations will be perceived as more valuable and credible.

3. **What specifically do I want to be done or changed?** Typically, military leaders are seeking consultation because they are looking for solutions to problems. Recommendations should be therefore practical and actionable. Military psychologists who can provide clear recommendations about "what to do" and "when to do it" are more likely to see their recommendations impact decisions.

THE PSYCHOLOGIST AS BEHAVIORAL HEALTH CONSULTANT

In many deployed locations, the military psychologist might be the sole mental health asset. Psychologists are therefore relied on by other medical professionals for assistance and consultation regarding the assessment and treatment of patients with suspected psychiatric health issues. This might include "on call" duties for behavioral health emergencies (e.g., psychotic or manic episodes, suicide risk, homicide risk), but it could also include requests for evaluation and/or treatment of suspected psychosocial health issues (e.g., impact of stress on headaches or high blood pressure). Although such consultation duties are familiar to many psychologists since similar services are typically provided in garrison as well, military psychologists must keep in mind the many contextual differences between deployed and garrison locations that can influence patient safety and the reasonableness of management and treatment strategies (Bryan, Kanzler, Durham,

West, & Greene, 2010). For example, because firearms are so easily accessible in combat zones, and in many places are required to be in the possession of service members at all times, suicide and homicide risk are increased exponentially, and common and otherwise highly effective risk management strategies such as means restriction are of limited utility due to easy access to other service members' weapons and other highly lethal means (e.g., explosives, chemicals, heavy machinery).

Perhaps less familiar to psychologists is more general, nonemergency behavioral health consultation to medical professionals. For example, psychologists might be asked to provide input and suggestions for the treatment of relatively common medical complaints such as gastrointestinal symptoms, headaches, sleep disturbance, and head injuries. The likelihood for such consultation requests increases for patients with recurrent health issues that do not remit or improve with standard medical interventions. Critically, military psychologists should be prepared to deliver brief, targeted assessments and interventions across a range of settings, whether in a troop medical clinic or a hospital ward. Psychologists with training in health psychology and experience working in integrated medical settings (e.g., primary care, internal medicine) are especially well prepared for meeting this need and supporting other deployed medical professionals.

THE PSYCHOLOGIST AS CLINICIAN

Of the many roles a military psychologist might fill while deployed, perhaps the most familiar is that of clinician. As deployed clinicians, military psychologists often provide traditional interventions for a wide range of psychosocial issues that mirror their work while in garrison ranging from occupational and/or family stress to adjustment problems to anxiety and depression. In addition to these "traditional" mental health concerns, military psychologists are often asked to provide psychoeducational classes for associated behavioral health issues such as tobacco cessation or sleep enhancement. In many ways, clinical work in deployed environments is very similar to clinical work in garrison; treatment is treatment, after all. However, as noted above, military psychologists must keep in mind contextual factors of the deployed environment that might influence treatment outcomes. For example, opportunities for behavioral activation (e.g., going to movies, engaging in hobbies) can be severely restricted, and common sources of social support (e.g., talking with a spouse, going to events with friends) might be limited or completely unavailable. A smaller "menu of options" for treatment can not only affect treatment planning, but it may also have an impact on clinical outcomes.

Combat-related PTSD is arguably the condition for which the issue of treatment effectiveness in deployed settings is of greatest relevance to military psychology. PTSD is one of the most frequent emotional consequences of exposure to violence and trauma, with estimates suggesting that around 15% of combat veterans suffer from this condition. Only 25–50% of these combat veterans are estimated to receive mental health treatment of any kind, however, most likely due primarily to pervasive fears and stigma about seeking out mental health care. Of this small treatment-seeking group, only around 30% will receive an empirically supported treatment such as prolonged exposure (PE) or cognitive processing therapy (CPT), which are the principal treatments for PTSD recommended by the Department of Veterans Affairs and the Department of Defense (2004) and the Institute of Medicine (2007), based on their consistent effectiveness in reducing symptoms and contributing to remission across dozens of clinical trials. Although the factors contributing to this low rate of providing PE or CPT for combat-related PTSD among service members are many, one that is particularly salient to the deployed military psychologist is clinician perceptions and beliefs. Many military psychologists have voiced concerns about providing PE or CPT in combat zones, the most common of which warrant discussion.

1. **Exposure-based therapies such as PE and CPT have not been shown to be effective for the reduction of combat-related PTSD among Active Duty personnel.** It is true that no randomized clinical trials with Active Duty military personnel have yet been published for either PE or CPT, but it is important to note that this limitation applies to almost all health or medical conditions and treatments, including many (arguably all) of our preferred mental health interventions, whether behavioral or pharmacologic. More specifically, none of the alternative treatments to PE and CPT that are widely used by military clinicians (e.g., supportive counseling, stress management classes, medications) have any scientific support with military populations either. Encouragingly, pilot data supports the effectiveness of exposure-based therapies when administered downrange with active duty military personnel suffering from combat-related PTSD (Cigrang, Peterson, & Schobitz, 2005). Given the absence of any controlled trials for any PTSD treatment among active duty military, military psychologists must turn to the next best available evidence, which are the civilian PTSD treatment studies, in which PE and CPT have shown clear and consistent advantages over other forms of treatment.

2. **Combat-related trauma should not (or cannot) be safely treated in a combat zone due to risk of retraumatization, insufficient resources to provide effective care and management, and/or the risk of clinical worsening in a high-risk operational context.** As above, it is correct that the safety of exposure-based therapies has never been explicitly tested in Active Duty samples, but this concern also applies to many common alternative treatments. Once again, military psychologists are forced to rely on the next best available evidence: civilian studies. In civilian studies, notable adverse event rates with PE or CPT have not been observed relative to comparison conditions including present-centered therapies, stress management skills training, or supportive counseling, supporting the safety of exposure-based therapies. On the contrary, significantly higher rates of improvement and recovery have been observed in CPT and PE relative to comparison treatments,

in some cases within two or three sessions of treatment initiation.

3. **Because many combat-related traumas occurred so recently, PTSD cannot yet be diagnosed. PE and CPT should therefore not be administered until full criteria for PTSD are met.** A unique aspect of clinical care for PTSD in the deployed setting is that the military psychologist can many times have direct contact with patients very soon after the traumatic event occurred. By definition, such patients cannot be diagnosed with PTSD, although many will meet criteria for a diagnosis of acute stress disorder. Critically, randomized clinical trials have supported exposure-based therapies as an effective early treatment for acute stress disorder, with decreased likelihood for subsequent onset of PTSD (Bryant, Moulds, & Nixon, 2003; Bryant et al., 2008). Military psychologists who administer these exposure-based therapies to service members who meet criteria for acute stress disorder in the period of time immediately following exposure to combat trauma are therefore well positioned to prevent the eventual development of PTSD and potential long-term psychiatric impairment.

4. **PTSD "symptoms" are adaptive or functional within combat zones. Treating these symptoms is therefore inappropriate and increases risk for service members.** Within combat zones, emotional suppression and increased vigilance are common and can actually function in an adaptive manner. For instance, being more alert or "on edge" when surrounded by potential sources of danger can increase the likelihood of detecting threat cues much sooner. Similarly, experiencing strong emotional responses (e.g., grief, fear) in response to life-threatening situations can jeopardize operational effectiveness and safety for an entire unit. Failing to recognize the social and environmental demands of the deployed environment can result in overpathologizing context-appropriate experiences and behaviors. Despite this, military psychologists must guard against underestimating or dismissing the signs and symptoms of PTSD as "normal" responses, as untreated

PTSD can result in occupational impairment that could escalate operational risks. Military psychologists must keep in mind the qualitative differences between expected combat stress reactions and symptoms of a clinical trauma reaction: insomnia is common when deployed, but sleep disturbances due to recurring nightmares of a traumatic event is not; emotional suppression may be an adaptive response, but pervasive avoidance of traumatic memories that contributes to functional impairment is not. Military psychologists must be prepared to identify and administer exposure-based therapies for deployed service members who exhibit these nonadaptive reactions to traumatic events.

Fortunately, several clinical trials are currently underway in which these treatments are explicitly being tested among active duty military personnel, so that concerns regarding their effectiveness and safety can be definitively addressed. Until these studies are complete, however, military psychologists should rely on the best available data and practice guidelines, which clearly support exposure-based therapies such as PE and CPT.

THE PSYCHOLOGIST AS SCIENTIST

Many military psychologists have graduated from programs identifying with the scientist-practitioner model of training. Although training programs certainly vary considerably in terms of the emphasis on this spectrum, all military psychologists have received at least some fundamental training in scientific methodology and principles. As alluded to above, our knowledge and understanding of many issues such as diagnosis, assessment, and treatment of conditions of particularly relevance to the military (e.g., PTSD, TBI, suicide) are directly hampered by the general lack of scientific studies conducted within the military, particularly within deployed settings. Military psychologists are therefore uniquely positioned to advance our understanding of many military-relevant issues and problems within deployed settings.

Unfortunately, there are many barriers to the easy and successful conduct of research within military settings. Approval for research is not only required from the appropriate regulatory and oversight bodies (e.g., institutional review boards), but also from local commanders, who must be willing to authorize or otherwise support the study with the personnel under their charge. The approval process is typically very slow and cumbersome, with multiple levels of approval required before a study can be initiated. Unfortunately, these barriers and layers of review and oversight are magnified exponentially within combat zones. Military psychologists must therefore possess and exercise a high level of patience and commitment if they decide to engage in research while deployed. Although difficult, the rewards for assuming a scientific role while deployed can be considerable, as very few have the opportunity to contribute to the knowledge base of our profession in this specific context. It is recommended that military psychologists interested in contributing to our profession through research in combat zones identify and collaborate with other researchers who have navigated the complex process of conducting research within military settings, especially downrange.

References

Bryan, C. J., Hernandez, A. M., Allison, S., Clemans, T., McNaughton, M., & Osman, A. (2012, June). *Combat exposure and suicidality*. Paper presented at the annual meeting of the DOD/VA Suicide Prevention Conference, Bethesda, MD.

Bryan, C. J., Kanzler, K. E., Durham, T. L., West, C. L., & Greene, E. (2010). Challenges and considerations for managing suicide risk in combat zones. *Military Medicine, 175*, 713–718.

Bryan, C. J., & Morrow, C. E. (2011). Circumventing mental health stigma by embracing the warrior culture: Feasibility and acceptability of the Defender's Edge Program. *Professional Psychology: Research and Practice, 42*, 16–23.

Bryant, R. A., Mastrodomenico, J., Felmingham, K. L., Hopwood, S., Kenny, L., Kandris, E., … Creamer, M. (2008). Treatment of acute stress disorder: A randomized controlled trial. *Archives of General Psychiatry, 65*, 659–667.

Bryant, R. A., Moulds, M. L., & Nixon, R. V. (2003). Cognitive behaviour therapy of acute stress disorder: A four-year follow-up. *Behaviour Research and Therapy, 41*, 489–494.

Cigrang, J. A., Peterson, A. L., & Schobitz, R. P. (2005). Three American troops in Iraq: Evaluation of a brief exposure therapy treatment for the secondary prevention of combat-related PTSD. *Pragmatic Case Studies in Psychotherapy, 1*(2), 1–25.

Department of Veterans Affairs & Department of Defense. (2004). *VA/DoD clinical practice guideline for the management of post-traumatic stress*. Washington, DC: Author.

Institute of Medicine. (2007). *Treatment of post-traumatic stress disorder: An assessment of the evidence*. Washington, DC: National Academies Press.

27 PROVISION OF MENTAL HEALTH SERVICES BY ENLISTED SERVICE MEMBERS

Richard Schobitz

The enlisted behavioral health technician is capable of providing a number of clinical and administrative services that are likely to expand behavioral health service capacity both in garrison and in deployed settings. This chapter will provide an overview of the training that behavioral health technicians receive prior to their first assignments, the roles that behavioral health technicians may play in different settings, and suggestions on how to supervise and support the development of behavioral health technicians.

TRAINING TO BECOME AN ENLISTED BEHAVIORAL HEALTH TECHNICIAN

Behavioral health technicians in training attend their service-specific basic training course and then the behavioral health technician course at the Medical Education and Training Campus at Fort Sam Houston, Texas. This course, approximately 5 months in duration, involves didactic training, practical experiences, and periodic testing which the candidate must pass in order to progress toward graduation.

The goal of the behavioral health technician course is to prepare service members to provide supervised clinical care in both garrison and deployed settings. It is expected that graduates of the behavioral health technician course are able to demonstrate skills in: conducting clinical interviews, basic counseling, knowledge of basic emergency medical treatment, and in assisting with the formulation and implementation of treatment recommendations and plans under the supervision of licensed providers. Training is provided on psychological testing, psychopathology and the use of the current edition of the *Diagnostic and Statistical Manual of Mental Disorders* (DSM), human growth and development, the use of relaxation techniques, and a number of other areas (US Army Medical Department Center and School, 2010).

This training is complemented by service-specific practicum training. Upon completing Phase I of the behavioral health technicians' course, Army and Navy trainees matriculate to Phase II, or the clinical portion of their training. Army trainees complete outpatient and inpatient practicum training (Phase II) at military and civilian sites around Fort Sam Houston upon completing the didactic portion (Phase I) of their training. Navy trainees receive their practicum training at various Naval clinical sites. The Air Force currently relies on on-the-job training for behavioral health technicians once they have graduated from the didactic portion of the behavioral health technician course.

SUPERVISION OF ENLISTED BEHAVIORAL HEALTH TECHNICIANS

Behavioral health technicians provide behavioral health services under the supervision of a licensed clinical psychologist, licensed clinical social worker, psychiatrist, or another licensed professional. Clinical supervision of technicians should be guided by the provider's code of ethics related to supervision. Information regarding the ethical standards for clinical supervision may be found in the National Association of Social Workers' Code of Ethics (2008) and the American Psychological Association's Ethics Code (2010).

It is important for providers working with behavioral health technicians to recognize that while all technicians receive the same didactic training, there will likely be variability among technicians in their training on direct clinical care. It is also important to note that upon graduating and throughout their careers, behavioral health technicians may be assigned to a variety of jobs, some of which are not clinically focused. Supervisors will need to develop an understanding of their technicians' relative strengths and weaknesses, and consider providing additional clinical training opportunities to assist in the development of the technician and to help to meet the mission.

In general, a new behavioral health technician coming directly from technical training will likely have limited experience in clinical settings. These technicians may benefit from opportunities to shadow clinicians and participate in assessments and treatment alongside their supervising clinician. In our clinics we often begin clinical exposure by providing junior behavioral health technicians with an opportunity to conduct clinical interviews with seasoned clinical staff. This allows for the newly trained technician to gain confidence in clinical settings, observe the work of other professionals, and then receive feedback and supervision after the completion of the session. Similarly, we often have behavioral health technicians in our clinics cofacilitate group therapy sessions.

As the technician gains experience and their skill level increases, it may be possible to expand their responsibilities and increase their autonomy while they continue to practice under supervision. Technicians may be provided opportunities to conduct initial evaluations in a model where they staff their cases with a licensed provider, conduct psychoeducational groups, and assist in testing if they have been provided with an opportunity to learn how to administer and score psychological assessments.

THE BEHAVIORAL HEALTH TECHNICIAN IN MENTAL HEALTH CLINICS

Behavioral health technicians can serve as an excellent resource to extend the clinical capacity of mental health clinics in garrison. Technicians whose clinical skill development has been supported may provide a variety of services under supervision. One area in which technicians are often involved is the triage process. The service members who are seen as clients at military behavioral health clinics are likely to have challenging schedules. These challenges often include time in the field, time spent on temporary duty away from home station, and time deployed to combat or operational missions. In order to make access to care as easy as possible, many behavioral health clinics on military instillations have adopted walk-in procedures. While this approach simplifies the process for the client, a clinic with walk-in procedures must dedicate clinical staff to be available. Behavioral health technicians are likely well suited to provide this support, as there is an emphasis on clinical interview and risk assessment procedures at the behavioral health technician course.

Depending on the population served, there may also be an added benefit to having behavioral health technicians provide the initial patient contact in military behavioral health clinics. For some service members, especially those very junior in rank who may have recently graduated from basic training, the opportunity to interact with officers is often limited. These service members may be more comfortable discussing their presenting problems with an enlisted technician as opposed to an officer. This process would mirror access to medical care for many service members, as they often enter the medical process by presenting to an enlisted medic or corpsman at sick call.

In addition to providing triage services, behavioral health technicians can extend the resources of a clinic by conducting intake assessments under the supervision of a licensed provider. In many clinics a standard intake interview is utilized in order to make sure that the interview is comprehensive and covers all of the information required. This practice can be particularly helpful when working with behavioral health technicians, especially those who are junior in their careers or those who have had nonclinical roles during previous assignments. Standard practices should be in place that allow for the appropriate level of staffing of the intake with a licensed provider.

Behavioral health technicians may also help increase access to clinical care by providing counseling in mental health clinics. The behavioral health technician course provides training on basic counseling techniques and theory, but the ability of technicians to practice counseling during the practicum portion of the course may have been limited. It will be important to work closely with the technician in order to identify their level of skill and comfort with counseling, and to provide additional training and support for technicians who will be providing counseling to patients. Providing a model for the technician to utilize in counseling, such as how to employ a solutions focused approach, may help the technician to develop both confidence and competence. In addition, providing cotherapy or counseling may help technicians to translate lessons learned from didactic training into the clinical setting.

In addition to outpatient behavioral health clinics, behavioral health technicians provide services in a number of settings. For example, military behavioral health technicians provide services in inpatient settings, partial hospitalization programs, and intensive outpatient programs focused both on the treatment of behavioral health disorders and on substance use disorders. Behavioral health technicians also provide administrative and clinical support for family advocacy programs. In addition,

behavioral health technicians can also be integrated into small clinics serving operational units, such as behavioral health clinics that are embedded into aid stations.

Advanced training is also available at the Medical Education and Training Campus through the Division of Behavioral Health Sciences to prepare technicians to provide additional services. Opportunities include: courses on the provision of individual and group treatment for substance abuse disorders, procedures for family advocacy cases, combat and operational stress control, the management of traumatic events, and others. Supervisors may find that these opportunities will increase the skill level of their technicians while also helping them to advance their careers. Additional information regarding training at Fort Sam Houston's Medical Education and Training Campus can be found at http://www.cs.amedd.army.mil.

THE ROLE OF BEHAVIORAL HEALTH
TECHNICIANS IN DEPLOYED SETTINGS

Enlisted behavioral health technicians are a key component of mental health care in deployed settings. Whether deployed in a remote combat zone or aboard a ship, resources to provide a mental health provider with back-up and support are limited. The deployed mental health provider may be faced with the challenge of providing mental health care as the sole provider for thousands of service members, and at the same time find that their only trained clinical support comes from their behavioral health technician. Furthermore, the deployed mental health provider may be tasked to provide mental health support to units that are geographically separated and travel may be both dangerous and a challenge to arrange. If this is the case, it may be helpful to split members of the team across two or more forward operating bases. This may increase access to behavioral health care while decreasing the need for patients to travel from base to base, which will likely reduce exposure to travel-related risks such as attacks on convoys.

While it varies between services, most uniformed behavioral health officers have little experience in field environments until they are deployed. Unless they have prior military experience in another role, most military behavioral health providers will spend most of their time providing clinical care in a medical setting while in garrison. This could make adapting to the deployed environment a challenge for many behavioral health officers. Enlisted behavioral health technicians, in many cases, will have much more skill and comfort in the field environment. One reason for this is that the basic training courses for enlisted service members tend to focus on these skills with the expectation that technical skills will be provided during follow-on training. Officer basic training courses, particularly those for medical officers, may not provide as much of an opportunity to develop skills in operational settings. When these training opportunities are provided, they may not be as comprehensive.

As part of a deployed combat and operational stress control team, the behavioral health technician will perform initial clinical interviews, conduct mental status exams, administer psychological tests, and provide counseling (US Army, 2006). The behavioral health technician also offers the perspective of an enlisted service member to the team. Because of this, behavioral health technicians may also have an easier time establishing rapport with other enlisted service members than the behavioral health officers.

One difference between providing behavioral health services in a deployed setting as opposed to providing those services in a garrison clinic is that the deployed mental health team is much more likely to take their services to the service member rather than remain in an office with all of their clients coming to them. One tool utilized is the concept of "therapy by walking around." This involves having members of the mental health team informally providing care to service members outside of the clinic, taking advantage of opportunities to provide support or education though spontaneous interaction. To do this, behavioral health providers and technicians need to be present in the environment where the service members spend their time.

Differences in rank may create a challenge for behavioral health providers who are interested in conducting therapy by walking around. The military culture, for many reasons, encourages a separation between junior enlisted service members, noncommission officers, and officers. This creates a culture where it would be unusual and uncomfortable for an officer to approach a group of enlisted service members for the purpose of simply spending time and being present and available to provide therapy by walking around. This will not be an issue for the enlisted behavioral health technician. It is more consistent with the norms of military culture for service members of similar rank to spend downtime socializing. This provides an opportunity for informal clinical intervention and support.

The behavioral health technician can serve as the front line of behavioral health support for units. Enlisted service members may be much more comfortable sharing with the behavioral health technician of similar rank. The information gathered by the behavioral technician can be useful as the behavioral health team develops plans to provide support to units and consultation to command. A technician who develops rapport and credibility with units may also be able to introduce other members of the behavioral health team to units or service members in need of assistance, reducing the barriers to care that may result from military cultural norms.

Behavioral health technicians can also be valuable extenders of services by assisting with the care provided by a behavioral health officer in the deployed setting. One example of this type of work is to include the behavioral health technician while working with service members who have mission-related anxiety. For instance, if a service member has anxiety about returning to convoy duty after witnessing an improvised explosive device impacting a vehicle, cognitive-behavioral techniques may be employed to treat the service member. This may include identifying anxiety provoking thoughts, replacing them with less distressing thoughts, and employing relaxation techniques such as diaphragmatic breathing in order to remain as calm as possible. In vivo exposure techniques may also be employed to help the service member to gradually return to convoy duty. A behavioral health technician can be of assistance by being present during the exposure process and reminding the client of the cognitive and behavioral techniques that have been taught. In many cases deployed behavioral health providers would not have the time away from their other duties to be present as the client returns to convoy duties, but behavioral health technicians may have more flexibility in their schedules.

Behavioral health technicians may also serve as a great resource through prevention efforts. This may include: classes to medics on the identification of behavioral health problems, training service members on buddy aid for service members showing signs of mental health risk, and assisting in the response to traumatic events. A trained member of the behavioral health team that is of similar rank to the service members they are either teaching or providing support after a traumatic event may again be an invaluable asset by reducing barriers that the behavioral health officers may inherently face due to their ranks.

In many cases the behavioral health technician in the deployed setting will work closely with their behavioral health provider in the same location. However, there may be times when it is beneficial to split the team up. For example, Army brigade behavioral health teams often provide coverage for battalions spread across a number of forward operating bases. Travel between forward operating bases is often dangerous, making access to behavioral health care challenging for service members on bases where the behavioral health team is not located. As mentioned earlier in this chapter, on these occasions it may be helpful to split team members across two or more locations in order to meet the mission.

When providers and technicians are geographically separated, a plan for supervision and support is needed. This plan will depend on factors such as ease and safety of travel between locations, access to communications, the skill level of the technician, and access to additional support at the location where the technician will be practicing. Ideally, the provider and technician will both have access to either telephonic or Internet-based communication, allowing for scheduled supervision on a regular basis as well as unscheduled consultation. In addition, developing a plan for on-site supervision from a non-mental-health medical provider such as a battalion physician assistant may be helpful. Finally, linking the technician with chaplains at their location may provide additional support while geographically separated from their supervising behavioral health provider. Developing a clear plan for supervision and support will help the technician to provide care in a safe manner that is consistent with their level of training.

References

American Psychological Association. (2010). *Ethical principles of psychologists and the code of conduct.* Retrieved from www.apa.org/ethics

National Association of Social Workers. (2008). *National Association of Social Workers code of ethics.* Retrieved from http://www.naswdc.org/pubs/code

US Army. (2006). *Combat and Operational Stress Control.* Field manual 4-02.51. Washington, DC: Headquarters, Department of the Army.

US Army Medical Department Center and School. (2010). Medical education and training campus behavioral health technician curriculum plan. Unpublished manuscript. Medical Education and Training Campus, Fort Sam Houston, Texas.

28 PROFESSIONAL BURNOUT

Charles C. Benight and Roman Cieslak

Researchers and practitioners have shown increasing interest in job burnout since the term was coined independently by Herbert J. Freudenberger and Christina Maslach in the late 1970s. As of May 2012 there were 3,682 publications recorded in the Web of Knowledge database that had job or work burnout in the topic. In 2010 there were 419, and in 2011 there were 493 such publications. These numbers show that job burnout is becoming one of the most popular fields of research in occupational health psychology.

The growing interest in job burnout has at least two sources. First, employees themselves have popularized the term "burnout" when describing their difficulties in dealing with intense work demands, challenging clients, and poor organizational resources. Second, occupational health psychologists have become increasingly focused on operationalizing the term, determining methods of assessment, validating different constructs, and applying theoretical systems to map burnout's trajectory. This has led to intriguing debates concerning identification of risk and protective factors linked to burnout in an attempt to generate a knowledge base for intervention strategies. Despite the popular use of the term, the scientific arena is emerging with significant gaps between what we understand intuitively and what we understand through theory and evidence related to job burnout.

DEFINITIONS AND MEASURES

There are many definitions and measures for job burnout. Job burnout is "a prolonged response to chronic emotional and interpersonal stressors on the job, and is defined by three dimensions of exhaustion, cynicism, and inefficacy" (Maslach, Schaufeli, & Leiter, 2001, p. 397). Although this definition is the most popular and was used for developing the frequently cited Maslach Burnout Inventory— General Survey (MBI-GS), it is not the only one. Three other definitions suggest that job burnout might be reduced to a single common experience: exhaustion. Each of these definitions has led to developing a different measure: Copenhagen Burnout Inventory (CBI), Burnout Measure (BM), Shirom-Melamed Burnout Measure (SMBM).

Demerouti and her colleagues proposed yet another conceptualization and measure of job burnout (Demerouti, Bakker, Vardakou, & Kantas, 2003). According to their conceptualization, job burnout consists of two dimensions: exhaustion and disengagement from work, which refers to "distancing oneself from one's work and experiencing negative attitude toward the work objects, work content, or one's work in general" (p. 14). Both dimensions are included in the Oldenburg Burnout Inventory (OLBI), an alternative to the MBI-GS. Conceptualization of exhaustion in the OLBI is broader than that in the Maslach measure, as it is seen as "a consequence of intensive physical, affective, and cognitive strain, i.e., as a long-term consequence of prolonged exposure to certain job demands" (p. 14).

In all of these alternatives to the MBI-GS conceptualizations and measures, professional inefficacy (a hypothetical third component of job-burnout) is consistently regarded as a separate construct. Across all definitions the overarching contributing factor to burnout has been intense prolonged exposure to significant job demands. Burnout might also arise from other less obvious sources.

Recently, job burnout has been also perceived as the consequence of indirect exposure to trauma in professionals working with traumatized clients (Stamm, 2010). Job burnout is understood here in a different way than in other conceptualizations, mentioned above. This type of burnout is "associated with feelings of hopelessness and difficulties in dealing with work or in doing your job effectively" (p. 13). Job burnout, along with secondary trauma reactions (e.g., posttraumatic stress symptoms) related to indirect trauma exposure, has important negative occupational and personal consequences including changes in cognitive beliefs about the self and the world.

ANTECEDENTS OF JOB BURNOUT

The list of job burnout antecedents is long, and includes both situational and individual factors. Two most frequently cited review papers on job burnout (Cordes & Dougherty, 1993, Maslach et al., 2001) indicated that job burnout might be caused or facilitated by work overload, time pressure, role conflict, role ambiguity, lack of social support, low control over work, low autonomy, and insufficient positive feedback. In addition to these job characteristics, important organizational, social, and cultural values that are not supported or realized through work are critical to consider. The following personality and individual difference factors also were found to be predictive of high job burnout: low hardiness (i.e., low commitment to job, low

job control, and tendency to appraise situation more like a threat than a challenge), external locus of control, passive or avoiding coping styles, low self-esteem, and low self-efficacy. Some demographic characteristics that contribute to job burnout include younger age or limited experience, being unmarried or single, and higher level of education.

THEORETICAL MODELS OF JOB BURNOUT

Along with the research aimed at testing the correlates (or antecedents) of job burnout, several theoretical models were proposed to explain processes and psychological mechanisms involved in developing job burnout. One of the popular theories is that job burnout is a prolonged response to chronic work stress. Although this thesis appeals to many practitioners and scientists, there are other symptoms that, along with the job burnout, may be considered the effect of prolonged exposure to chronic job-related stress such as depression and work dissatisfaction. This theory is not specific enough to explain processes that are unique to job burnout.

Other theoretical approaches, so-called developmental models, concentrate on developmental trajectories of job burnout over time. In these approaches, job burnout is not a static constellation of symptoms but a process that, for example, may start from emotional exhaustion leading to cynicism, which finally affects perception of inefficacy at work.

The job demands-resources (JD-R) model is currently the most influential theoretical approach to understand job burnout (Demerouti & Bakker, 2011). According to this model, when defining risk and protective factors for job burnout one should consider the occupational setting. These factors, different for various work settings, can be categorized into two broad categories: job demands and job resources. Job demands refer to those aspects of the job that require effort or skills and therefore lead to some physiological and psychological costs. Job resources relate to components of the job that are helpful in (1) achieving work-related goals, (2) reducing job demands and costs associated with these demands, and (3) stimulating personal development (Demerouti & Bakker, 2011). Through health impairment and motivational processes, job demands and resources directly, or in interaction with each other, affect job burnout and ultimately affect work engagement. The JD-R model shows that from organizational and individual perspectives it is important to know what factors lead to a negative outcome, such as job burnout. At the same time, however, knowledge about factors promoting positive outcomes, such as work engagement, is also necessary.

WORK ENGAGEMENT

Work engagement is sometimes perceived as the opposite end of the job burnout dimension and therefore is characterized by high energy, involvement, and perceived efficacy at work (Maslach et al., 2001). Another conceptualization of work engagement is of an independent construct, which is negatively correlated with job burnout and defined by three symptoms: vigor (e.g., a high level of energy and persistence), dedication (e.g., involvement and a sense of significance of the job), and absorption (e.g., concentration on a job to the extent that one has a sense of time passing quickly; Bakker, Schaufeli, Leiter, & Taris, 2008). Work engagement is often measured with the Utrecht Work Engagement Scale (UWES, 17- or 9-item version).

JOB BURNOUT AND WORK ENGAGEMENT AMONG PSYCHOLOGISTS

For practitioners, the notion that work engagement is separate from the job burnout phenomenon has important implications. Those practitioners who want to optimize their functioning at work and improve work-related well-being should not only take some actions to prevent job burnout, but also take some, probably different, actions to increase work engagement. In thinking about ways to foster

work engagement and reduce job burnout, one must consider both contributing factors of resources and demands. Generating increased resources such as social support may influence work engagement but not reduce burnout. Whereas reducing job demands might positively impact burnout, it may not increase work engagement. Importantly, studies among practicing psychologists have shown that work-home conflict and home-work conflict are positively related to job burnout and that these types of conflicts may mediate the effects of job demands and resources on job burnout (Rupert, Stevanovic, & Hunley, 2009). Thus, determining an appropriate balance between personal and professional demands and resources is an important challenge for all psychologists.

CONSEQUENCES OF JOB BURNOUT

Job burnout has significant consequences (see Maslach et al., 2001 for review). Most of them relate to job performance and subjective well-being or health. Interestingly, the same outcomes are included in studies on consequences of work stress. This indicates possible connections or overlaps between work stress and burnout processes. In terms of job performance, high job burnout is related to higher absenteeism, higher turnover or intention to quit the job, lower effectiveness at work, and low job or organizational commitment. It may also affect organizational standards and culture, making burned out individuals less focused on high quality performance and respecting human values in day-to-day operations.

Discussion of health-related outcomes of job burnout should be contextualized in the existing diagnostic categories and diagnostic systems. Job burnout symptomatology partially reassembles diagnostic criteria for neurasthenia, described in the World Health Organization's International Classification of Diseases (ICD-10) under code F48, "other neurotic disorders." The term "burn-out," defined as a "state of vital exhaustion," may also be found under code Z73.0 in "problems related to life-management difficulty." Job burnout is not recognized in the *Diagnostic and Statistical Manual* (DSM-IV-TR) but, in the current proposal for the DSM revision, it might be classified under category G 05 "trauma- or stressor-related disorder not elsewhere classified."

Physiological correlates of job burnout are typical of the effects of prolonged exposure to stress and include more frequent and stronger somatic complaints (e.g., headaches, chest pains, nausea, and gastrointestinal symptoms). People with high job stress are also at risk for developing depression and anxiety, but the causality of this relationship is not clear, as both anxiety and depression may also contribute to the development of job burnout.

SPILLOVER AND CROSSOVER EFFECTS OF JOB BURNOUT

Most definitions assume that job burnout is related to only one domain of human functioning (i.e., work and job-related activities). However, the consequences of job burnout may be experienced in other domains of life, such as family life. This interdomain transmission of the effects is called spillover. The example of negative spillover effect might be a situation when family roles or activities are disrupted due to job burnout. Positive spillover may take place when resources from one domain (e.g., family life) are used as a protective factor, acting against developing job burnout or reducing its negative consequences. For example, fulfilled family life and satisfactory family relationships may protect from emotional exhaustion and cynicism.

Whereas spillover is an intrapersonal transfer of consequences across different domains of functioning, crossover is an interpersonal transmutation of consequences. For example, an employee's burnout has an effect on a spouse's burnout and in that indirect way reduces life satisfaction of the spouse (Demerouti, Bakker, & Schaufeli, 2005). These are critical implications to consider in developing new interventions related to burnout.

PREDICTORS OF JOB BURNOUT AMONG MILITARY PSYCHOLOGISTS

There is limited evidence for the prevalence of job burnout and its risk factors among military mental health providers. Ballenger-Browning et al. (2011) showed that in a nonrepresentative sample of 97 providers, 27.8% reported high levels of emotional exhaustion, 18.6% had high levels of depersonalization, and 4.1% had indicated low levels of personal accomplishment, measured with the MBI version for human services (MBI-HSS). The intensity of job burnout among military mental health providers was compared to burnout levels among 730 civilian mental health providers. The results showed that military providers had lower depersonalization and higher personal accomplishment (Ballenger-Browning et al., 2011). The same study showed that risk factors for emotional exhaustion were: being a psychiatrist (comparing to other mental health professions), working long hours, and being female. High depersonalization was predicted by having a high percentage of patients with personality disorders and low percentage of patients with traumatic brain injury in providers' caseloads. Low personal accomplishment was reported more often by those who were not psychologists, were seeing a high number of patients per week, indicated low support from work and reported fewer years of clinical experience.

RECOMMENDATIONS FOR MILITARY PSYCHOLOGISTS

Recommendations for job burnout prevention among military psychologists are difficult to provide given the limited data in this area. However, the general (i.e., useful for a majority of working population) or specific (i.e., unique for job demands in that profession) interventions can focus on the individual or the organization. Given the unique nature of the military hierarchical environment, organizational interventions become more complex. However, efforts should be made to increase workload control, work flexibility, and enhancement of peer and supervisory support. Individual interventions that promote individual resource development (self-care strategies, work/home balance, symptom processing), professional skill promotion, and social resource enhancement (peer support, friends, etc.) prove to be effective in many cases. Military psychologists (Linnerooth, Mrdjenovich, and Moore, 2011) shared the professional experiences that helped them to cope with job burnout. Although the job demands were different for the predeployment, deployment, and postdeployment phases, the coping mechanisms were similar across these phases and included investment in individual resources (e.g., military and professional trainings), developing social network (family and professional relations), and acting proactively with the awareness that ethics standards and self-care are important parts of military psychologists' jobs. There is more work to be done to help determine the most beneficial methods to assist military psychologists.

References

Bakker, A. B., Schaufeli, W. B., Leiter, M. P., & Taris, T. W. (2008). Work engagement: An emerging concept in occupational health psychology. *Work and Stress, 22*(3), 187–200. doi:10.1080/02678370802393649

Ballenger-Browning, K. K., Schmitz, K. J., Rothacker, J. A., Hammer, P. S., Webb-Murphy, J. A., & Johnson, D. C. (2011). Predictors of burnout among military mental health providers. *Military Medicine, 176*(3), 253–260.

Cordes, C. L., & Dougherty, T. W. (1993). A review and an integration of research on job burnout. *Academy of Management Review, 18*(4), 621–656. doi:10.5465/AMR.1993.9402210153

Demerouti, E., & Bakker, A. B. (2011). The job demands–resources model: Challenges for future research. *SA Journal of Industrial Psychology, 37*(2), 1–9. doi:10.4102/sajip.v37i2.974

Demerouti, E., Bakker, A. B., & Schaufeli, W. B. (2005). Spillover and crossover of exhaustion and life satisfaction among dual-earner parents. *Journal of Vocational Behavior, 67*(2), 266–289. doi:10.1016/j.jvb.2004.07.001

Demerouti, E., Bakker, A. B., Vardakou, I., & Kantas, A. (2003). The convergent validity of two burnout instruments: A multitrait-multimethod analysis. *European Journal of Psychological Assessment, 19*(1), 12–23. doi:10.1027//1015-5759.19.1.12

Linnerooth, P. J., Mrdjenovich, A. J., & Moore, B. A. (2011). Professional burnout in clinical military psychologists: Recommendations before, during, and after deployment. *Professional Psychology: Research and Practice, 42*(1), 87–93. doi:10.1037/a0022295

Maslach, C., Schaufeli, W. B., & Leiter, M. P. (2001). Job burnout. *Annual Review of Psychology, 52*(1), 397–422. doi:10.1146/annurev.psych.52.1.397

Rupert, P. A., Stevanovic, P., & Hunley, H. A. (2009). Work-family conflict and burnout among practicing psychologists. *Professional Psychology: Research and Practice, 40*(1), 54–61. doi:10.1037/a0012538

Stamm, B. H. (2010). *The concise ProQOL manual* (2nd ed.). Pocatello, ID: ProQOL.org.

29   SUICIDE IN THE MILITARY

M. David Rudd

Over the course of the wars in Iraq and Afghanistan, suicide has emerged as arguably the most challenging mental health problem in the military today. Tragically, it has been the second leading cause of death in the US military for the last several years, with deaths by suicide actually exceeding combat-related losses in Iraq and Afghanistan (Ritchie, Keppler, & Rothberg, 2003; US Department of Defense, 2010). In 2008 active-duty military suicide rates surpassed those of comparable-age civilians for the first time in modern history (Kang & Bullman, 2009). Prior to Operation Iraqi Freedom (OIF) and Operation Enduring Freedom (OEF), active military duty served as a protective factor for suicide risk, with Active Duty service members experiencing dramatically lower suicide rates than comparable-age civilians. Prior to the Global War on Terror, the belief was that military service was protective given the prominence of unit cohesion, a common and clearly identified mission, and related social support, all elements that are generally consistent with the core tenets of the interpersonal-psychological theory of suicide (Joiner, Witte, Van Orden, & Rudd, 2009).

The largest branch of the Department of Defense, the US Army, has experienced the greatest increases, with the suicide rate more than doubling from 9.6 per 100,000 in 2004 to 21.9 per 100,000 in 2009. The rate is projected to climb to 24.1 per 100,000 in 2012 (given the 2-year lag time in reporting, these data are not yet available for review and analysis). Marine Corps rates have, for the most part, paralleled Army rates. Air Force and Navy suicide rates have been noticeably lower. Given the variable missions across service branches, with the US Army and Marines having the primary responsibility for ground combat, these numbers have raised questions about the specific role of combat exposure and post-trauma symptoms as suicide risk factors. Preferred methods of suicide in the military are comparable to those of the general US population, with the majority using firearms (50.6%), followed by hanging/suffocation (24%), and poisoning (18%) (US Department of Defense, 2012).

THE UNIQUE NATURE OF RISK IN MILITARY POPULATIONS

Consistent with the extant literature, available data indicate that psychiatric illness has played a major role in the sharp increases in military suicides over the course of the last 10 years, with major depression, posttraumatic stress disorder (PTSD), and substance abuse having been found to elevate risk for suicidal thinking, suicide attempts, and death by suicide (Jakupcak et al., 2009). It is estimated that well over 90% suffer from a diagnosable mental illness at the time of death, and many are characterized by marked comorbidity. It is important to note that these findings are generally consistent with those from the Vietnam era. With more than two million service members having deployed as part of the Global War on Terror, the number of troops serving in war zones and exposed to combat is the largest since World War II.

Naturally, the role of deployment and combat exposure as risk factors for suicide among military personnel has been studied and debated. Most studies indicate that deployment and combat exposure are important variables and elevate overall risk, but the precise mechanism of action is unclear. As mentioned above, both the Army and Marine Corps have experienced dramatic increases in suicide rates and also have the highest volume of soldiers exposed to combat relative to the Navy and Air Force. As suicide rates started to increase noticeably in 2005, it is important to recognize that approximately 40% of those dying by suicide were not deployed personnel. Since 2009, however, data have emerged suggesting that multiple deployments and severity of combat exposure may well play significant and specific roles.

Previous studies have linked combat experience to the emergence, persistence, and severity of suicidal thinking but not death by suicide. A recent Institute of Medicine (2007) report concluded that there is a link between combat exposure and enduring suicide risk after returning home (particularly among veterans), but the nature of the relationship and related risk is unclear. A number of investigators have explored the link between "moral injuries" (i.e., "killing" or "failing to prevent injury or death") in combat and suicide risk. More specifically, some have speculated that one of the unique characteristics of combat exposure can be captured by targeting resultant guilt and shame rather than specific psychiatric diagnosis. Some studies have found differential susceptibility to psychiatric illness across genders following combat exposure, with women evidencing increased vulnerability relative to men, particularly with respect to major depression and posttraumatic stress disorder. A recent study found a clear link between the severity of combat exposure, risk for psychiatric illness, and subsequent suicide

risk (Rudd, 2012). More specifically, 93% of those with "heavy" combat exposure qualified for a diagnosis of posttraumatic stress disorder and more than two-thirds exceeded the cutoff score for significant suicide risk.

Identifiable precipitants and stressors among military personnel are not dissimilar from those found in the general population, aside from the unique domains of deployment and combat exposure noted above. Among the most frequently cited stressors associated with military suicides are the following: work-related problems, relationship stress, legal problems, having been a victim of abuse (broadly defined), and financial problems. Of interest, it has been found that 84% of suicide attempts were related to work stress (i.e., job loss, coworker issues, poor work performance, work-related hazing) (US Department of Defense, 2012) and 60% were precipitated by a failed relationship (individuals can have multiple precipitants). It is important to note that legal issues in the military can have unique characteristics as they include courts-martial, administrative separation actions, nonselection for promotion, and disciplinary actions (e.g., Article 15). Although there has been some discussion in the literature of suicide "triggers" (i.e., an event known to be the motivation for death), current methodologies do not allow for such precision in understanding the associated stressors in the vast majority of cases. Less than half (44%) of those dying by suicide in the military were involved in treatment at the time of their death, despite considerable investment by the Department of Defense in suicide prevention and treatment promotion campaigns. It has been estimated that 37% of those dying by suicide were evaluated at a military treatment facility within 30 days of their death. These data suggest that cultural issues and stigma are profound challenges in a military environment.

Although clinical markers of suicide risk in the military do not appear to differ in any significant way from those in the general population, there are some potentially profound cultural differences, differences that may make it more difficult and a unique challenge to compel those struggling with psychiatric symptoms to pursue clinical care. In what has been referred to as a "Warrior Culture," psychiatric illness and psychological injuries from combat are often viewed as a personal failing or weakness rather than an illness or injury. It is also important to emphasize that easy access to firearms in a military environment may play a role in increased rates and difficulty in effective prevention and safety planning, as firearms continue to be the primary method used in suicide.

The role of prescription medications in military suicides has received considerable popular attention, with data revealing 29% of those dying had a known history of medication use, including antidepressants (22%), anxiolytics (10%), antipsychotics (5%), and anticonvulsants (3%). Not surprising, and consistent with general population findings, substance abuse has been found to be a major contributor to suicide risk among military personnel, with 28% of those dying having been engaged in active abuse (across a broad range of substances). In contrast to suicide in the general population, however, the majority of those dying by suicide in the military (67%) did not communicate their intent. Again, this may be related to some unique cultural variables, but targeted research is needed to answer this question.

As reservists (USAR) and National Guard (ARNG) service members have assumed increasing responsibility for the Global War on Terror, and over 40% of deployed soldiers fall into their ranks, they have faced and presented a number of unique suicide risk factors. Perhaps foremost have been the pressures created by an economy in recession. As national unemployment figures have continued to hover around 9%, and underemployment being reported as high as 18.5%, USAR and ARNG service members have been hit uncharacteristically hard. To a large degree, Active Duty soldiers have been insulated from the economic downturn. Repeated deployments and stints of Active Duty service have proven a unique challenge and stressor for USAR and ARNG, as they move in and out of jobs after fulfilling their military commitments. This translates to higher unemployment rates for veterans, with those aged 18–24 experiencing rates as high as 21%. It should be no surprise

that USAR and ARNG soldiers had significant increases in both suicides attempts and deaths by suicide over the last 2 years of available data. National data for the general population have demonstrated a clear link between the economic downturn and the rise in suicides between 2005 and 2010 (moving from under 11.0/100,000 to 12.0/100,000).

ENDURING SUICIDE RISK
IN MILITARY VETERANS

There is little debate that veterans are at significantly higher risk for suicide when compared to the general population. According to the Department of Veterans Affairs (DoVA), 20% of all suicides are veterans, with a tragic 18 veteran suicides per day. It is estimated that less than one-third were in active treatment with the DoVA at the time of their suicide. This again hints at the potential role of stigma and the unique characteristics of a warrior culture. Suicide rates among OIF/OEF veterans receiving VA care were noticeably higher than Active Duty service members, with rates as high as 28 to 38/100,000 (over a 3-year span). The general population rates for comparable age groups range from 17 to 19/100,000 (over a 3-year span). Male veterans from this cohort have experienced remarkably high suicide rates ranging from 30 to 43/100,000. Clearly, OIF/OEF veterans are at significantly higher risk for suicide than comparable-age civilians. Overall, the same clinical risk factors seem be involved, with the single greatest problem being untreated or undertreated psychiatric injury/illness (and related comorbidity). Confirming the magnitude of the problem for veterans, the Dole-Shalala Commission reported that "56 percent of active duty, 60 percent of reservists, and 76 percent of veterans" acknowledged mental health symptoms to a health care provider (Dole & Shalala, 2007, p. 15).

Psychological Injuries in Combat and Suicide

It is estimated that between 20 and 25% of those engaged in combat experience psychological injuries, with posttraumatic stress disorder (PTSD) and depression being the most common. Some studies have estimated that up to one-third of OIF/OEF veterans could suffer from a psychological injury. Traumatic brain injury (TBI) has been referred to as the signature wound among OIF/OEF veterans, with mild to moderate TBI posing significant diagnostic challenges, particularly given the high rates of comorbidity with PTSD and depression. As noted above, the overwhelming majority of those dying by suicide suffer a diagnosable mental illness at the time of death. It is also important to recognize that large numbers of veterans will suffer from "subclinical" symptoms (i.e., those that fall short of meeting full diagnostic criteria) with some estimates suggesting that an additional 20% will fall into this category.

Accurate differential diagnosis has been a major challenge for military clinicians and has complicated treatment efforts. There is considerable overlap in the symptom constellations for PTSD, major depression, and TBI, with mood changes, concentration problems, and sleep disturbance being most prominent. Accurate diagnosis is critical, since it drives subsequent treatment efforts. From this constellation, sleep disturbance has emerged as one of the most promising early warning signs, a symptom that cuts across many diagnostic categories including PTSD, major depression, TBI, and a range of anxiety disorders.

The connection between psychological injury and disciplinary problems in the military environment has received some attention and support. It has been found that soldiers with psychological injuries are twice as likely to experience disciplinary problems as their counterparts. Similarly, the burden on family members is profound, with noticeable increases in female soldier divorce rates over the past decade. It is also estimated that more than 2 million children have had a parent deploy during the Global War on Terror, with evidence of increased rates of behavioral and health problems when compared to those families in which a parent has not deployed. Clearly, the strain on military families has been considerable over a decade of war.

THE PROMISE OF TREATMENT

Over the course of the last 5 years, in particular, the Department of Defense has provided historic levels of funding for intervention and treatment research, including projects targeting PTSD specifically, as well as projects geared entirely toward suicidal behavior regardless of diagnosis. For the first time in history, randomized clinical trials are underway providing treatment for PTSD in Active Duty military service members prior to them being identified as "disabled" and treated in the DoVA. Initial findings from cognitive processing therapy (CPT) and prolonged exposure (PE) trials are quite promising, with several publications soon to appear in the literature. The military's PTSD clinical trials are organized under an initiative called Strong Star (http://delta.uthscsa.edu/strongstar/). The link provided includes a summary of all current PTSD treatment trials. As data become available they will be posted and accessible from the site.

In addition to the randomized clinical trials on PTSD, there are one half-dozen trials underway exploring the utility of a range of psychotherapeutic approaches for the treatment of suicidal behavior (including suicide attempts and suicidal ideation). Although none of the studies is yet completed, several are nearing completion, and data will soon be available. Preliminary results are promising. The Military Suicide Research Consortium was developed to provide up-to-date information about these studies: https://msrc.fsu.edu/.

The challenge of military suicide is a significant one. Aside from the unique contextual variables of deployment and combat, clinical markers of risk are remarkably consistent with findings from the general population, with untreated or undertreated psychiatric injury/illness arguably the central problem. Hopefully, as the high rates of psychological injury in combat become more broadly understood and accepted it will help facilitate a willingness to pursue clinical care and undermine some of the formidable barriers in a warrior culture. In many ways, though, the cultural barriers presented by the military are similar to traditional masculine values (particularly for young males) found in the general population, values that impact males' willingness to pursue any type of health care regardless of the nature of the problem. Health care providers have tried for decades to convince men to recognize early warning signs for a range of health problems and seek treatment before the problems escalate, all with limited success. In short, the issue of stigma and psychological injury and illness is a problem that is certainly not unique to the military. There is, however, much to engender hopefulness. Active clinical trials hold considerable promise for effective day-to-day management and treatment for suicidal behavior, with pending results likely to shape the nature of clinical care for decades to come.

References

Dole, B., & Shalala, D. (2007). *Serve, support, and simplify: Report of the President's Commission on Care for America's Returning Wounded Warriors.* Washington, DC: US Government Accountability Office.

Institute of Medicine. (2007). *Gulf War and health: Vol. 6. Deployment-related stress and health outcomes.* Washington, DC: National Academies Press.

Jakupcak, M., Vannoy, S., Imel, Z., Cook, J. W., Fontana, A., Rosenheck, R., & McFall, M. (2010). Does PTSD moderate the relationship between social support and suicide risk in Iraq and Afghanistan war veterans seeking mental health treatment? *Depression and Anxiety, 27*(11), 1001–1005.

Joiner, T. E., Witte, T., VanOrden, K., & Rudd, M. D. (2009). *Clinical work with suicidal patients: The interpersonal-psychological theory of suicidality as guide.* Washington, DC: American Psychological Association.

Kang, H. K., & Bullman, T. A. (2008). Risk of suicide among US veterans after returning from the Iraq or Afghanistan war zones. *JAMA: Journal of the American Medical Association, 300*(6), 652–653.

Ritchie, E. C., Keppler, W. C., & Rothberg, J. M. (2003). Suicidal admissions in the United States military. *Military Medicine, 168*(3), 177–181.

Rudd, M. D. (2012). *Severity of combat exposure, psychological symptoms, social support and suicide risk in OEF/OIF Veterans.* Manuscript under review.

US Department of Defense, Task Force on the Prevention of Suicide by Members of the Armed Forces. (2010). *The challenge and the promise: Strengthening the force, suicide and saving lives: Final report of the Department of Defense task force on the prevention of suicide by members of the armed forces.* Washington, DC: Department of Defense.

US Department of Defense, Task Force on the Prevention of Suicide by Members of the Armed Forces. (2012). *Army 2020: Generating health and discipline in the force ahead of the strategic reset.* Washington, DC: Department of Defense.

30 WOMEN IN COMBAT

Dawne Vogt and Amy E. Street

Women's presence in war zones, as well as their roles in the US military, has changed considerably over the years. The number of women in the military has grown substantially since the 2% cap on women's participation in the Active Duty military was lifted in 1968. While about 7,000 women were deployed to Vietnam, almost six times this number (40,000) were deployed in support of the 1990–1991 Gulf War, and over 200,000 women have been deployed for the wars in Afghanistan (Operation Enduring Freedom, OEF) and Iraq (Operation Iraqi Freedom, OIF; and Operation New Dawn, OND).

The increase in the number of women participating in combat operations has been accompanied by changes in women's roles in these operations. While the primary role of female service members during the Vietnam War was health care related or clerical in nature, women served in a much broader range of roles during the 1990–1991 Gulf War. During this war, women served in a variety of combat support roles that placed them in much closer proximity to enemy positions than was typical in prior wars. In recognition of women's contributions in the military, more than 90% of military occupations were opened to female service members following the Gulf War. Though women are still officially barred from serving in direct combat positions (e.g., Marine and Army infantry), female service members have served in a variety of positions during the wars in OEF/OIF/OND that involve leaving military bases, working side-by-side with male combat soldiers, and coming under direct fire. For example, female service members have been brought in to search Muslim women in suspected insurgents' homes, exposing them to the same dangers that male infantry face during these missions to find and remove weapons. Similarly, women in combat support positions have been assigned to drive trucks in supply convoys, placing them at considerable risk if those convoys are attacked by insurgents. In early 2012 additional changes were made to military policy regarding women's role in combat, opening even more combat-related positions to women and eliminating some rules that limit women's exposure to combat. This was followed by the historic announcement in January 2013 that the U. S. military was lifting its ban on female service members serving in combat roles. These policy changes are likely to translate into even greater similarity in the military roles of female and male service members in future combat deployments.

Another factor that has contributed to women's increased risk for combat in more recent deployments is the changing nature of modern warfare. Whereas earlier wars were characterized by a more defined front line, more recent combat operations such as OEF and OIF have involved guerrilla fighting in which combat and noncombat roles are less distinguishable. With this type of warfare, the threat of combat exposure is not restricted to service members in combat roles and service members in non-combat roles may also be exposed to combat. As a consequence, most female service members deployed to the wars in Afghanistan and Iraq report at least some combat exposure. For example, in a national survey of US OEF/OIF veterans, 77% of female service members (compared to 85% of men) reported exposure to at least one combat experience (Vogt et al., 2011). While nearly as many women as men report exposure to combat, women are still less likely to be exposed to high-intensity combat experiences. For example, another study found that among active duty soldiers, 31% of women (compared to 66% of men) reported exposure to death, 9% of women (compared to 45% of men) reported witnessing killing, and 4% of women (compared to 36% of men) reported having killed a combatant (Maguen, Luxton, Skopp, & Madden, 2012).

UNIQUE CONTEXT SURROUNDING WOMEN'S EXPOSURE TO COMBAT

Importantly, the context within which female service members experience combat may be quite different than it is for men. While both women and men are exposed to a range of potentially stressful events during deployment, some stressors are more salient for women. For example, women often report unique health and hygiene concerns related to living and working in the war zone (Trego, 2012). Lack of privacy is a concern that comes with being in the gender minority in the war zone. Most shower and bathroom facilities are intended for men and finding private facilities can be a daunting challenge for female service members. Difficulty maintaining menstrual hygiene is another frequently reported concern for women. Due to the often unsanitary living conditions, as well as the lack of consistent access to bathrooms and shower facilities, deployed female service members are at risk for both urinary tract and vaginal infections. Challenges associated with accessing gynecological care in the war zone may compound these problems by increasing female service members' risk for negative health sequelae, such as kidney or pelvic infections. Another concern that is unique to female personnel is the risk of unintended pregnancies associated with inconsistent access to, and/or use of, birth control methods during deployment.

In addition to these more basic concerns related to the difficult living and working environment, women face a number of interpersonal stressors in the war zone. One stressor that has received a great deal of attention recently is exposure to sexual harassment and sexual assault during deployment. While men can also be victims of sexual harassment and sexual assault during military service, these experiences are much more common among women (Street et al., 2011). Recent surveillance data focusing on victimization experiences in a 1-year period found that 4.4% of military women reported some form of sexual assault (DMDC 2010 Sexual Assault Report), 8% of women reported quid pro quo sexual coercion (e.g., offers of special treatment for sexual cooperation), and 22% of women reported some other form of unwanted sexual attention (Rock, Lipari, Cook, & Hale, 2011). These experiences of sexual harassment and assault are not limited to the peacetime military but may also occur during combat deployments, compounding the potential mental health sequelae of already stressful deployments for victims of sexual trauma.

Another deployment stressor that may be of particular concern for female service members is gender harassment. Gender harassment involves hostile behaviors that are used to enforce traditional gender roles or in response to violations of those roles. These behaviors are not aimed at sexual cooperation, but instead convey insulting or degrading attitudes about women. Research findings indicate that gender harassment occurs more frequently than sexually based harassment, with 43% of military women reporting experiencing sexist behavior consistent with gender harassment in the last year (Rock et al., 2011). Like experiences of sexual harassment, exposure to gender harassment may contribute to a more stressful overall deployment experience for deployed women.

Importantly, female service members may experience combat and other deployment stressors in a context in which they have less social support available to them from military peers and superiors, as female military personnel consistently report less unit social support than their male counterparts. This is particularly concerning given that unit social support may protect against the negative mental health effects of combat exposure. Moreover, lack of access to social support may be especially problematic for female service members, as there is some evidence that social support may be more protective for female compared to male service members (e.g., Vogt et al., 2005).

Female service members may also be more likely to experience stress related to prolonged family separations during deployment. Though female and male service members are about equally likely to be parents, women are more likely than men to serve as children's primary caregivers and to have added responsibilities related to caring for extended family members like aging parents, which may contribute to increased stress in the face of family separations. Moreover, military mothers are more likely than men to be single parents or to be married to other service members (introducing the possibility that both service members may be deployed at the same time), and thus, women's deployments may be associated with more disrupted and/or less stable child care arrangements. Given these family dynamics, female service members may experience elevated family stress when compared to their male counterparts. This is particularly concerning given that family stress has been shown to have more detrimental effects on the postdeployment mental health of female service members when compared to their male counterparts (Vogt et al., 2005).

WOMEN'S PSYCHOLOGICAL RESPONSE
TO COMBAT

As a consequence of their exposure to combat and other associated deployment stressors, female service members may experience the same postdeployment adjustment issues that are common among their male counterparts, including mental health conditions like posttraumatic stress disorder (PTSD), depression, and substance abuse. Though much of the prior research on the prevalence and predictors of postdeployment mental health conditions has focused primarily on male service members, an increasing number of studies address women's adjustment following deployment. A question that has received a great deal of attention in recent research is whether the mental health consequences of OEF/OIF/OND deployment are similar or different for women and men. While studies based on the broader population indicate that women are about twice as likely as men to develop PTSD following trauma exposure, evidence for gender differences in PTSD among the most recent cohort of OEF/OIF/OND veterans has been more limited. While not all investigations support the idea that women and men are similarly impacted by service in the war zone, most studies have revealed fairly comparable rates of probable PTSD for returning female and male OEF/OIF/OND service members, averaging between about 10% and 15% (Street, Vogt, & Dutra, 2009). In contrast, findings regarding gender differences in both major depression and substance abuse in this veteran cohort have been generally consistent with the broader literature, with women at higher risk for depression and men at higher risk for substance abuse (Seal et al., 2009). Findings also support high levels of comorbidity for both female and male OEF/OIF veterans, with many veterans who experience one mental health condition also meeting criteria for other conditions.

Though these findings shed light on the overall impact of deployment on women's and men's mental health, they do not address whether there are gender differences in the effect of particular deployment experiences, such as combat exposure. While there is some evidence for gender differences in the impact of specific high-intensity combat experiences on the postdeployment mental health of this veteran cohort (e.g., Maguen et al., 2012), study findings available at this point do not appear to support gender differences in the overall effect of combat exposure on postdeployment mental health. Instead, recent findings suggest that female OEF/OIF service members may be as resilient to combat-related stress as men (e.g., Vogt et al., 2011). The conclusion that gender differences in the impact of combat-related stressors on mental health are minimal is consistent with comments offered by Hoge, Clark, and Castro (2007) in their recent commentary on women in combat, in which they suggested that combat duty may be a great equalizer of risk due to its persistent level of threat. More generally, it may be that the increasing similarity in women's and men's military experiences contributes to similar outcomes after deployment. Additional research is needed to better understand gender-specific effects with respect to service members' exposure to combat-related stressors and associated consequences for their postdeployment mental health. Studies on this topic are critically important to ensure that the unique postdeployment needs of female service members are adequately addressed.

IMPLICATIONS FOR CLINICAL CARE

The finding that the vast majority of female service members deployed in support of OEF/OIF/OND have had at least some exposure to combat highlights the need for increased attention to combat-related stress in the assessment of returning female veterans. In particular, the fact that many female service members report exposure to combat despite their exclusion from ground combat roles underscores the importance of asking about specific combat experiences rather than relying on reported occupational roles during deployment (e.g., combat role vs. service support role) in the assessment of deployment stressors. In addition, the finding that many female veterans report exposure to a range of other stressors in the war zone, including especially interpersonal stressors such as sexual harassment and concerns related to family members at home, suggests that these factors must also be addressed as part of a comprehensive assessment. A broad assessment of deployment stress among returning female veterans, including both combat and interpersonal stressors, can provide a more integrated clinical picture and facilitate more effective and targeted treatment plans.

As described in this chapter, findings available to date suggest that combat exposure may have similar implications for the postdeployment mental health of returning OEF/OIF/OND veterans. Fortunately, there are a number of evidence-based cognitive behavioral psychotherapies and pharmacotherapies that can be used to treat the most common mental health conditions experienced by female and male veterans returning from deployment, including PTSD, depression, and substance abuse. Information on these treatments and their application among female and male veterans is available in the VA/DoD clinical practice guidelines (http://www.healthquality.va.gov/). The absence of any evidence to date that a particular treatment is more effective for one gender suggests that clinical decision-making regarding treatment selection should be similar for male and female Veterans.

Finally, a comprehensive assessment and subsequent treatment planning with all veterans should include an assessment of potential comorbidities, or the presence of co-occurring mental disorders. Comorbidities, which are common among returning veterans of both genders, can present unique treatment complications for mental health conditions such as PTSD. As reviewed above, female veterans are more likely than their male counterparts to report symptoms consistent with depression (and conversely, male veterans are more likely than their female counterparts to report substance abuse issues). Thus, assessments of female veterans should focus particularly on the possibility of comorbid depression.

AUTHOR NOTE

This research was supported, in part, by a Department of Veterans Affairs Health Sciences Research and Development Service grant (DHI 09-086; PI: Dawne Vogt, PhD).

References

Hoge, C. W., Clark, J. C., & Castro, C. A. (2007). Commentary: Women in combat and the risk of post-traumatic stress disorder and depression. *International Journal of Epidemiology*, 36, 327–329.

Maguen, S., Luxton, D., Skopp, N., & Madden, E. (2012). Gender differences in traumatic experiences and mental health in active duty soldiers redeployed from Iraq and Afghanistan. *Journal of Psychiatric Research*, 46, 311–316.

Rock, L. M., Lipari, R. N., Cook, P. J., & Hale, A. D. (2011). *2010 Workplace and gender relations survey of active duty members: Overview reports on sexual harassment and sexual assault* (DMDC Report No. 2010-025 March 2011; DMDC Report No. 2011-023 April 2011). Arlington, VA: Defense Manpower Data Center, Human Resources Strategic Assessment Program.

Seal, K., Metzler, T., Gima, K., Bertenthal, D., Maguen, S., & Marmar, C. (2009). Trends and risk factors for mental health diagnoses among Iraq and Afghanistan veterans using Department of Veterans Affairs health care, 2002–2008. *American Journal of Public Health*, 99, 1651–1658.

Street, A. E., Vogt, D., & Dutra, L. (2009). A new generation of women veterans: Stressors faced by women deployed to Iraq and Afghanistan. *Clinical Psychology Review*, 29, 685–694.

Street, A. E., Kimerling, R. Bell, M. E., & Pavao, J. (2011). Sexual harassment and sexual assault during military service. In: J. I. Ruzek, P. P. Schnurr, J. J. Vasterling, & M. J. Friedman (Eds.), *Caring for veterans with deployment-related stress disorders* (pp. 131–150). Washington, DC: American Psychological Association.

Trego, L. (2012). Prevention is the key to maintaining gynecological health during deployment. *Journal of Obstetric, Gynecologic, and Neonatal Nursing*, 41, 283–292.

Vogt, D., Pless, A., King, L., & King, D. (2005). Deployment stressors, gender, and mental health outcomes among Gulf War I veterans. *Journal of Traumatic Stress*, 18(3), 272–284.

Vogt, D., Vaughn, R., Glickman, M., Schultz, M., Drainoni, M., Elwy, R., & Eisen, S. (2011). Gender differences in combat-related stress exposure and postdeployment mental health in a nationally representative sample of U.S. OEF/OIF veterans. *Journal of Abnormal Psychology*, 120(4), 797–806.

31 PSYCHOTHERAPY WITH LESBIAN, GAY, AND BISEXUAL MILITARY SERVICE MEMBERS

Matthew C. Porter and Veronica Gutierrez

In September 2011 the US military's ban on lesbian, gay, and bisexual (but not transgendered) service members was lifted. The repeal of Don't Ask Don't Tell (DADT) was a landmark step in the advancement of civil rights for sexual minorities in the United States. One of its effects was to open the door for military psychologists to begin to provide culturally sensitive services to a segment of the US military population whose treatment needs have until now been difficult to meet. The absolute number of Active Duty service members who are lesbian, gay, bisexual, or questioning their sexual orientation (LGBQSMs) is currently unknown. While estimations will by necessity wait for the Department of Defense's relevant census initiatives, anecdotal reports suggest that the numbers are likely substantial and, due to the repeal of DADT, may be growing (e.g., National Defense Research Institute, 2010; Porter & Gutierrez, 2011).

HISTORICAL SUMMARY

Prerepeal

Extensive literature already documents the long history of sexual minorities serving in the US military (e.g., Berube, 1990; Herek, 1993, National Research Institute, 2010). Essentially, gay men, lesbians, and bisexuals have always served in the US military. Until September 2011, these service members had been obliged to keep their identities, behaviors, and lifestyles as secret as possible. Failure to do so resulted in harassment, prosecution, or discharge from military service. The rationale for this has varied over the years. Before the 1940s, homosexual sexual behavior had been considered a criminal offense under the Uniform Code of Military Justice (UCMJ). During World War II, a more complete adoption of the medical model shifted the overall approach of US military toward homosexuals from the prior focus on criminal behavior to a newer focus on abnormal (or pathological) identity. This lasted until 1993, when with the passing of DADT the military's approach shifted again. This time, suppositions of both criminality and pathology were explicitly avoided, in favor of a stated concern about a destabilizing, antisocial impact that the open presence of active duty LGBQSMs within the US military might possibly exert on unit cohesion and morale. Though DADT placed ostensible constraints on social or administrative enquiry into service members' sexual behavior and orientation, sexual-orientation-related discharges escalated, totaling almost 14,000 during this 18-year period (National Defense Research Institute, 2010).

Postrepeal

The September 2011 repeal of DADT allows LGMSMs to manage their privacy, including the concealment or disclosure of behaviors, identities, or attitudes related to sexual minority status, more freely than ever before in history. Yet the absence of administrative constraints on coming out has not yet clearly created an unequivocally affirming environment for sexual minorities serving within the US military.

While some generalizations can be made about the tendency toward socially conservative, heteronormative attitudes in US military culture, each unit or work environment is characterized by its own social norms and prevailing attitudes. At worst, disclosure (intentional or inadvertent) in socially dangerous environments within the military is believed to have already led to overt discrimination: increased targeting of LGBQSMs for verbal harassment, violence, military sexual trauma (MST), and even murder. At best, disclosure in a socially accepting environment can lead to improvements in mental health, social cohesion, task cohesion, commitment to the organization, and overall well-being. Somewhere in between lies the pernicious and difficult-to-identify territory of covert discrimination: disclosure that leads to social or professional exclusion (including, potentially, reduced opportunities for professional advancement or desirable work assignments) and loss of potential access to cultural capital, military camaraderie, and solidarity. In fact, as history has already demonstrated in the case of women and African Americans, integration of new sociocultural elements into US military culture takes time. DADT's repeal has opened the door to a host of psychosocial concerns that may impinge on the mental health of LGBQSMs until true and universal acceptance is reached.

Furthermore, while the Department of Defense's efforts to reduce discrimination of LGBQSMs are, over time, likely to reduce overt discrimination, they are less likely to be successful at reducing covert discrimination or at improving frank cultural acceptance of sexual minorities. This is particularly unfortunate, as social inclusion, access to cultural capital, and military camaraderie can exert important buffering effects on the high levels of stress and trauma often involved in military service. Even partial exclusion from these important aspects of military life may place some LGBQSMs at greater risk for developing symptoms as a result of military service, deployment-related stress, or trauma exposure.

CONTENT CONSIDERATIONS: CULTURALLY SALIENT AREAS OF POTENTIAL CLINICAL CONCERN

The sociohistorical and current heteronormativity within the US military discussed earlier is likely to impact the willingness and readiness of the LGBQSM client to discuss sensitive clinical issues in psychotherapy. Psychologists (military and otherwise) working with LGBQSMs (active duty and otherwise) need to be aware of several important areas of potential clinical concern.

Clinicians working with this population can learn to balance an appreciation of the client's developing readiness to discuss sensitive issues with the therapeutic goal of including culturally important issues in the treatment. Other than in cases of potential high risk, prematurely introducing sensitive LGB-relevant topics may risk rupturing a fragile therapeutic relationship. Our recommendation is that psychologists focus first on developing trust and rapport, allowing the client an opportunity to develop his/her disclosure progressively and cumulatively within the treatment. Nevertheless, when appropriate, demonstrating knowledge and concern about culturally relevant areas of potential concern can strengthen the LGBQSM client's confidence in both the psychologist and the treatment, improving both working alliance and treatment outcomes. Such dialogues can unroll fruitfully over the entire course of the treatment, returned to as needed, given the ongoing nature of most LGBQSMs' negotiation of the following issues.

Identity Development

For the estimated 50% of potential LGBQSM clients who are between the ages of 17 and 24, issues concerning identity development are likely to need attention. For many sexual minorities, this is the age when sexual or affective tendencies toward same-sex relationships may be first noticed, or when tendencies noticed previously become too strong to continue ignoring. Furthermore, for younger LGBQSMs, this may be their first extended period of time away from the influence of a potentially constraining family environment. Positive development of mature gender and sexual orientation identities during this age range is crucial for a successful transition to adulthood and sustainable, long-term mental health.

For young recruits, military service is a precious opportunity to become adult men and women. This process may be more complicated for the young LGBQSM, for whom healthy development of an adult gender role identity is likely to be constrained on one end by the military's valuation of stereotypical masculinity and on the other by their recognition of their own inherent differences from that stereotype. Certainly, not all men who are interested affectively or sexually in other men necessarily exhibit other stereotypically feminine behaviors or interests, nor do all women interested in women exhibit other stereotypically masculine behaviors or interests. Nonetheless, affective or sexual tendencies toward same-sex relationships transgress mainstream gender norms. Furthermore, outside the military, many lesbian and/or gay communities establish cultural capital on the basis of transgression or exaggeration of mainstream gender stereotypes. Young LGBQSMs may become confused or distressed by these conflicting social forces. This can lead to isolation, maladaptive coping styles, psychopathology, and, in a minority of cases, high-risk behaviors or suicide.

Feelings of discomfort regarding one's tendencies toward affective or sexual relationships with same-sex partners are a normal part of adjusting to a potential change in self-concept and its related social implications. The mere presence of such feelings does not indicate the use of "reparative therapy" to attempt to change the tendencies. Such therapies remain ethically controversial, lack empirical support, and are generally repudiated by mainstream psychological and psychiatric associations. Rather, psychotherapists can help clients investigate a range of profound and crucial identity developmental questions that are likely to underlie any immediate social or professional concerns brought to treatment by the LGBQSM. These may include variants on "What kind of man/woman am I?", "If I have 'fooled around' with other men in the service, does it mean that I am 'gay'?", "How can I feel good about being an adult man/woman when I transgress core elements of normative masculinity/femininity?" and "What kinds of future can I envision for myself within the US military, as I increase my autonomy, personal power, and cultural capital as a LGB person?" Through investigation and clarification of these questions, the psychotherapy will assist in authentic self-development and self-determination, rather than proscriptive behavioral changes of dubious durability or psychological merit.

Managing Disclosure and Concealment

Concealing sexual identity at work has generally been linked to lower job satisfaction, work cohesion, and task cohesion, as well as higher turnover rates and greater overall distress for sexual minorities. However, disclosing sexual identity (inadvertently or intentionally) in a hostile environment can also lead to difficulties. Fortunately, managing concealment and disclosure need not be an all-or-nothing affair. With the help of the clinician, LGBQSMs may be able to identify certain friendships, work relationships, or environments that seem likely to support full disclosure, and others that do not. If indicated, psychotherapists can begin a dialogue with their LGBQSM clients about the costs and benefits of coming out (disclosing their identity), as well as about what might be its potential gradations (partial or selective disclosure; intentional or inadvertent dissemination through social networks, including social media; managing sexual behavior on and off base).

Potential High-Risk Clinical Issues

1. Substance misuse: Extraordinary care should be taken in ongoing treatment to evaluate LGBQSMs for potential substance misuse. Both military service members and LGBQ people are at increased risk for substance misuse compared to the general population. LGBQSMs, subject to the stressors and social norms of both groups, may represent an even higher risk. Though the US military's zero tolerance policy for illicit substance use is often effective at prevention, misuse of alcohol is likely to be high among LGBQSMs seeking treatment, and can lead to other problems, such as risky sexual behavior. Clinicians working in the United States should be aware of the high rates of methamphetamine use among gay men in the general population, and its potential combination with risky sexual behavior and promiscuity.

2. Unsafe sex and HIV/AIDS: Psychotherapists could help their LGBQSM clients, particularly sexually active men, by developing comfort around discussing the specifics of safer and risky sexual behavior, assessing their clients' health behavior in this domain on an ongoing basis. Over 50% of the US military is currently under the age of 25. In this younger age bracket, which did not directly experience the most socially traumatic years of the AIDS epidemic, rates of risky sexual behavior are increasing. Where appropriate within the treatment, exploration of the client's thoughts and feelings about HIV/AIDS, including the way it can impact careers within the US military (e.g., constraint on deployments) can be used to facilitate greater awareness and improve health behavior. LGBQSMs who are seropositive may be at higher risk for isolation and depression, and will need additional support within the therapy related to any potential feelings of shame, guilt, or fear regarding their condition.

3. Military sexual trauma (MST): Both lesbian and gay male service members may be at high risk of being targeted for MST,

principally by male service members (e.g., Burks, 2011). LGBQSM clients should be screened for potential MST histories and treated as appropriate. Additionally, regardless of exposure history, the possibility of future MST may be a safety concern for LGBQSM clients, adding to their overall level of distress, and impacting the client's management of concealment and disclosure.

4. Suicide: Both LGBQ and military populations are at higher risk for suicide compared to members of the general population. Sadly, suicide rates in the US Armed Forces have been increasing since 2008, a trend understood to be in part a function of distress related to multiple deployments to combat zones. Further, LGBQ people, particularly youth and young adults, have ended their own lives at high rates in response to social disenfranchisement and harassment and as a result of conditions of depression, low self-acceptance, and challenges within identity development. An initial screening for suicide risk should be supplemented by ongoing, informal assessments and support, when indicated.

TECHNICAL CONSIDERATIONS: DIAGNOSIS, TREATMENT APPROACH, ETHICS

Diagnosis

Beyond the various categories for general Axis I psychopathology available in the *Diagnostic and Statistical Manual of Mental Disorders* (4th ed., text rev.; DSM–IV–TR; American Psychiatric Association, 2000), any of which may apply to the LGBQSM client, current US psychiatric nosology offers little nuance with which to characterize sexual orientation identity-related distress. Adjustment Disorder (309.X) captures emotional and behavioral presentations of distress related to living as a sexual minority within a heternormative environment, such as the US military even after the repeal of DADT. Identity Problem (313.82) indicates significant distress related to self-questioning about sexual orientation and behavior. Though Sexual Disorder Not Otherwise Specified (302.9) can be used for cases of significant, long-lasting distress regarding sexual orientation, it risks pathologizing the orientation or behavior, rather than the heteronormative environment. This is an arguably unethical step as it could potentially cause further harm to an already fragile client.

Treatment Approach

Using the content areas provided above as touchstones in the treatment, psychotherapists of LGBQSM clients can work effectively within any modality; no theoretical orientation is clearly better than any. Regardless of modality or theory, the approach to treatment can benefit from the following considerations: creating an LGBQ-affirming space, cultivating awareness of one's own homophobia or heterosexism, and avoiding pigeonholing.

1. Creating an LGBQ-affirming space: When the LGBQSM client walks into the office, he or she scans it and looks for evidence that it is safe. In turn, the psychotherapist who has items, books, or objects that clearly are LGBQ-affirming communicates from before the first word is uttered that he or she is open-minded, welcoming of him or her, and accepting of who he or she is as a person.

2. Cultivating vigilance against one's own homophobia: Avoiding heteronormative assumptions and using neutral language are steps toward greater inclusivity. Using terms such as "partner" or "significant other" rather than "girlfriend" or "spouse" can signal an openness to working with people of various sexual orientations and/or relationship arrangements. This stance requires constant attention and self-reflection. The APA has provided guidelines that can help psychotherapists understand their roles and have reasonable expectations in psychotherapy with LGBQSM clients (APA Div. 44, 2000). The authors of this chapter have also written on couple therapy with this population (Porter & Gutierrez, 2011).

3. Avoiding pigeonholing: When LGBQSMs seek psychotherapy, their concerns do not necessarily hinge on their sexual or affective tendencies. Sexual orientation might be disclosed only later in the treatment, depending on the client's readiness. Again, some feelings of discomfort regarding a potential shift in self-concept or social identity are developmentally normal, and do not indicate therapeutic attempts at "reparation."

Ethical Considerations

Psychotherapist lapses in cultural competence can result in therapeutic ruptures, diminished client engagement, or an impoverished alliance, any of which could lead to a potentially high-risk client prematurely terminating treatment. Psychotherapists newer to this population are advised to assess their own knowledge base, skills, attitudes, and awareness of issues related to sexual minorities using published self-report measures of cultural competency (e.g., Bidell, 2005). Given the high-risk behaviors (including suicide) in this population, psychotherapists with substantial heteronormative biases of their own should consider referring LGBQSM clients to someone able to provide more appropriate care. Yet, with appropriate training, clinical competence, and sensitivity, military psychologists can provide valuable and needed mental health services to LGBQSMs.

References

American Psychiatric Association. (2000). *Diagnostic and statistical manual of mental disorders* (4th ed., text rev.). Washington, DC: Author.

American Psychological Association Division 44, Committee on Lesbian, Gay, and Bisexual Concerns Task Force. (2000). Guidelines for psychotherapy with lesbian, gay, and bisexual clients. *American Psychologist, 55*, 1440–1451.

Berube, A. (1990). *Coming out under fire: The history of gay men and women in World War Two*. New York, NY: Free Press.

Bidell, M. P. (2005). The Sexual Orientation Counselor Competency Scale: Assessing attitudes, skills, and knowledge of counselors working with lesbian/gay/bisexual clients. *Counselor Education and Supervision, 44*(4), 267–279.

Burks, D. J. (2011). Lesbian, gay, and bisexual victimization in the military: An unintended consequence of "Don't Ask, Don't Tell"? *American Psychologist, 66*, 604–613.

Herek, G. M. (1993). Sexual orientation and military service: A social science perspective. *American Psychologist, 48*, 538–549.

National Defense Research Institute. (2010). *Sexual orientation and U.S. military personnel policy: An update of RAND's 1993 study*. Santa Monica, CA: RAND Corporation. Retrieved from http://www.rand.org/pubs/monographs/MG1056

Porter, M., & Gutierrez, V. (2011). Enhancing resilience with culturally competent therapy for same-sex military couples. In B. Moore (Ed.), *Handbook of counseling military couples* (pp. 295–320). New York, NY: Routledge.

32 MILITARY PSYCHOLOGISTS' ROLES IN INTERROGATION

Larry C. James and Lewis Pulley

Military psychologists' role in interrogation continues to be an area of complexity and controversy. Due to previous missteps by a minute number of psychologists, many psychologists cried out against behavioral health professionals' involvement in interrogation in any form. While interrogation consultation is not a specialty compatible with all military psychologists' skill sets, it is nevertheless a critical role that must be filled by some. Military psychologists have a moral imperative not only to protect detained persons but also to protect the public at large (Greene & Banks, 2009; James, 2008). Moreover, the challenges of being deployed and the effective role of the psychologist in operational settings to manage these challenges are well documented (e.g., Adler, Bliese, & Castro, 2010; Kennedy & Zillmer, 2012).

While the authors oppose missteps or misconduct in this area, they also take issue with the notion of psychologists being uninvolved in the interrogation process. Greene and Banks (2009) agree: "we categorically state that to run away from an area where we can help both the country and the individuals in detention is simply wrong. We believe that to do nothing when we have the knowledge to do good is to run away from preventing evil" (p. 30). Suedfeld (2007) similarly noted that psychologists' involvement helps to prevent interrogations from deviating into dangerous waters.

Prior to the Global War on Terror, detention facilities and intelligence operations used indigenous assets to accomplish their missions. Often, these intelligence units lacked the resources of a well-trained operational psychologist. In fact, most intelligence units did not have a psychologist assigned to its unit. However, as the operational tempo in Iraqi and Afghanistan escalated, an increase in the number of detainees and military detention facilities increased as well. As a result, within a short period of time, the need for psychologists from other parts of the DoD to serve as Behavioral Science Consultants (BSCs) drastically increased. Given the national security emergency, the demand for BSC services and reported abuses at detention facilities, the American Psychological Association convened a Task Force to develop guidelines for psychologists serving in intelligence units and detention facilities (Presidential Task Force on Psychological Ethics and National Security, 2005). To this day, this Task Force report serves as the standard for all psychologists who serve in detention facilities and intelligence operations around the world. Even though military members perceive the report as "the" guide for those working in the intelligence community, it has received criticism.

As a result of guidelines put forth by the PENS report and well-organized training programs developed by the US Army, BSCs function as consultants who possess psychological knowledge and skills to directly aid commanders in performing ethical, legal, and effective intelligence interrogations, detention operations, and detainee debriefing operations (Ritchie, 2011). They also provide direct consultation to the interrogation team itself (Staal & Stephenson, 2006).

PRIMARY ROLES FOR BSCS

Ritchie (2011) asserted that two main objectives comprise Behavior Science Consultants' mission. First, they offer psychological proficiency in the areas of watching, consulting, and giving feedback on the detention facility to aid the command in ensuring a humane environment for detainees and providing for the well-being of US service members. Second, BSC and BSC Teams (BSCTs) use psychological prowess in assessing persons and their surroundings in order to offer effective insights to enhance interrogations, debriefings, and detention proceedings. Staal and Stephenson (2006) added that operational psychologists aid in the formulation of interrogation planning, strategy formulation, and coaching interrogators in how to use cultural factors and personality features to establish rapport with subjects. While fulfilling this mission, they may function as special staff to the commander of both detainee and interrogation operations. Ideally, BSCs report solely and directly to the commander rather than to the detention or interrogation debriefing center commander. This reporting structure best facilitates the BSC's overall consultation role (Ritchie, 2011).

Interrogation places considerable mental, emotional, and physical stress on not just detainees, but also the interrogators themselves. (Alexander, 2008; McCauley, 2007; Soufan, 2011). Consequently, the BSC safeguards the interrogation process by monitoring the interrogators and providing oversight for the process as a whole. There are several potentially problematic phenomena that BSCs are particularly suited to identify. Behavioral drift is one of these. Ritchie (2011) defined behavioral drift as "the continual reestablishment of new, often unstated, and unofficial standards in

an unintended direction" (p. 699). She added that this drift often occurs in environments where one set of people has control over the actions and living conditions of others. Social influence and various psychological factors contribute to psychological drift's occurrence. It manifests itself in conditions of unenforced policy, unclear guidance, poor supervision, and inappropriate training (Ritchie, 2011). Moral disengagement often follows behavioral drift and results in protective guidelines being disregarded supposedly in favor of a higher moral purpose (Dunivin, Banks, Staal, & Stephenson, 2011). Said differently, moral disengagement is akin to the erroneous axiom that "the end justifies the means." Providing countermeasures for these factors falls well within the expertise of experienced behavioral consultants.

Staal and Stephenson (2006) maintained that the operational psychologist's toolbox must contain "tools" from several psychological subdisciplines including clinical, social, industrial/organizational, and forensic psychology. Skills from these areas aid in conceptualizing enemy subjects, illuminate options to find said subjects, or contribute in recovering information when those subjects are apprehended. Social psychology illuminates the notions of diffusion of responsibility, groupthink, moral disengagement, human influence, and social attribution theory. Other essential skills that BSCs bring to the fight are their knowledge of organizational behavior and culture, group dynamics, sources of motivation, and leadership principles. Finally, BSCs use their assessment expertise to create target profiles or aid in choosing service members who possess special skills for particular missions.

ETHICAL ISSUES

James (2008) has discussed many ethical issues and concerns in the interrogation and prison settings. For example, in these settings serving as a doctor and a consultant to interrogations at the same time, using medical records to enhance interrogations, not working as an advocate for the prisoner and serving in a combatant role can lead to ethical conflicts. Another concern is the blurring of roles for the psychologist in interrogation support. In other words, is the psychologist a health care professional or a clearly defined combatant? The current policy is for the psychologist to be "noncredentialed" and serve as a combatant consultant to their command. However, the psychologist is still a licensed health care professional and must abide by the ethical principals of the American Psychological Association and the laws mandated by the state license.

OTHER ISSUES

Prisons in Europe, particularly some found in France and Spain, have become centers for Islamic radicalization or even formulation of terrorist plots in some cases (Silke, 2011). While these prisons may offer some differences from the detention centers used by the US Military, BSCs are an extra set of "eyes and ears" to monitor these possible threats and assist detention personnel in deterring or reacting to problematic trends.

Toye and Smith (2011) relayed that detainees commonly strive to gain control, secure certain privileges, or manage boredom using aberrant methods. This is often observed in detainees' aggression, atypical behavior such as washing in excrement, and suicidal statements or parasuicidal gestures. Such activities are usually attempts at manipulation or tests of behavioral health and prison personnel. Toye and Smith (2011) added that a detainee hunger strike offers particular assessment, intervention, and consultation challenges for the behavioral health psychologist. Detainees may use hunger strikes as a way to court international attention, since they have historically been used to protest political establishments. For more in-depth details on this challenge, see Toye and Smith (2011).

Behavioral Science Consultants play an integral role in the war on terror and augmenting our military's efforts therein. "Above all else, the operational psychologists' primary objective is to ensure a safe, legal, ethical, and effective interrogation and detention process" (Staal & Stephenson, 2006, p. 269).

This simultaneously entails safeguarding the detainees in our custody and protecting the country that we serve. Although most of the national attention has been placed on the role of the BSCT in the interrogation process, the Behavioral Science Consultant has the added mission of ensuring the welfare of the service members in the unit as well as the detainees. Additionally, enhancing the combat effectiveness of the intelligence unit through consultations and training is another critical role for the BSCT. One should be mindful of the fact that the majority of enlisted service men and women range in age from 18 to 26, and their experience levels vary greatly. Thus, the BSCT can provide valuable assistance in the proper manner of interviewing a detainee.

Often, in a combat zone, sleep deprivation, behavioral drift, and behaviors inconsistent with good order and disciple can occur. The Behavioral Science Consultant's "watchful eye" can serve the INTEL unit well in preventing abuses and ensuring that all members of the command comply with the standard operating procedures and all applicable laws.

References

Adler, A. B., Bliese, P. D., & Castro, C. A. (Eds.). (2010). *Deployment psychology: Evidenced-based strategies to promote mental health in the military.* Washington, DC: American Psychological Association.

Alexander, M. (2008). *How to break a terrorist: The U.S. interrogators who used brains, not brutality, to take down the deadliest man in Iraq.* New York, NY: Simon and Schuster.

Dunivin, D. L., Banks, L. M., Staal, M. A., & Stephenson, J. A. (2011). Behavioral science consultation to interrogation and debriefing operations: Ethical considerations. In C. H. Kennedy & T. J. Williams (Eds.), *Ethical practice in operational psychology* (pp. 85–106). Washington, DC: American Psychological Association.

Greene, C. H., & Banks, L. M. (2009). Ethical guideline evolution in psychological support to interrogation operations. *Consulting Psychology Journal: Practice and Research, 61*(1), 25–32.

James, L. C. (2008). *Fixing hell*. New York, NY: Grand Central.

Kennedy, C. H., & Zillmer, E. A. (2012). *Military psychology: Clinical and operational applications* (2nd ed.). New York, NY: Guilford.

McCauley, C. (2007). Toward a social psychology of professional military interrogation. *Peace and Conflict: Journal of Peace Psychology, 13*(4), 399–410. doi:10.1080/10781910701665576

Presidential Task Force on Psychological Ethics and National Security. (2005). Report of the American Psychological Association Presidential Task Force on Psychological Ethics and National Security. Washington, DC: American Psychological Association. Retrieved from http://www.apa.org/pubs/info/reports/pens.pdf

Ritchie, E. C. (2011). Military forensic mental health. In E. C. Ritchie (Ed.), *Combat and operational behavioral health* (pp. 693–702). Falls Church, VA: Office of the Surgeon General, US Army; Fort Detrick, MD: Borden Institute.

Silke, A. (2011). Terrorists and extremists in prison: Psychological issues in management and reform. In A. Silke (Ed.), *The psychology of counterterrorism* (pp. 123–134). New York, NY: Routledge.

Soufan, A. H. (2011). *The black banners: The inside story of 9/11 and the war against al-Qaeda*. New York, NY: W.W. Norton.

Staal, M. A., & Stephenson, J. A. (2006). Operational psychology: An emerging sub discipline. *Military Psychology, 18*(4), 269–282.

Suedfeld, P. (2007). Torture, interrogation, security, and psychology: Absolutistic versus complex thinking. *Analyses of Social Issues and Public Policy, 7*(1), 55–63.

Toye. R., & Smith, M. (2011). Behavioral health issues and detained individuals. In E. C. Ritchie (Ed.), *Combat and operational behavioral health* (pp. 645–656). Falls Church, VA: Office of the Surgeon General, US Army; Fort Detrick, MD: Borden Institute.

33 INTERACTING WITH THE MEDIA

Nancy A. McGarrah and Diana L. Struski

WHY TALK TO THE MEDIA?

Why speak to the media? Military psychologists have an opportunity to tell their story as a health care leader and subject matter expert. The media offer opportunites to deliver key messages to the community and to educate the public on topics relevant to them. Sharing one's expertise with the media can help ensure that accurate information is provided to the public and it can help enhance the image of the profession. Media psychology is still an emerging field and constantly faces interaction with a multitude of established as well as new technologies (Luskin & Friedland, 1998). This

is a field that many psychologists are nervous to enter, and media interviews can be challenging and stress-inducing even for experienced communicators. Doing some homework before jumping into an interview will go a long way toward easing this stress and ensuring effective communications with the media.

FIRST STEPS

When asked by the media to provide an interview, the psychologist should first contact the unit's public affairs officer for assistance. Types of assistance include: obtaining clearance for the media if necessary; understanding the "angle" of the story and the deadline; helping the psychologist prepare for the interview and possibly being present for the interview; and obtaining written consent from patients if necessary to protect patient confidentiality.

Media representatives will often ask the psychologist for patient examples either before or during an interview. Discussing specific patients raises issues of confidentiality, and it is better to speak about therapy populations and treatments, not about individual patients. If an interview is to feature a particular patient, the patient's consent must be obtained in advance. In addition, reporters often ask the psychologist for a response to an item in the news, and it is important to talk in generalities and be clear that the person in the news has not been evaluated by the psychologist being interviewed. Standard 5.04, Media Presentations, of the APA Ethics Code, states that psychologists must ensure that their statements are based on "their professional knowledge, training, or experience in accord with appropriate psychological literature and practice...and do not indicate that a professional relationship has been established with the recipient" (American Psychological Association [APA], 2010, p. 8).

To prepare for the interview, it is necessary to know if the interview will be conducted live for radio or television, recorded for a later news broadcast/news magazine, or produced for print medium, the Web, or other technology. Prior to participating in the interview, it is important to first establish ground rules with the media representative, such as:

- When the interview will occur
- The length and location of the interview
- The topics covered
- Any resources the reporter can send ahead of time (such as a breaking news story or research study)
- How you will be addressed (e.g., as Doctor, by your rank, etc.)
- If you will have the opportunity to edit or review the completed interview prior to it being aired, released, or published.

At times, a psychologist may be contacted to participate in an interview the same day as the request is received. When the interview is requested to obtain comments and opinions about a current event or news item there may be pressure to respond right away. The military psychologist must not give in to pressure and skip important steps in preparing for the interview. It is vital that the psychologist both contact the installation's public affairs officer before agreeing to provide an interview and do the research and preparation necessary to provide a credible and competent interview.

It is often possible to preview an interview for a print medium, but this may be more difficult in other situations. Even if the psychologist is unable to preview the interview, it is important to review it after publication to check for accuracy. The psychologist should contact the media representative if there are any problems. Doing so is consistent with Standard 1.01 of the APA Ethics Code (APA, 2010), Misuse of Psychologists' Work, which requires psychologists to "take reasonable steps to correct or minimize the misuse or misrepresentation" of their work (p. 4).

COMPETENCE

Once the psychologist knows the topic of the interview, the ethical standard of competence must be considered. Reporters will ask for comments on a wide variety of subjects, for most of which the psychologist may have no specific expertise. However, the military psychologist knows about mental health and the military, and the reporter will rely on this expertise. The psychologist is often competent to do the interview after reviewing any information provided by the media representative and doing their own literature search. Sources for literature searches include APA Journals, Google Scholar (http://scholar.google.com/), and PubMed (http://www.ncbi.nlm.nih.gov/pubmed/). Professional listservs can also be useful tools for acquiring fast responses to requests for research on different topics. However, if the psychologist then determines that the topic is not in his or her area of expertise, a referral should be made to another psychologist who has the needed relevant experience and expertise.

When interacting with the media, it is important that military psychologists keep in mind that they are representing the profession of psychology and their command, branch of service, and the US Military. It is thus extremely important to only answer questions and make statements within the psychologist's areas of expertise, consistent with Standard 5.04, Media Presentations, of the APA Ethics Code (APA, 2010) as has been highlighted.

Key Message

When preparing to start the interview, the psychologist should think of certain key messages that are critical to impart to the audience. It is essential that military psychologists answer questions in complete, easy to understand sentences, without using jargon or acronyms. It is best to speak in sound bites, which are memorable quotes and stand alone answers, and to provide succinct responses. In this way, when editing occurs, the psychologist's comments will be aired or printed in a way that makes sense. If a reporter asks a question that includes misinformation, it is an opportunity for the psychologist to respond professionally and provide positive and correct information.

The Four C's

Some common principles can make the military psychologist's presentations to the media more effective. These principles, referred to as the four C's, are: commercial, control, credibility, and cosmetics. The commercial refers to the most critical element of any interview, which the psychologist wants to be sure to get across to the audience. This critical element is usually summed up in two or three main messages about the topic. Psychologists use control to

keep the key messages in front of the intended audience. "Bridges" are often used as a means of keeping control of an interview, and can move from one aspect of an issue to another. Verbal bridges allow the military psychologist to steer a reporter back to relevant topics and key messages if the reporter loses focus or moves onto an unimportant topic. The "bridge" can be conceptualized as an ABC process, with "A" being the acknowledgment of the reporter's question; "B" is bridging the response; and "C" is the delivery of the commercial (key message). Examples of this technique are:

- "Yes...(the answer), "and in addition to that..." (the bridge).
- "No, let me explain..."
- "That's the way it used to be...here's what we do now."
- "What's most important is..."
- "The bottom line is..."
- "That's not my area of expertise, but I do know that..."
- "What the research tells us is...".

The most important quality of a spokesperson is credibility. The military psychologist should never speak beyond his or her realm of expertise. In the face of adverse or hostile questioning, the psychologist must remain professional. Regardless of the approach taken by the interviewer, it is important to remain calm and to never become antagonistic or defensive. The psychologist should never be afraid to say, "I don't know." Credibility is critical to the success of an interview, to the reputation of the military psychologist, and to the future working relationship with the media. To add credibility to the military organization, the psychologist can utilize various types of information to support the message. Techniques most commonly used to support statements include personal experience, human interest story (with careful consideration of informed consent and confidentiality issues), facts, statistics, quoting authorities/experts, and analogies or comparisons.

Finally, appearance is important when conducting an interview, since nonverbal communications are important, especially when the medium is television. The media representative can be helpful in suggesting appropriate attire for the location of the interview. The filming usually includes a significant number of bright lights, which are difficult to become accustomed to when trying to appear cool, calm, and collected.

TRAINING OPPORTUNITIES

Training in media psychology is very helpful in developing competence and awareness of ethical challenges. There are ongoing and emerging issues in the field of media psychology, and reporters are calling on psychologists at increasing rates. New technologies such as Web-based outlets are also used for media interviews. Opportunities for training in media work can be found at the psychologist's state psychological association, at their military installation especially through their installation's Public Affairs Office, and at the American Psychological Association's Division of Media Psychology (http://www.apa.org/divisions/div46/). Consider joining the Division of Media Psychology and joining its electronic mailing list (for more ideas, see McGarrah, Alvord, Martin, & Haldeman, 2009). APA also has a very active Public Education Office, which can offer advice regarding working with the media. For more specific help with tricky ethical issues, psychologists can contact the APA Ethics office, their state psychological organization ethics committee, or their risk management advisor at their malpractice insurance company.

RECOMMENDATIONS

As a spokesperson, the military psychologist interacting with the media should:

- Know the subject matter of the interview, who is conducting the interview, where it is to take place, and the allotted time.
- Have a reasonable amount of time to research the issue and develop key messages and supporting facts and information.
- Coordinate all interview requests with the installation's Public Affairs Office and consider having a public affairs officer or other organization representative present during the interview.
- Determine whether audio- or videotaping of the interview is appropriate.
- Understand in advance whether other guests will appear, and, if so, know their identities and positions.
- Ask if there will be an opportunity to review or edit the interview before it is printed or aired.
- Determine the details of the publication of the interview (live, taped, etc.).

References

American Psychological Association. (2010). *Ethical principles of psychologists and code of conduct.* Retrieved from www.apa.org/ethics

Luskin, B. J., & Friedland, L. (1998). *Task force report: Media psychology and new technologies.* Washington, DC: Division of Media Psychology, Division 46 of the American Psychological Association.

McGarrah, N., Alvord, M., Martin, J., & Haldeman, D. (2009). In the public eye: The ethical practice of media psychology. *Professional Psychology: Research and Practice, 40*(2), 172–180.

34 PREPARATION AND TRAINING AS A MILITARY PSYCHOLOGIST

Peter J. N. Linnerooth[†] and Brock A. McNabb

DECIDING TO BECOME A MILITARY PSYCHOLOGIST

Military psychology is a difficult but interesting, rewarding, and fulfilling career. Future military psychologists (FMPs)—doctoral students considering military practice (and their families, if any)—must be willing to serve anywhere, in any assigned role, during the psychologist's service obligation of 4–6 years. Duty may include lengthy periods of time away from family, practicing under isolated, environmentally deprived, or dangerous conditions (e.g., combat).

To make an educated decision, the FMP should seek advice from current or former military psychologists. To make the right decision, the FMP should include their spouse or partner. The military is a "greedy institution" (Coser, 1974). Military duties can often consume all the time and energy the MP has and still demand more. This is challenging for marital and family relationships, so it is crucial to consider more than just the financial incentives and travel offered by the military. The FMP and his/her family must be sure that the military is a lifestyle they can accept, as it is a uniquely demanding context in which to practice and live as a family.

Some of the roles a psychologist might play in the US Army (USA) include: serving with a Combat Stress Control unit that might deploy

to support any area of conflict doing prevention, assessment, and treatment for all units fighting in that region; serving as a brigade psychologist, who trains and then deploys with a specific brigade combat team caring for its 5,000–8,000 fighting men and women; or practicing in a clinic setting, from a small troop medical clinic, to a community hospital, or in a major medical center such as the well-known Walter Reed Army Medical Center. In the US Air Force (USAF), a psychologist might be attached to the medical group that cares for a USAF Wing; be a flight psychologist, evaluating and/or treating the unique problems of pilots and aircrew; become a SERE psychologist providing psychological "inoculation" to service members (SMs) learning to "survive, evade, resist, and escape" when operating in enemy territory; and, as a SERE psychologist, also provide treatment, helping repatriate personnel who have spent time as prisoners of war. USAF psychologists are also sometimes assigned to support USA units in combat. In the United States Navy (USN) one could be assigned duty on a ship or serve "shore duty" as a clinician in a USN clinic or hospital, stateside, or abroad. USN psychologists also serve with Marine combat units, and, like USAF MPs, may even directly supporting USA units in combat.

Typically, an MP begins as a generalist, assessing and treating all presenting problems within a given combat unit or military community. This, although challenging, is one of the true delights of practice as an

[†]Peter J.N. Linnerooth unfortunately passed away before this book was completed.

165

MP. The MP is often the only mental health expert on scene, and does everything in their power to effect valid assessment or effective prevention or treatment. But, there are also military-sponsored postdoctoral fellowships. These are offered at major military medical centers, and include neuropsychology and child, pediatric, and health psychology.

MILITARY PSYCHOLOGISTS' PERSONAL SAFETY

FMPs should also consider personal safety when making their decision. Military psychology may require training and working in high-risk environments. Some MPs receive special, potentially dangerous training such as airborne (parachute) qualification. Others are frequently "in the field" or "with the fleet" (at sea) with their units, environments that can pose dangers from heavy equipment, adverse environments, and accidents. Finally, every FMP must consider and be ready to serve in combat. Although MPs are not combat arms (frontline, fighting SMs), combat environments are hazardous at any location. And circumstances, and the need for their skills, may force health care professionals into dangerous circumstances (cf. Jadick & Hayden, 2007).

CULTIVATE MILITARY KNOWLEDGE AND MILITARY BEARING

Once the FMP's decision is made, it is advisable to cultivate military knowledge, appearance, and behavior ("military bearing"). At first, a goal like good military appearance may seem superficial, but in fact, is not. One aspect of military appearance, physical fitness, for example, is believed to have a role in preventing PTSD (Taylor et al., 2008). Likewise, appearance, knowledge of military customs and courtesies, such as how to salute, wear the uniform, and speak to subordinates and superiors can help to gain the trust of patients, their commanders, and others in the military community who may become referral sources.

Understanding military culture is very important. The Ethical Principles of Psychologists and Code of Conduct (American Psychological Association [APA], 2010) would not condone psychologists serving a culture of which they have no knowledge. Each military branch has its own history, language, and behaviors. Therefore we believe it is necessary to seek cultural competence. For example, FMPs could learn important cultural subtleties like never verbalizing excuses for failure or ignorance to a superior officer. FMPs might also chafe at the idea of following orders or other military behaviors that can be inscrutable to a new officer. Some of this cultural competence will come from formal military training ("Officer Basic Course," Internship in a military hospital). However, turning to current MPs or other psychologists who work with SMs or veterans is also useful preparation.

ACQUIRE CLINICAL SKILLS RELEVANT TO MILITARY PRACTICE

As an MP, one may have to handle "everything that comes through the door." Military practice is likely to present new mental disorders, as well as unique presentations of PTSD, depression, and marital problems, different from what the FMP saw in his/her civilian training. To prepare, we suggest three skills a MP should develop, the abilities to:

- Rapidly assess complex behavioral problems with minimal assistance/assessment tools
- Rapidly make high stakes decisions (e.g., danger to self/others, unfit for duty)
- Remain calm, confident, and able to communicate effectively with patients and the chain of command

USEFUL TRAINING CONTEXTS

We believe there are three very useful training contexts in which a FMP could start gaining these three core skills before entering active duty:

- The Emergency Department in a busy hospital (or another entity specializing in crisis services)

- A general psychiatric facility serving both acute and chronic inpatients
- A [nonacute] Veterans Administration (VA) mental health outpatient clinic or PTSD treatment program

In the Emergency Department, the FMP could gain experience dealing with suicidal and homicidal crisis intervention and other emergent situations requiring rapid, quality assessment. In the psychiatric facility, the FMP could gain experience recognizing and working with a variety of serious and/or chronic and persistent mental disorders. For example, it is relatively common for young SMs to experience their first psychotic episode or to present with delirium. Such presentations tend to make military leaders uncomfortable. It is important that they not surprise the military psychologist and that the MP's initial case conceptualization be quick and accurate. Training in a VA outpatient clinic, or other PTSD treatment program, will provide the FMP with key clinical skills for treating combat related PTSD and depression, and provide the context FMPs need to understand what is unique about SMs, the traumas they survive, and their treatment. For example, military operations, whether combat or repeated accidents (especially in USN environments) can present multiple, perhaps dozens, of traumatic events. As FMPs hear the stories of SMs from a spectrum of conflicts and operations they can learn that SMs experience and cope with traumas somewhat differently than survivors of civilian traumas.

In all three training environments, the prospective military psychologist could train with a variety of skilled supervisors and colleagues. Real-world learning, relevant to future military practice, under mentors from various disciplines could be very helpful. It could provide exposure to some of the same types of cases the FMP will soon be expected to handle independently. And the FMP could learn the language and skills of other disciplines. Once on active duty, MPs will regularly require the assistance of their physical medicine colleagues. For example, in a combat theater, "chemical restraint" of an agitated patient with haloperidol can

be lifesaving. Also, simple interventions such as rest, induced/enforced by benzodiazepine treatment, can be a powerful short-term intervention for acute combat exhaustion. Once on active duty there will be too much work and too few providers for interdisciplinary squabbling. FMPs must be ready to communicate and cooperate with military mental/health care providers from all disciplines.

OTHER TRAINING

Other potentially useful skills and experiences an FMP might seek include facilitating group therapy and exposure to pastoral counseling. Regarding group treatment, data from our own clinic suggested that patient visits were 10 times higher in 2005–2006 than in prewar 2000–2001. While that might not be generalizable, it suggests that any skill that helps the MP cope with a high workload may be very valuable. FMPs may also consider seeking exposure to pastoral counseling (counseling provided by religious leaders). In our experience, ability to work synergistically with our 8 battalion chaplains was critical to providing adequate services for the roughly 20,000 soldiers in our catchment area in Iraq. Our clinic was manned by 1 MP and 2 psychology technicians. The clinic was open 106 hours/week for appointments, and saw an average of 400 patient visits per month. Our brigade chaplain was also required to keep detailed records, and so was able to report with certainty that his 8 chaplains saw another 400 visits per month. These were low severity mental health problems, not spiritual counseling cases. Again, while not necessarily generalizable, our experience suggests knowledge of pastoral counseling and the ability to work with religious leaders may be very helpful to an MP.

Not all of an FMP's training need be directly relevant to the military or PTSD. SMs and their families need all the types of expertise that a civilian community would require. Background, for example, in marital and family therapy, behavioral pediatrics, parent training, or addictions would be welcome in a military community. In Iraq, for example, we began offering "stop smoking" classes. Surprisingly, these were well attended, and not insignificantly, provided soldiers with a way to establish trust with the providers without a negative "mental health" stigma. Thus, FMPs still in the decision process should not feel constrained to study only what is suggested here, or only what seems military-relevant. Solid skills in data-based assessment and empirically supported treatment will transfer readily from civilian training to military contexts and problems.

CARING FOR SELF AND PEERS: VICARIOUS TRAUMA AND PROVIDER BURNOUT

It may also be useful for the FMP to concentrate a part of their training on vicarious traumatization and provider burnout. Vicarious traumatic stimuli can affect all SMs. Even "noncombat" SMs working far from the battlefront, including military health care providers, can suffer from vicarious traumatization as well. While they were not at the scene of injury/death, they are repeatedly exposed to the product of those situations as they treat wounded SMs. Thus, knowledge of how to recognize, assess, and treat provider burnout may be critical skills for many MPs. The intense workload of military providers, the adverse conditions under which they may practice, and possibly vicarious traumatization can cause burnout. One of our initiatives in Iraq, for example, was to measure provider burnout with a psychometrically sound, anonymous, self-report instrument, brief command on the results, and institute rotation of personnel to a "nontrauma aid station," with less severe cases, for periods of recovery.

CASE EXAMPLES

To close the chapter, it may be useful to present some of the assessment questions/ treatment cases we faced as MPs. These are authentic, with only identifying details eliminated to maintain confidentiality. There is no one "right answer" in these cases. As FMPs consider a career as MPs, especially as they reach the end of their doctoral training, it might be useful to consider cases such as these (and others in the literature) as thought-provoking scenarios. For each of these scenarios, you could ask yourself questions such as: How would I assess the SM(s) problems(s)? What would be my initial treatment plan? How would I follow up over time? What might I say to educate the patient's command? What would I advise the commander to do to help the SM?

- An SM has reported suicidal or homicidal ideation with a vague plan. He'll contract for safety, but only for 24 hours, until your next session.
- An SM commits suicide in a private location, but very near where his platoon is training. The platoon is in shock, and angry. They are willing to sit down together, "but only once, and only if speaking is optional."
- An NCO brings in "one of his best, smartest soldiers," a young man who has done nothing wrong until he was found crying inconsolably after a battle. The NCO asks, "How can we trust this formerly excellent soldier in combat?"
- Across 1 month, 3 SMs are brought to the aid station in a moderate to severe delirium. The NCOs who brought them in all reported that these are healthy SMs, all from different units, all support troops not involved in combat, and are all reported to be extremely conscientious, very hardworking, meticulous SMs, perhaps the best in their platoons. None has ingested any substance. Each is slightly dehydrated, very sleep deprived, and has been eating a little poorly.
- An SM is showing increasing signs of PTSD, but has just started his/her second combat deployment. A brief treatment of a few days off combat patrols, rest, good diet, and a chance to talk seems to have made symptoms worse.
- An SM presents with a complaint of ADHD and severe memory problems. The SM sought treatment immediately after losing an extremely valuable and tactically

sensitive piece of equipment such as night vision goggles.

- An SM was previously diagnosed with an Axis II disorder (usually considered incompatible with further military service). The SM has frequently been disciplined in garrison for minor crimes. But the SM has close friends, and in combat was cited for repeatedly returning to an extremely hazardous position to rescue his living comrades and retrieve the remains of those killed in action.

- An SM is brought to the aid station on a litter after "shutting down" toward the end of an intense firefight. He fought hard, but suddenly froze at his machine gun, mute and unmoving. He arrives uninjured and healthy, but has a very rigid posture, does not respond to verbal commands and cannot speak at all. The final words he spoke, while still in battle, were to tell his team that he had accidentally killed a fellow member of the unit (and good friend), literally "sawed him in half" with machine gun fire. No wounded or dead soldiers have been brought in, or even reported.

- An SM presents with a self-inflicted gunshot wound to thigh. The command wants to rule out a suicide attempt, then discipline the SM for self-injury to avoid duty. The SM claims it was an accident, and continues to go on all his unit's missions, despite severe pain from his serious wound.

This chapter provided decision guidance for potential future military psychologists (FMPs) and suggestions for how FMPs might acquire military cultural competence. It outlined three core skills potentially useful to every MP, and proposed three training contexts to supply that skill training. It also provided suggestions for legitimizing patients' psychological treatment, and provided ideas for good, preparatory training of a FMP, both in content knowledge and in developing supportive and synergistic relationships with fellow mental/health care providers. Throughout, the chapter highlighted issues important to military practice, including combat deployment and gave case examples for thought and professional development.

References

American Psychological Association. (2010). *Ethical principles of psychologists and code of conduct.* Retrieved from www.apa.org/ethics

Coser, L. A. (1974). *Greedy institutions: Patterns of undivided commitment.* New York, NY: Free Press.

Jadick, R., & Hayden, T. (2007). *On call in Hell: A doctor's Iraq war story.* New York, NY: NAL Caliber.

Taylor, M. K., Markham, A. E., Reis, J. P., Padilla, G. A., Potterat, E. G., Drummond, S. P. A., & Mujica-Parodi, L. R. (2008). Physical fitness influences stress reactions to extreme military training. *Military Medicine, 173,* 738–742.

35 THE IMPACT OF LEADERSHIP ON MENTAL HEALTH

Richard L. Dixon Jr.

"Toxic leadership" can create a negative and highly stressful environment in which subordinates feel helpless in dealing with their situation; it can contribute to a number of psychological and substance use disorders. In extreme cases, an individual feeling no escape from a toxic leadership environment can resort to suicide.

Results from a Mental Health Assessment Team Report from Iraq showed a disturbing relationship between soldier and marine rates of mental health problems and the perception of their leaders. Depending on the intensity of the combat experienced, soldiers and marines having an unfavorable opinion of their leaders screened positive for mental health problems at two to three times the rate of those having a favorable opinion of their leaders (US Army Office of the Surgeon Multinational Force Iraq, 2006). A first-of-its-kind 2007 Canadian study involving 8,441 Active Duty military personnel examined the relationship between deployment-related experiences and mental health problems. The study found 14.9% of the participants had been assessed for mental health issues in the past year, and 23.2% self-assessed that they needed help (Sareen et al., 2007). The DSM-IV mental health disorders detected were: major depressive disorder, posttraumatic stress disorder, generalized anxiety disorder, panic disorder, social phobia, and alcohol dependence. Using the criterions of DSM diagnosis and perceived need or service use, more than 30% of the population surveyed would be considered in need of mental health care (Sareen et al., 2007). The study also found that soldiers reported "attitudinal barriers" to mental health care due to the perception that they would be seen as "weak" and that "my unit leadership might treat me differently" (Sareen et al., 2007, p. 844).

TOXIC LEADERSHIP

The "toxic leader" has a lack of concern for subordinates, interpersonal skills that negatively affect the command climate, and a primary motivation of self-interest. Toxic leadership has long existed in our military, and, according to Colonel Denise Williams's 2005 Army War College research paper, "the paradoxical nature of military leadership" is that it tends to reinforce some toxic leadership traits—"busy, rigid, in control, enforcing, confident, and street fighter"—and that these "may be characteristics the Army values in a leader" (Williams, 2005, p. 14). A recent Army survey of over 22,630 soldiers from the ranks of E-5 through O-6 and Army civilians (pay grade unknown), found that roughly one in five believed the person they worked for was "toxic and unethical." Such leadership may not only contribute to psychological problems, but it can create an environment inconducive to seeking mental health treatment. Recent news stories of hazing in the Army and Marines and a court-martial for attempted suicide underscore the importance of positive-leadership-driven environments. The best approach to preventing suicide is proactively treating psychiatric conditions; the right leadership environment makes this possible.

When a service member works within a leadership environment that does not value the individual nor place an emphasis on mental health, why doesn't the individual just leave? Unfortunately, the nature of the military mission does not permit the freedom of movement quite often enjoyed within the private sector. Beyond this explanation, however, Jean Lipman-Blumen has identified psychological, existential, financial, political, and social barriers that not only prevent individuals from escaping their toxic leadership environments but also make people more predisposed to tolerate such leadership. He refers to these as "Six Aspects of the Human Condition That Make Us Susceptible to Toxic Leaders" (Lipman-Blumen, 2005, pp. 3–4):

1. *Existential anxiety*—Humans grapple with the idea that death is certain but living is uncertain. For those service members that have endured multiple deployments, a fatalistic outlook can develop in which the feeling is you endure your situation (learned helplessness) until you die or make it home from deployment ("suck it up and drive on," "it don't mean nuth'n," "same shit, different day," "BOHICA"—Bend over here it comes again, etc.)

2. *Psychological need*—In accordance with Maslow's Hierarchy of Needs, leaders (good or bad) fulfill many human needs, including group membership and rewards. Depending on an individual's self-esteem, they may be more likely to self-identify with a toxic leader, and/or they may feel powerless to escape.

3. *Situational fear*—Crisis, change, and uncertainty can elicit fear in individuals. This can manifest in anyone from a young recruit in basic training to a seasoned veteran in a combat environment. A leader (good or bad) provides direction in the turbulence, and subordinates fall in line to survive.

4. *Historical time frame*—In every generation some type of fear or challenge faces the populace, and people gravitate to those leaders perceived to have the ability to guide them through the tough times (Global War on Terror, bad economy, etc.). For many Americans, 9/11 was a defining moment in their lives and spurred military service or redefined it. Individual sacrifice in the name of fighting terrorism is seen as for the greater good.

5. *Hope for the future*—The idea that the future holds unlimited possibilities and the desire for individuals to define the meaning of their lives lends itself to following leaders who promise better futures.

6. *Neverending knowledge*—The world will always have untold secrets and discoveries and inventions waiting to be made. These types of challenges require leadership to make them possible.

Lipman-Blumen doesn't speak specifically to service members, but it is easy to see how his "Six Aspects" can hold true in the structured environment of the military. The fears and anxieties relating to fitting into a unit and participating in its respective subculture, and not wanting to fail the group or disappoint the unit's leadership, are powerful forces that act on the individual. Toxic leadership can act on these fears and anxieties and can ultimately lead to a variety of psychological disorders, whether an individual is exposed to combat or not.

GENERATION Y

In addressing mental illness in today's military, the ultimate concern is for those individuals contemplating suicide. According to a recent *Army Times* article (2011), "the majority of service members who commit suicide have never been deployed or seen combat" (Kime, 2011, para. 13). The article went on to list the percentages by service of those that had committed suicide and had no deployment history: Air Force 68%, Marines 20%, Army 70% (one or zero deployments), and no correlation for the Navy. One possible explanation for this is the presence of Generation Y, also known as the Millennial Generation (born between 1980 and 2000) within the military.

Generation Y members are described as having had pampered upbringings and are, according to associate managerial science professor Jordan Kaplan, "much less likely to respond to the traditional command-and-control type of management still popular in much of today's workforce" (Armour, 2005, para. 5). They have high expectations of themselves and others, and were "brought up in the most child-centered generation ever. They've been programmed and nurtured" (Armour, 2005, "Conflicts", para. 5). This need for constant feedback and the desire to offer input versus the traditionally directive leadership nature of the military appears to be in direct conflict with Generation Y's expectations and possibly their coping skills.

Whether Generation Y service members are dealing with the stressors of military service in general, combat experience, or the new phenomena of budget cuts and involuntary separations, leaders need to be aware that this generation may not have the hardy coping skills required to deal with these life challenges. It is important that leaders not only mentor these individuals but also promote an environment in which mental health counseling is seen as a viable option for self-development.

FUTURE IMPLICATIONS

With across-the-board reductions in military manpower already being enacted, thousands of service members will be facing the stress of going before retention boards and many of them will be forced out of the military. This will occur despite previously signed enlistment contracts "guaranteeing" a set number of years of service. It will occur despite some service members having stellar records and the recommendations of their superiors for continued service. It will end the careers of service members that had a life-long goal of retiring from Active Duty military service, and now face the unknown prospect of civilian employment in a poor economy.

To a civilian, the aforementioned paragraph describes an understandably disappointing situation, but it may not capture the devastation and sense of betrayal felt by someone being forced to leave a career they love—being in the service of their country. Less than 1% (0.45%) of the US population has served in the military since 9/11, and many veterans feel the public does not understand what they and their families have gone through (Pew Research Group, 2011). According to the Pew Research Report, 44% of post-9/11 veterans report difficulties readjusting to civilian life, versus 25% from previous wars. Whether this is a product of the public lack of a common reference point for veterans or of the coping skills of Generation Y veterans requires more research. At issue will be many veterans feeling anger, betrayal, resentment, guilt, and depression from leadership that made the decision to prematurely end their military service. These unchecked emotions have the potential to develop into psychological issues.

These psychological issues have the potential to worsen when service members transition from active duty to the Veterans Administration for mental health care. According to testimony before the US Senate by representatives from the Wounded Warrior Project, they found in a survey of more than 935 veterans that 62% tried to get mental health counseling from the VA; two in five reported difficulty in getting the treatment and more than 40% reported not getting the treatment at all (Wounded Warrior Project, 2011). With the current proposed reduction in forces, 100,000 ground troops between the Army and Marines will be eliminated from active duty within the next 5 years (Herb, 2012). With no budget plan in place to significantly increase the VA budget to meet the onslaught of newly discharged service members, mental health care is not likely to improve for our veterans.

THE IMPORTANCE OF LEADERSHIP

Whether it is perceived poor leadership in a combat environment, toxic leadership in a garrison/nondeployed setting, or the stigma of seeking mental health counseling in any situation, leadership has likely played a fundamental role in which veterans present for treatment. Leadership will continue to play a role in veterans' ability to continue counseling and the current environment in which the individual interacts in the work setting. For those service members facing a Reduction in Forces (RIF), it will be the leadership that determines how those individuals are treated, counseled, and prepared for civilian life.

For mental health professionals who may provide treatment to military service members and veterans, it will be important not only to help these service members in a clinical capacity but also to be a patient advocate in an environment where institutional pressures encourage suboptimal care or a lesser diagnosis in the name of cost savings. Ultimately, leadership is the moral courage to stand up for what is right, whether it is working silently behind the scenes to foster an environment conducive to service members seeking mental health treatment or taking a more visible and potentially unpopular standpoint with your peers or supervisors.

In order to be a proper health care advocate, the provider will need to be an instrument of change, promoting an environment in which mental health counseling is encouraged rather than stigmatized. This may involve dealing with toxic leaders or just uninformed leaders, providing educational outreach that can reverberate throughout a military organization. As part of the culture in the military or any bureaucracy, a "check the box" mentality exists in which an activity is done to meet a requirement established by higher authority. No one really takes the requirement seriously, but they know they need to complete it to stay out of trouble. Mandatory pre- and postdeployment mental health questionnaires often fall into this checklist mentality. As a mental health advocate, it will be incumbent on you to reach out to military leadership and service members to help change these and other perceptions, helping military leaders to become mental health advocates as well. Assisting them to see the value of mental health services and how they promote combat readiness and mission effectiveness is an important aspect of this role. Conducting seminars (not presentations) in which service members are encouraged to participate in topics such as mental health issues, stress resilience, and the importance of seeking help without repercussions, puts a face on what you do and who you are. This can be conducted at the unit level for active duty

members and even incorporated into a weekend drill schedule for reservists. Interacting with service members outside of a clinical setting could make all the difference in someone seeking mental health services when they are needed.

References

Armour, S. (2005, November 6). Generation Y: They've arrived at work with a new attitude. *USA Today*. Retrieved from http://www.usatoday.com/money/workplace/2005-11-06-gen-y_x.htm

Herb, J. (2012, January 26). "Tough budget choices": Pentagon cuts will shrink military, reduce ground forces. *DEFCON Hill*. Retrieved from http://thehill.com/blogs/defcon-hill/policy-and-strategy/206865-pentagon-budget-cuts-will-shrink-military-reduce-ground-forces

Kime, P. (2011, September 9). Services still grappling with suicide trends. *Army Times*. Retrieved from http://www.armytimes.com/news/2011/09/military-suicide-prevention-services-090911w/

Lipman-Blumen, J. (2005). Toxic leadership: When grand illusions masquerade as noble visions. *Leader to Leader, 3–4*. Retrieved from http://www.connectiveleadership.com/articles/when_grand_illusions_masquerade_as_noble_visions.pdf

Pew Research Group. (2011, October 5). War and sacrifice in the post-9/11 era. Retrieved from http://www.pewsocialtrends.org/2011/10/05/war-and-sacrifice-in-the-post-911-era/?src=prc-headline

Sareen, J., Cox, B., Afifi, T., Stein, M., Belik, S., Meadows, G., & Asmundson, G. (2007). Combat and peacekeeping operations in relation to prevalence of mental disorders and perceived need for mental health care. *Archives of General Psychiatry, 64*(7), 843–852. Retrieved from http://arc,hpsyc.ama-assn.org/cgi/content/full/64/7/843#otherarticles

Williams, D. (2005). *Toxic leadership in the U.S. Army*. Retrieved from http://usawc.sirsi.net/uhtbin/cgisirsi/x/0/0/5?searchdata1=288536

Wounded Warrior Project. (2011, November 30). VA mental health care: Addressing wait times and access to care. (Testimony before the US Senate Committee on Veterans' Affairs). Retrieved from the Wounded Warrior website http://www.woundedwarriorproject.org/programs/policy-government-affairs/wwp-testimony.aspx

36 TRAINING INITIATIVES FOR EVIDENCE-BASED PSYCHOTHERAPIES

Jeanne M. Gabriele and Judith A. Lyons

IDENTIFYING EVIDENCE-BASED PSYCHOTHERAPIES

Two approaches have been used to identify evidence-based psychotherapies. The first involves training providers in the evidence-based practice process (McHugh & Barlow, 2012). Evidence-based practice involves making clinical decisions by integrating the best available evidence with practitioner expertise and resources, and with the characteristics, state, needs, values, and preferences of those who will be affected (Council for Training in Evidence-Based Practice, 2008). The evidence-based practice process consists of five steps (Council for Training in Evidence-Based Practice, 2008):

- Ask client-oriented, relevant questions to inform treatment decisions.
- Acquire the best available evidence to answer the question.
- Appraise the evidence on quality, applicability, and meaningfulness.
- Apply shared decision making.
- Analyze and adjust.

EBBP.org, a website funded by the Office of Behavioral and Social Sciences Research at the National Institutes of Health, provides online training modules and tools for providers to gain knowledge and skills in the evidence-based practice process.

The second approach of identifying evidence-based psychotherapies involves systematically reviewing literature and developing treatment or practice guidelines (McHugh & Barlow, 2012). These guidelines are typically created by governmental agencies or professional organizations. For example, in the area of PTSD, treatment guidelines have been created by Veterans Affairs/Department of Defense (VA/DoD), American Psychiatric Association, United Kingdom National Institute of Health and Clinical Excellence, Australian National Health and Medical Research Council, and the International Society for Traumatic Stress Studies (Institute of Medicine, 2012). Within the VA/DoD, an Evidence-Based Practice Workgroup coordinates evidence-based reviews to support recommendations for care, as well as to maintain and update VA/DoD evidence-based practice guidelines. A list of current VA/DoD practice guidelines can be found at www.healthquality.va.gov.

TRAINING PROVIDERS IN EVIDENCE-BASED PSYCHOTHERAPIES

Training programs in evidence-based psychotherapy differ widely in the amount of didactic training and competency training they provide. Didactic training aims to increase knowledge about an intervention so clinicians know why, how, and when to use an intervention. Didactic training focuses on procedural elements of an evidence-based psychotherapy, timing and structure of the treatment, how to identify appropriate patients for the psychotherapy, and strategies for problem-solving complications and barriers to implementation. Workshops are the most commonly used format for didactic training. Such workshops typically involve a lecture and slide presentation along with handouts, a review of a treatment manual, and role-plays or experiential exercises. In addition to workshops, didactic training may also include completing Web-based training, reading treatment manuals, and watching videos (McHugh & Barlow, 2012).

Competency training, the "procedural learning for the application of knowledge to a clinical encounter" (McHugh & Barlow, 2012, p. 44), is essential for skill acquisition and incorporation of evidence-based psychotherapies into sustained clinical practice. Competency training involves supervision in administering the intervention. Supervision is typically from an advanced, trained clinician and may be conducted in an individual or group format. Supervision sessions may occur through in-person sessions with the provider(s), phone sessions, or Web conference. Supervision methods for competency training vary from clinician self-report to observing or listening to encounters. These can occur through in-person observation, video-taping, or audio-taping. Supervision sessions allow clinicians to discuss challenges faced and receive feedback and suggestions to assist in overcoming these challenges (Martino, 2010; McHugh & Barlow, 2012).

Quantitative feedback on the provider's implementation of the psychotherapy, such as ratings of adherence and competency, is an important component of the consultation process. Self-evaluations of provider competence are available; however, these evaluations can be misleading, as providers often overestimate skills. Consequently, methods in which the supervisor rates the trainee's interactions and skills in implementing treatment strategies using a reliable and valid scale are preferred. Feedback is typically needed on 10–15 samples before the provider can consistently implement

the treatment with competence (Stirman et al., 2010).

Client outcomes can also be used to enrich the consultation process and evaluate the provider's implementation of treatment. Measures such as the Beck Depression Inventory-II, PTSD Checklist, and Insomnia Severity Index can be used to regularly monitor patient symptoms during treatment and evaluate whether the provider's implementation of the EBP leads to positive behavior change and symptom improvement.

THE ACCESS MODEL

The Access Model (Stirman et al., 2010) provides a framework for disseminating evidence-based psychotherapies, taking into account issues at the system/institutional level as well as at the level of the individual clinician. This model has six steps: assess and adapt, convey basics, consult, evaluate, study outcomes, and sustain. The first step involves assessing the day-to-day operations, available resources, potential constraints, and readiness for change of the agency; engaging stake-holders in the development of a training plan; and assessing and adapting training for the agency's mission, clientele, and clinicians attitudes, needs, and skills. The second step (i.e., convey the basics) is the didactic training component and aims to increase knowledge about the evidence-based psychotherapy. The third and fourth steps (consult and evaluate work samples) are the competency-training components that aim to translate basic learning to sustained practice. The consult step involves coaching and follow-up support. This includes assisting the provider in selecting appropriate cases, reviewing and discussing treatment sessions, discussing concerns, identifying barriers and problem solving to overcome these barriers, and conducting experiential exercises. Evaluating work samples includes rating multiple sessions with a competency rating scale, assessing conceptualization and treatment-planning skills, delivering feedback, developing goals, and identifying strengths and challenges. The fifth step (study the outcomes) consists of program evaluation or research on the impact of the training program to assist stakeholders in making informed decisions about the future of the training program. The final step (sustain) involves maintenance of change and prevention of drift. This may include fidelity monitoring and provision of support and information after the training is complete.

VETERANS HEALTH ADMINISTRATION EVIDENCE-BASED PSYCHOTHERAPY INITIATIVE

The Veterans Health Administration (VHA) is transforming its mental health system to increase delivery of evidence-based psychotherapies. Funding has been allocated to support the development and implementation of national EBP training programs. The original goal of these efforts was to ensure that all veterans with PTSD, depression, and severe mental illness have access to EBPs for these conditions. Over time, the training programs have expanded to address other needs and conditions common to veterans (McHugh & Barlow, 2012). Table 36.1 provides information on the current evidence-based psychotherapies being implemented within VHA.

The VHA EBP training programs use a combination of didactic training and competency training. Providers selected to attend a training program receive didactic training through a three to four-day in-person workshop. For some training programs, such as the cognitive processing therapy training, providers are asked to complete online trainings prior to attending the workshop. At the workshop, providers are provided with power point presentations, treatment manuals, and other resources to facilitate the learning process. Role-play and group experiential exercises are used to practice skills.

After the workshop, participants return to their institution and engage in a 6-month consultation process (see Karlin et al., 2010). Providers are assigned to a training consultant and speak to this consultant weekly through group and/or individual phone supervision sessions. Most training programs require audio-recordings of sessions and use a rating system to assess competency and fidelity to the treatment protocol. For example, with the cognitive-behavioral therapy training program, providers' sessions are evaluated using the Cognitive Therapy Scale. To complete the consultation process, providers are required to attend calls, complete the protocol with a certain number of veterans specified by the training program, and achieve certain ratings on the assessment of their audiotapes.

After completing the training programs, providers continue to have support for the

TABLE 36.1. Current EBP Training Initiatives

EBP	Targeted Condition or Population
Cognitive-Behavioral Therapy (CBT)	Depression
Acceptance and Commitment Therapy (ACT)	Depression and Anxiety
Cognitive Processing Therapy (CPT)	PTSD
Prolonged Exposure (PE)	PTSD
Social Skills Training (SST)	Severe Mental Illness
Behavioral Family Therapy (BFT)	Severe Mental Illness
Multifamily Group Therapy	Severe Mental Illness
Integrative Behavioral Couples Therapy (IBCT)	Couples with adjustment difficulties
Cognitive Behavioral Therapy for Insomnia	Insomnia
Problem-Solving Therapy	Depression
Cognitive-Behavioral Therapy for Chronic Pain Management	Chronic Pain Management
Motivational Enhancement Therapy	Individuals in need of, but ambivalent regarding, addiction treatment
Motivational Interviewing	Individuals ambivalent about treatment, medication management, or making change

implementation of the therapy. Videos that show the implementation of the skills are readily available to providers. Share-point drives have been established for providers to have easy access to useful information. For some programs, monthly webinars are offered to provide further information on the evidence-based psychotherapy such as overcoming barriers, delivering the therapy to specific populations, or refining skills. In addition, regular group conference calls are available for any provider to dial in to seek consultation as needed.

The VHA uses a train-the-trainer model to increase the number of both trainers/workshop leaders and consultants (individuals who provide supervision on telephone consultation calls). This model increases the capacity to provide training to more providers. Training consultants will nominate individuals who exhibit strong skills in implementing the treatment and show good interpersonal skills. If a provider is interested in becoming a trainer or consultant, the individual attends a 5-day consultation training program. In this training, the provider receives additional training on the treatment to ensure understanding of the model, refines skills in delivering the therapy, and learns about the consultation process.

CHALLENGES AND RECOMMENDATIONS

Challenges occur at the level of the organization, the clinician, and the person served. Agencies may face pressure from influential groups or persuasive marketers to confer "evidence-based" status to therapies based on limited data or data that may or may not generalize to that agency's population of persons served. Organizationally, significant resources are required to train, evaluate, and certify clinicians and to then monitor sustained implementation of evidence-based protocols. In systems in which a different model had been in place (e.g., psychotherapy that was longer-term, less focused, more supportive), the transition to time-limited evidence-based protocols can evoke push-back from supervisors, clinicians,

and clients. Persons served may feel services have been "cut off" after a protocol ends or if they are discharged from the protocol for not completing assignments that require more active treatment engagement than that to which the client was previously accustomed. The standard protocols may lack flexibility to take into account the client's available time and resources (e.g., 12 weekly group sessions may not be feasible for a client whose job involves travel/rotating shifts or who lives far from the clinic). Clinicians may similarly feel that broader longer-term care is needed and may resist the extra level of planning and organization that some protocols require. Clinicians who already practice similar therapies may view the standard protocols and certification process as restrictive or redundant. Seasoned clinicians may view being critiqued by junior trainers/consultants with some trepidation or resentment. Team leaders, supervisors, and administrators face the challenge of carving out staff time for training/consultation, deciding which clinicians to train first and how to shift other workload assignments to both allow for and reinforce the adoption of EBPs.

As with the roll-out of any institutional change, engaging key individuals who have social influence is often key, and success breeds success. This makes the initial stages of dissemination particularly important. Securing the voluntary engagement of respected clinicians who can administer the treatment effectively increases the likelihood of word-of-mouth support among clinicians as well as across the client pool, as clients see other persons served benefiting from the new treatment. Devotion of time to EBPs can be reinforced by prorating other workload reductions. Parallel to transition of time-limited EBPs, it is important to identify/develop options to fill the long-term-support gap that may result from diversion of resources toward time-limited EBPs and away from less structured supportive therapy. The recovery model of care (e.g., Jacobson & Greenley, 2001), with its emphasis on peer support and community integration, complements EBP initiatives in this respect.

References

Council for Training in Evidence-Based Practice. (2008). Definition and competencies for evidence-based behavioral practice (EBBP). Retrieved from http://www.ebbp.org/documents/EBBP_Competencies.pdf

Institute of Medicine. (2012). *Treatment for posttraumatic stress disorder in military and veteran populations: Initial assessment.* Washington, DC: National Academies Press.

Jacobson, N., & Greenley, D. (2001). What is recovery? A conceptual model and explication. *Psychiatric Services, 52*(4), 482–485.

Karlin, B. E., Ruzek, J. I., Chard, K., Eftekhari, A., Monsoon, C. M., Hembree, E. A.,...Foa, E. B. (2010). Dissemination of evidence-based psychological treatments for posttraumatic stress disorder in the Veterans Health Administration. *Journal of Traumatic Stress, 23*(6), 663–673.

Martino, S. (2010). Strategies for training counselors in evidence-based treatments. *Addiction Science and Clinical Practice, 5*(2), 30–39.

McHugh, R. K., & Barlow, D. H. (2012). *Dissemination and implementation of evidence-based psychological interventions.* New York, NY: Oxford University Press.

Stirman, S. W., Bhar, S. S., Spokas, M., Brown, G. K., Creed, T. A., Perivoliotis, D.,...& Beck, A. T. (2010). Training and consultation in evidence-based psychosocial treatments in public mental health settings: The ACCESS Model. *Professional Psychology: Research and Practice, 41*(1), 48–56.

37 UNIQUE CHALLENGES FACED BY THE NATIONAL GUARD AND RESERVE

Michael Crabtree, Elizabeth A. Bennett, and Mary E. Schaffer

The US Military is currently deployed in more than 150 countries around the world. Many of these overseas military personnel are in combat zones. Since October 2001, the US Military has seen over 1.7 million military personnel deployed in support of Operation New Dawn (OND; Iraq), Operation Enduring Freedom (OEF; Afghanistan), Operation Iraqi Freedom (OIF; Iraq), and Operation Noble Eagle (ONE; homeland security). While most research has focused on Active Duty veterans, National Guard and Reserve troops make up about half of those actually fighting the current conflicts (US Department of Defense, 2012). This chapter examines some of the distinctive characteristics and needs of these troops and their families.

WHO ARE THE RESERVE COMPONENT?

The term Reserve Component, or RC, is used to refer collectively to the seven individual Reserve Components of the armed forces:

• The Army National Guard of the United States
• The Army Reserve
• The Navy Reserve

- The Marine Corps Reserve
- The Air National Guard of the United States
- The Air Force Reserve
- The Coast Guard Reserve

The role of these seven RCs, as codified in law by 10 USC (United States Code) § 10102, is to "provide trained units and qualified persons available for active duty in the armed forces, in time of war or national emergency, and at such other times as the national security may require, to fill the needs of the armed forces whenever…more units and persons are needed than are in the regular components." Though RC troops and their families exhibit many of the same strengths and needs as their Active Duty counterparts, they also have unique characteristics because they alternate between their military lives and responsibilities and their civilian lives and responsibilities. For this reason we will also use the term Citizen Warrior to refer to RC troops, because we feel this label captures their overlapping identities in two worlds.

The concept of a Citizen Warrior as a function of military strategy goes back to pre-revolutionary war times with the colonial militia. During the cold war, the RCs were a rarely activated strategic force. From 1945 RCs were activated an average of only once per decade during times of sudden national emergency. During this time period the RC was generally not deployed, deployed to noncombat theaters, or deployed to combat theaters primarily in combat support or combat service support roles. September 11th had a pivotal effect on utilization of the RC.

CONTEMPORARY ROLE AND IDENTITY
OF THE RC

The past decade has seen the RC move from being a strategic reserve force to an operational, combat-ready, tactical component used to address current international conflicts. In the decade since 9/11/01, the frequency and length of RC deployments has increased, and the time between deployments (commonly called "dwell time") has decreased at rates never before experienced. When not participating in training, activated, or deployed, today's RC troops are called on to participate in frequent communications, responsibilities, and leadership tasks related to their RC status. Their codified role as an emergency ancillary to the Active Duty components has not been revised to reflect this change, but the demands on Citizen Warriors are more intense and more frequent than they have been at any other time in history.

As of July 17, 2012, there were 63,498 RC troops on Active Duty status, and 789,252 total RC veterans who had been deactivated from Active Duty status since 9/11/02 (US Department of Defense, 2012). This number of deactivated Citizen Warriors is important because it reflects the substantial probability that civilian medical and behavioral health practitioners will encounter combat veterans, or the family members of combat veterans, among their civilian patients. Combat veterans are our neighbors, our coworkers, our teachers, and for clinical practitioners, our patients. Many RC service members do not fit a stereotypical image of a warrior as presented in the media (e.g., being young and unmarried), increasing the chances their civilian providers may not recognize them as such (Bennett et al., 2012).

Even when the military experience of RC troops is recognized, their circumstances and needs cannot be generalized from their Active Duty military counterparts; these Citizen Warriors are different from their active counterparts in several important ways prior to, during, and after deployment. RC personnel tend to be older and have additional responsibilities outside the military, such as full-time employment. In a survey of returning RC members, Britt et al. (2011) found 45% were over the age of 45 and 57.9% had children. Enlisted RC troops also tend to have higher education levels than their Active Duty counterparts. Additionally, many RC personnel come from rural locations where access to behavioral health care can be limited (Schaffer, Crabtree, Bennett, McNally, & Okel, 2011).

The RC often embraces a different and more multifaceted professional identity than their Active Duty counterparts, viewing their military commitment as a part-time job and their civilian occupation as their full-time professional identity. This dual identity requires reservists to negotiate values and attitudes from both cultures.

In each environment, the ability to adapt and integrate may be complicated by RC status. The RC are sometimes viewed as "weekend warriors," a pejorative term implying less commitment, experience, and training. RC troops typically serve and train for 39 days each year, participating in one drill weekend a month and two weeks of annual training. Their Active Duty counterparts serve year-round and typically participate in more frequent and longer training exercises. While the RC do have extended predeployment training, and many RC members have active duty experience, the intermittent training afforded by the schedule of the RC may result in them being somewhat less prepared for the emotional and physical rigors of battle, and less experienced in utilizing specialized skills. At times, the RC has less access to the most updated equipment than Active Duty components. Less frequent training and less updated equipment for the RC can lead their Active Duty counterparts to have lower expectations of, and confidence in, RC personnel, even when they are side by side on the battlefield. Perceptions also exist that Active Duty units and personnel are given preference and status over the RC in the overall military structure.

Despite this lack of complete functional and cultural integration, today's RC troops appear to be more integrated into actual military operations than their Cold War predecessors. RC troops are currently just as likely to experience combat as their Active Duty counterparts, and to be wounded or be killed as Active Duty troops. Since 9/11/01, many RC troops have completed repeated, prolonged deployments, experiencing the same frequency and intensity of combat exposure as their Active Duty counterparts. Because certain types of units and skills are predominately found in the RC, many RC troops have experienced an even greater frequency and intensity of combat exposure than some Active Duty troops. A US General Accounting Office (2004) report noted that RC troops are "fighting…side by side with their active duty counterparts, facing the same dangers and making the same sacrifices" (p. 1).

Their civilian lives are also affected in unique ways by RC status. Social support has been shown to be a protective factor in both military and civilian populations for depression, PTSD, and other behavioral health concerns. When Active Duty troops return from a combat deployment, they typically return to a supportive military installation where many individuals share similar experiences and culture, and where there is an array of accessible military-specific services. In contrast, RC troops have no such buffer; it is not unusual for them to experience the streets of Kabul one week as a warrior and the next week experience their hometown as a civilian. Often, they return home to a well-meaning community exhibiting little or no actual understanding of the RC's warrior experience. Current Department of Defense policy guarantees RC troops returning from deployment a full 60 days off from their military duties, often increasing separation from military support and those who have similar experiences. Schaffer et al. (2011) found that reestablishing previous civilian social supports and ties can be difficult and can take time for Citizen Warriors. This is compounded by RC troops from the same unit frequently coming from varying locations. When the unit leaves for deployment, open positions may be filled by troops from other areas. When these "cross-leveled" troops return, they lack shared deployment experiences with other unit members; this denies them an important form of social support.

CIVILIAN BEHAVIORAL HEALTH PROVIDERS AND THE RC

There are many reasons civilian providers should be alert for behavioral health concerns in the deactivated RC community. Compared to their Active Duty counterparts, RC troops are more likely to experience behavioral health concerns during and after deployment.

Researchers have found 20.3% of Active Duty, compared to 42.4% of the RC, screened positive for PTSD, alcohol misuse, major depression, anxiety, or other mental health problems (Milliken, Auchterlonie, & Hoge 2007). Multiple factors may explain these higher rates of mental health problems in RC troops versus Active Duty troops who have deployed to a combat zone. In a study of National Guard soldiers, Riviere, Kendall-Robbins, McGurk, Castro, and Hoge (2011) found correlations between financial hardship, civilian job loss, lack of civilian employer support, and perceptions of negative effect of deployment absence on civilian coworkers with rates of depression and posttraumatic stress disorder (PTSD).

Many behavioral heath problems such as PTSD do not manifest during or immediately following deployment, instead presenting months or even years later. The Post-Deployment Health Reassessment (PDHRA) is completed 90 to 180 days after all troops return from deployment. However, when symptoms appear after the 90- to 180-day window, they may go unrecognized in RC veterans who are back in their civilian communities. This increases the burden on civilian providers to be aware of, and screen for, combat-stress-related concerns.

In Bennett et al.'s (2012) research on civilian providers who may treat members of the RC, the researchers found providers who routinely asked their clients if they had experience in the military recognized a significantly higher percentage of their patients had been in combat in Iraq or Afghanistan than those who did not ask. These providers who were asking also believed that a significantly higher percentage of their patients who had been in combat would experience symptoms. Asking about service and combat is pivotal.

Even with alert providers, Citizen Warriors and their families face obstacles in getting treatment. As of 2008, 41% of all veterans enrolled in the Veterans Administration (VA) resided in rural or highly rural areas (US Department of Veterans Affairs, 2010). Physical and cultural barriers to mental health care for rural RC veterans include distance from and limited access to services, a shortage of rural mental health

services, a need to feel self-sufficient, a social stigma associated with seeking mental health treatment, a belief system often incongruent with care-seeking behaviors, and rural community stigma. Bennett, Crabtree, Schaffer, and Britt (2011) found rural RC veterans cited as the primary reasons for not seeking treatment their beliefs that "others were worse off" and they would be prescribed medication that would negatively impact their military and civilian careers. Additionally, rural providers may be primarily trained in models developed from the treatment of civilians living in urban areas, and not specifically in the needs of rural veterans. Bennett et al. (2012) found many of these rural providers were alarmingly uninformed and even misinformed about evidenced-based psychosocial and pharmacological treatments for PTSD.

BEHAVIORAL HEALTH RESEARCH AND THE RC

Little research has been conducted specifically on the characteristics, needs, and strengths of RC families. These families are often separated from families who share their experience and also from military support services. In an analysis of focus groups of family members of RC troops, Schaffer et al. (2011) found RC family members expressed concerns about the following issues:

- Not being prepared for the changes in their RC members
- Inaccurate perceptions by their community and the media of their RC member's experience
- The development of family and marital conflict
- Difficulty accessing a confusing array of military services often located a considerable distance from the families
- Difficulty renegotiating household roles and responsibilities
- Difficulty helping their RC members reestablish bonds with spouses and children
- Concern about stigma associated with behavioral health concerns

- Seeking and finding support for those concerns in both civilian and military cultures
- Concern about an array of behavioral heath symptoms not previously present.

Concerns over behavioral health symptoms are compounded by the alarming rates of attempted and completed suicide among military personnel. The suicide completion rate among the military has historically been lower than in the civilian population; however, recently it has risen to exceed that in the civilian population. As of spring 2012, statistics indicate more American troops are dying due to suicide than direct combat. According to spring 2012 figures from the Pentagon, Active Duty troops are completing suicide at the rate of one per day. Non-active-duty reservists are not counted among these statistics by the military. Returning veterans under Veterans Administration care are also omitted, as are returning veterans who are not enrolled in the VA system. Figures are not available for RC personnel, or for those who have left active duty, but many estimates are that the suicide rate is even higher for those who are not on active duty. Reports do indicate that most of the suicides within both the Army Reserve and Army National Guard occur when the soldier is in civilian, rather than military, status. This means the actual figures for the number of suicides may be much higher than what is currently reported (*Daily Mail Reporter*, 2012).

RECOMMENDATIONS

It is important to keep in mind that most returning troops, both RC and Active Duty, are resilient, readjust, and exhibit no diagnosable behavioral health disorders. However, the incidence of behavioral heath concerns is significant. Military and civilian psychologists will best serve their RC clients by increasing awareness of their unique needs and concerns, and by helping Citizen Warriors and their families, whenever possible, to overcome the obstacles to their treatment. Based on our research, several recommendations have been developed

to assist practitioners in effectively treating members of the RC (Bennett et al., 2012).

- Ask every client at intake if he or she, or a family member, has been in the military, regardless of the probability of an affirmative response. Because you can't identify RC personnel based on appearance, this will eliminate missing critical concerns of RC clients who do not fit the stereotype of a warrior.
- If you are a civilian practitioner, become familiar with military culture and values. However, always ask questions about the RC member's individual experience in the military and their feelings and thoughts about that experience.
- Ask if the client or a family member has been in combat or in a combat zone even if they do not acknowledge military service (many civilians with no military experience have experienced the current conflicts as contractors).
- Post placards in your waiting area welcoming veterans and encouraging clients to let staff know if they, or a family member, has served.
- Use a brief, validated screener for PTSD, such as (http://www.ptsd.va.gov/professional/pages/screening-and-referral.asp) in order to assist in identifying symptoms that may underlie the client's presenting concerns.
- When a diagnosis of PTSD is appropriate, facilitate effective treatment using an evidence-based treatment. This may involve a referral if the provider is not trained in one of these techniques.
- Always screen every client for suicidal ideation and intent.

References

Bennett, E. A., Crabtree, M., Schaffer, M. E., & Britt, T. W. (2011). Mental health status and perceived barriers to seeking treatment in rural reserve component veterans. *Journal of Rural Social Sciences, 26*(3), 74–100.

Bennett, E. A., Schaffer, M. E., Wynn, G. H., Oliver, K., Pury, C., Crabtree, M., & Britt, T. W. (2012). *Your first step in identifying and diagnosing combat stress in returning reserve component veterans: Just ask*. Unpublished manuscript, Combat Stress Intervention Program, Washington and Jefferson College, Washington, PA.

Britt, T. W., Bennett, E. A., Crabtree, M., Haugh, C., Oliver, K., & McFadden, A. (2011). Using the theory of planned behavior to predict whether reserve component veterans seek treatment for a psychological problem. *Military Psychology*, 23(1), 82–96.

Daily Mail Reporter. (2012). Shocking figures show one U.S. soldier commits suicide every day. Retrieved from http://www.dailymail.co.uk/news/article-2156250/Suicides-U-S-troops-hit-DAY.html

Milliken, C. S., Auchterlonie, J. L., & Hoge, C. W. (2007). Longitudinal assessment of mental health problems among active and reserve component soldiers returning from the Iraq war. *Journal of the American Medical Association*, 298(18), 2141–2148.

Riviere, L., Kendall-Robbins, A., McGurk, D., Castro, C., & Hoge, C. (2011). Coming home may hurt: Risk factors for mental ill health in US reservists after deployment in Iraq. *British Journal of Psychiatry*, 198, 136–142.

Schaffer, M. E., Crabtree, M., Bennett, F., McNally, M., & Okel, A. (2011). Identifying barriers to treatment for PTSD among reserve component veterans in rural Pennsylvania: An analysis of five focus groups. *Journal of Rural Community Psychology*, E14(1). Accessed at http://www.marshall.edu/jrcp/volume14_1.htm

US Department of Veterans Affairs. (2010). *Veterans Health Administration Office of Rural Health Strategic Plan 2010–2014*. Washington, DC: Author.

US General Accounting Office. (2004). *Reserve forces: Observations on recent National Guard use in overseas and homeland missions and future challenges* (Publication No. GAO-04-670T). Washington, DC: Author.

US Department of Defense. (2012). Reserve components Noble Eagle/Enduring Freedom. Retrieved from http://www.defense.gov

PART IV
Clinical Theory, Research, and Practice

38 PREVALENCE OF MENTAL HEALTH PROBLEMS AMONG MILITARY POPULATIONS

Sherrie L. Wilcox, Kimberly Finney, and
Julie A. Cederbaum

The mission of the Department of Defense (DoD) is to maintain forces to deter war and protect the security of the United States. Thus, military personnel are often called on as a first line of defense for conflict or disasters. As part of training and preparation for war or conflict, military personnel engage in physically and mentally strenuous activities, often involving separation from family members. In addition to separation related to training, military personnel face the possibility of being deployed. Since the start of the post-9/11 wars (i.e., Overseas Contingency Operations), more than 2 million service members have been deployed, many of whom have experienced repeated deployments. Deployment, particularly to a combat zone, creates added stress for service members and their families.

Mental health problems that military personnel experience may be a direct effect of combat or deployment, or a result of the compounding factors associated with the military lifestyle, including relationship conflict, parenting difficulties, work stress, and reintegration challenges (Hazle, Wilcox, & Hassan, 2012). Premilitary service events also predispose individuals to mental health problems. Those factors associated with highest/increased risk for mental health stressors include: (1) deployment to a combat zone, (2) younger age, (3) a

history of hospitalization, (4) Reservist and National Guard membership, and (5) reintegration challenges for military retirees. Men and women appear to experience psychological stressors at similar rates.

This chapter focuses on the commonly reported mental health problems, including posttraumatic stress disorder (PTSD), major depressive disorder (MDD), and generalized anxiety disorder (GAD). Other psychiatric disorders, such as schizophrenia or bipolar disorder, are much less common among military personnel, due to screenings completed before entering military service. Such severe psychiatric problems can impact readiness, which influences mission accomplishment and could inadvertently lead to injury or death of unit members.

PREDEPLOYMENT MENTAL HEALTH STATUS AMONG MILITARY PERSONNEL

Despite the DoD's implementation of predeployment mental health assessments in 1998, there are few studies that assess or report mental health problems among military personnel prior to deployment. Among the limited studies, findings indicate that some military personnel report mental health symptoms before being deployed. The rates of reported predeployment mental health problems range from 5 to 16% among samples of Active Duty post-9/11 veterans (Hoge et al., 2004). Specifically, the reported rates of predeployment PTSD (9.4%) and GAD (15.5%) in the sample of the 2,530 post-9/11 service members were higher than the lifetime prevalence of PTSD (6.8%) and GAD (5.7%) in the general population (Hoge et al., 2004; Kessler, Chiu, Demler, & Walters, 2005). The reported rate of predeployment MDD (11.4%) in the sample was also elevated, but slightly lower than the lifetime prevalence of MDD (16.6%) in the general population.

According to the July 2012 Armed Forces Health Surveillance Center's Deployment Health Assessment Report (AFHSC-DHA), none of the 214,912 service members who completed the predeployment mental health assessment reported any symptoms of depression or more than two symptoms of PTSD. Predeployment GAD symptoms were not presented in the report. In addition to the low levels of PTSD and depression symptoms, there were low levels of mental health referrals (1.1% among active component, 0.4% among reserve component).

It is likely that the underreporting may be due to fear of being unable to deploy with a predeployment mental heath problem. Additionally, the underreporting may be due to a lack of military-specific PTSD or depressive symptoms. Overall, military personnel have reported predeployment mental health problems, which have the potential to become exacerbated by deployment and increase the probability of deployment-related mental health problems. The predeployment mental health problems may be due to traumatic experiences not related to the military, to intense military-related training before deploying, to anxiety related to an upcoming deployment, or to other general life challenges and stressors, such as divorce.

POSTDEPLOYMENT MENTAL HEALTH STATUS AMONG MILITARY PERSONNEL

Not surprisingly, military personnel returning from deployment are significantly more likely to report mental health problems than those in predeployment. Additionally, the rates of postdeployment mental health problems among military personnel are also higher than lifetime prevalence of mental health problems among the general population. A sample of 815 marines and 2,856 soldiers reported rates of postdeployment PTSD, MDD, and GAD with ranges of 6–13%, 14–15%, and 6–8%, respectively. However, the rates of any postdeployment mental health disorder ranged from 11.2 to 29.2% (Hoge et al., 2004). Among the sample, rates of marines and soldiers who received professional help in the past month ranged from 6.6 to 11.4%, indicating that a number of marines and soldiers are not receiving treatment for mental health problems.

The rates of postdeployment PTSD and depression symptoms vary across branches of the military. According to the July 2012 AFHSC-DHA Report (Armed Forces Health Surveillance Center [AFHSC], July 2012), baseline postdeployment reports of two or more PTSD symptoms ranged from 2.7 to 12.0% and any depression symptoms ranged from 11.6 to 35%, depending on the branch of service. The Army and Marine Corps tend to have the highest rates, particularly among the reserve component. Additionally, the rates tend to increase with time across all branches, as the 3- to 6-month rates were significantly higher than the initial postdeployment rates. Postdeployment GAD symptoms were not presented in the report. Table 38.1 presents the mental health symptoms by branch for active and reserve components (AFHSC, July 2012).

In addition to the high rates of PTSD and depression symptoms, an average of 7.1% of the Active Duty members and 5.6% of Reserve component members received a mental health referral at the initial postdeployment assessment. This rate increased slightly at the 3- to 6-month follow-up for the Active Duty members (7.3%), while it tripled for the Reserve component members (17.8%). Moreover, while over 96% of Active Duty members made a medical visit after referral, only 43.8% of Reserve component members made a medical visit within 6 months after referral (AFHSC, July 2012).

The effects of untreated mental health problems may be most damaging. Undiagnosed and untreated mental health problems can impact

TABLE 38.1. July 2012 Postdeployment Mental Health Symptoms

	Army		Navy		Air Force		Marine Corps	
	Post-deployment n = 124,038 (%)	Reassessment n = 131,321 (%)	Post-deployment n = 14,540 (%)	Reassessment n = 19,179 (%)	Post-deployment n = 52,022 (%)	Reassessment n = 4 5,429 (%)	Post-deployment n = 37,348 (%)	Reassessment n = 35,931 (%)
Active Component								
PTSD symptoms	12.0	13.8	6.3	8.7	2.7	3.1	7.1	12.0
Depression symptoms	29.5	31.3	20.7	27.6	11.6	13.0	27.0	32.6
Reserve Component	Post-deployment	Reassessment	Post-deployment	Reassessment	Post-deployment	Reassessment	Post-deployment	Reassessment
PTSD symptoms	10.6	18.0	6.8	13.6	3.1	3.8	10.8	21.1
Depression symptoms	28.2	33.0	22.7	31.5	13.2	15.2	35.0	38.8

both the service member and the family in multiple ways. Military personnel with untreated mental health problems are at increased likelihood of engaging in unethical behavior (i.e., injuring noncombatants or destroying property), substance abuse, and homelessness. For example, a study by the Department of Veterans Affairs found that PTSD alone does not increase homelessness, but rather the personal and economic consequences of untreated PTSD, including social isolation and violent behavior, are what contribute to homelessness (Williamson & Mulhall, 2009). Suicide is also of concern; rates of suicide among military personnel have increased annually since 2004. While definitive number of suicides are difficult to track, veterans make up about 13% of the US population, they account for 20% of suicides (Williamson & Mulhall, 2009). Last, untreated mental health issues lend to family stressors including marital strain, divorce, and family reintegration.

Overall, the rates of mental health problems surpassed those among the general population and were greater 3–6 months post deployment versus immediately post deployment. The effects of war are cumulative, and therefore rates of mental health symptoms and meeting of diagnostic criteria increase with time. The postdeployment rates of mental health problems are especially troubling in the reserve components, particularly around 3–6 months post deployment. This is often the time when postdeployment reintegration challenges are surfacing (Hazle et al., 2012). It is also important to note that the rates of reported mental health symptoms are grossly underreported and should be viewed with caution and an understanding that actual rates may be higher than reported rates. In addition to higher rates of mental health symptoms, the low rates of medical visits after a referral in the reserve component underscore challenges that reserve members face compared to active duty members.

MENTAL HEALTH AMONG MILITARY FAMILIES

Military Spouses and Partners

Military families experience a different side of deployment and their own deployment-related stressors. Military separations can lead to a disconnection between service members and their families; children have aged while the service member was away and family roles have shifted during the deployment. For the military families who remain "home" during a deployment, the stress of deployment can lead to mental health strain for the nondeployed spouse or partner. Being married has been found to be both a protective factor by providing a dedicated source of support, as well as a risk factor and leading to greater family distress.

Divorce is also a concern among military families. Divorce has been found to be strongly related to mental health problems and suicide (Hyman, Ireland, Frost, & Cottrell, 2012). In 2007 the Air Force (7.44%) had the highest rate of divorce, followed by the Army (6.34%), the Navy (3.67%), and the Marine Corps (3.39%). Among those who completed suicide, the Air Force (8.11%), again, had the highest rate of divorce, followed by the Navy (5.13%), the Marine Corps (3.03%), and the Army (2.63%) (Hyman et al., 2012). It is important to note that the divorce rates fluctuate and the military divorce rate is generally similar to that of the civilian population. Despite similar rates to that of the civilian population, divorces add an extra layer of distress to military families and can negatively influence service member readiness and mission accomplishment.

Military Children

Distress and conflict among military parents can also directly impact mental health among military children. Parent distress is predictive of child depression, externalizing symptoms, and well-being (e.g., academic engagement, family functioning). Children in military families tend to have high rates of mental health problems. A study of 307,520 Army children found that boys (19.6%) and girls (15.7%) with a deployed parent were more likely to have any mental health diagnosis compared to boys (16.3%) and girls (13.6%) without a deployed parent (Mansfield, Kaufman, Engel, & Gaynes, 2011). Deployments are also associated with behavioral problems in military children. Nearly 50% of military children report conduct problems or emotional problems.

FACTORS INFLUENCING RATE FLUCTUATION

For military personnel, reporting mental health problems is a difficult decision that could likely impact their career. Rates of reported mental health problems vary between samples within published studies and samples within DoD reports. Knowledge of the regulations regarding confidentiality in the military may deter service members from seeking treatment, due to fear of lost privacy or lost privileges, and thus reduce the likeliness of honestly reporting symptoms of mental health problems. Research suggests that allowing service members to anonymously report mental health problems increases honesty in reporting symptoms (Warner et al., 2011). However, anonymous reporting will not help those with a mental health problem get referrals for mental health services. This section briefly describes factors that can influence the reported rates of mental health problems among military personnel.

Reduced Confidentiality

Confidentiality applies to communication between a patient and a clinician made for the purpose of facilitating diagnosis or treatment of a patient's mental condition. Information discussed in this process should be protected from unauthorized disclosure. Military mental health clinics have unique regulations regarding standards of confidentiality and informed consent. Per military doctrine, a confidential communication will be disclosed to those with a proper and legitimate need for information and who are authorized by law or regulation to receive it, unless it is evidentiary privilege. Moreover, in order to receive time away from work to go to an appointment, the service member must request the time away and may be required to state the reason for the appointment to the commanding officer. Thus, before even receiving a diagnosis, a military leader must be informed that there may be a problem, and once the appointment is made and a problem is reported, military leaders are again informed of the extent of the problem. This also infers that if a service member believes that he or she has a mental health problem, they might avoid reporting symptoms and seeking treatment at a usual appointment because want to maintain a level of privacy regarding their mental health status.

Negative Job Impact

Military mental health clinicians must practice within the guidelines of DoD Directive 6490.4 and DoD Instruction 6490.1, which outline policy and administration actions, clinical evaluations of imminently dangerous service members, recommendations to commanders, and other responsibilities of military clinicians and commanders. These documents also outline the protection rights of service members against improper referrals of service members for evaluation and treatment by their commanders. For example, a commander can give a legal order for a service member to have a mental health assessment if he or she believes that the service member has a mental health problem that is impacting the service member's ability to perform their military duties. However, the commander who orders the mental health evaluation must be able to articulate reasons that the service member should be evaluated. Failure of the service member to comply can result in judicial punishment. Additionally, although the service member has the right to accept or refuse treatment, if treatment is refused and job performance is hindered, then the command can exercise their right to discharge the service member from the military, as the service member will be unfit to continue to serve in the military.

Another policy worth mentioning is DoD Instruction 5200, the Information Security Program. All military personnel are required to have a security clearance, which is deemed necessary for the safety and security of the country. This policy addresses behavioral requirements that must be demonstrated in order to maintain a security clearance. This Instruction states that military personnel must be free of mental illness, drug and alcohol addictions, and demonstrate financial responsibility, among other criteria aimed to decrease government personnel from being a security liability. This Instruction also outlines the service member's responsibility as it relates to restoring their mental wellness following treatment, as it relates to their personal security clearance and national security.

Personal Gain

Although most active duty service members tend to prefer to underreport mental health problems, some service members may falsely report problems. Malingering refers to fabricating or exaggerating symptoms for secondary motives, including financial compensation or avoiding work. Malingering is occasionally a concern when service members seek mental health treatment and are placed on duty restriction or are not allowed to deploy. Ultimately, the clinician is responsible for correctly identifying that a service member has a mental health problem or is malingering.

The rates of mental health problems among military personnel, as well as the related problems among military families, underscore the potential impact that service in the military can have on both service and family members. In general, rates of mental health problems are higher among military personnel than in the general population, and even higher in reserve component members. Moreover, reserve members are less likely to receive needed treatment for mental health problems, likely due to fear of a potentially negative impact on their job. These rates highlight the need for both preventive interventions and quality treatment. Treatment is the best choice when an illness or disorder has been identified. However, this is not always an easy choice for service members, who tend to underreport mental health symptoms. The road to treatment for many service members has many curves and twists and it is best traveled with a mental health clinician who is aware of the implications of each treatment choice.

References

Armed Forces Health Surveillance Center. (2012, July). *Deployment health assessments U.S. Armed Forces: July 2012*. Washington, DC: Defense Medical Surveillance System.

Hazle, M., Wilcox, S. L., & Hassan, A. M. (2012). Helping veterans and their families fight on! *Advances in Social Work, 13*(1), 229–242.

Hoge, C. W., Castro, C. A., Messer, S. C., McGurk, D., Cotting, D. I., & Koffman, R. L. (2004). Combat duty in Iraq and Afghanistan, mental health problems, and barriers to care. *New England Journal of Medicine, 351*(1), 13–22.

Hyman, J., Ireland, R., Frost, L., & Cottrell, L. (2012). Suicide incidence and risk factors in an active duty US military population. *American Journal of Public Health, 102*(S1), S138–S146. doi:10.2105/ajph.2011.300484

Kessler, R. C., Chiu, W. T., Demler, O., & Walters, E. E. (2005). Prevalence, severity, and comorbidity of 12-month DSM-IV disorders in the National Comorbidity Survey Replication. *Archives of General Psychiatry, 62*(6), 617–627. doi:10.1001/archpsyc.62.6.617

Mansfield, A. J., Kaufman, J. S., Engel, C. C., & Gaynes, B. N. (2011). Deployment and mental health diagnoses among children of US Army personnel. *Archives of Pediatric and Adolescent Medicine, 165*(11), 999–1005. doi:10.1001/archpediatrics.2011.123

Warner, C. H., Appenzeller, G. N., Grieger, T., Belenkiy, S., Breitbach, J., Parker, J.,...Hoge, C. (2011). Importance of anonymity to encourage honest reporting in mental health screening after combat deployment. *Archives of General Psychiatry, 68*(10), 1065–1071. doi:10.1001/archgenpsychiatry.2011.112

Williamson, V., & Mulhall, E. (2009). *Invisible wounds: Psychological and neurological injuries confront a new generation of veterans.* Retrieved from http://iava.org/files/IAVA_invisible_wounds_0.pdf

39 CHALLENGES AND THREATS OF COMBAT DEPLOYMENT

Heidi S. Kraft

The recent long wars, which were fought in Iraq and continue to rage in Afghanistan, have changed the landscape for many who choose to serve their country today. Military psychologists are no exception. After over a decade at war, psychologists in uniform understand the demands placed on them during their service, and how they differ from those who preceded them.

Concerns about the emotional readiness and psychological resilience of today's combat troops continue to make headlines and demand the attention of military leaders. The treatment of the mental health of combat operating forces remains a top priority for military medicine leaders. Uniformed psychologists enter service during this time with the certainty that they will deploy to a combat zone as soon as they are licensed.

There has been some recent attention in the literature about the nature of the high-risk work that will necessarily define these psychologists' careers, the ethical challenges they might face in these situations, and the need for specific supervision to prepare them for these deployments (Johnson, 2008; Johnson & Kennedy, 2010; Kraft, 2011; Moore & Reger, 2006). Others have written recent personal accounts of the unique experiences faced by mental health, medical, and religious personnel who serve alongside combat troops on the battlefield, with descriptions of some of the potential risks to their lives, safety, and emotional health (e.g., Kraft, 2007).

Drawing from the limited literature about psychologists in combat during recent conflicts, as well as from many personal accounts relayed through correspondence and other forms of storytelling-based communication, challenges facing combat psychologists are becoming better defined. Of course, some trials include those experiences faced by nearly everyone on wartime deployments, including extended family separation, exposure to harsh elements, austere living conditions, chronic sleep deprivation, and possible leadership/command culture issues (Reger & Moore, 2009).

In addition to deployment-related general and chronic stressors, further risks of concern facing combat psychologists might be conceptualized in four categories: (1) threat to physical safety and well-being; (2) exposure to medical trauma, and therefore seriously injured or deceased service members and/or civilians; (3) vicarious trauma through the treatment they provide for acute combat stress injury and grief; and (4) unique and often surprising ethical situations in which the answer is not obvious, for even the most experienced clinician.

PEOPLE ARE ACTUALLY TRYING TO KILL YOU

In his acclaimed work *On Killing*, Lieutenant Colonel David Grossman wrote, "Killing is what war is all about, and killing in combat, by its very nature, causes deep wounds of pain and guilt." (Grossman, 1995, p. 92). Most psychologists depart on wartime deployments with the understanding that they will be asked to provide treatment for those invisible wounds in their warrior patients. These same psychologists likely also deploy with genuine hope—despite the knowledge that killing is a part of war—they will never need to use those side arms they are issued before they leave, or put into practice that convoy egress training they had to repeat over and over. In many cases, however, the truth about current combat experiences is quickly revealed. Because modern conflicts and their counterinsurgency missions frequently do not have a defined frontline, medical, religious, and mental health providers—including those assigned to ostensibly "safe" combat hospitals—can find themselves in very hostile conditions. Even without being embedded with line units, these individuals can experience many of the same wartime conditions as those lived by their patients. They can receive indirect fire on forward operating bases, experience hostile fire from small arms and various other weapons while passengers on convoys and in aircraft, or live through explosions from improvised explosive devices—among other combat-related exposure.

Psychologists might be faced with their own mortality for the first time in their lives. They might experience stark fear of injury or capture. And they might come face to face with the grief and guilt that accompanies the loss of comrades in combat. These experiences are described as frightening, exciting, and surreal—but most psychologists' personal accounts agree that combat exposure provides perspective on their patients' stories and traumas in a way they never imagined before their deployments (e.g., Kraft, 2007). Above all, contact with the horror of combat may be inescapable, and in order to provide quality mental health care in a war zone, psychologists will need a mechanism to survive its effects on their own emotional well-being.

THERE WAS A REASON YOU DECIDED NOT TO BE A SURGEON

Different aspects of helping professions appeal to different people. Although there are some psychologists who train in a surgical or trauma environment, others potentially select their profession partly because their workplace of choice is other than an operating room. Mental health professionals are not typically accustomed to spending quality time with patients in trauma bays, intensive care units, or surgery recovery suites. And yet, because of the very nature of the forward-deployed field combat hospital, some find themselves washing blood off their own boots—along with nurses, surgeons, and corpsmen.

During a mass casualty in combat, everyone is expected to assist in any way possible. This can include carrying litters, holding IV bags, assessing mental status of grievously wounded service members, and even helping in the OR if necessary. Some combat psychologists may also be called to perform roles they never imagined—providing support for dying patients, sitting bedside with injured service members before or after surgery, or providing notification of death or injury to friends, leaders, or comrades.

The type and severity of injuries caused in combat are often shocking, even for those with experience in trauma. The extent of patients' pain is difficult to tolerate as a bystander, especially one without a scalpel. Graphic memories of those severely wounded patients may be some of the same intrusive images that invade thoughts even years later. And yet, despite the potentially horrible sights of combat trauma triage bays or operating suites, psychologists can discover rare opportunities to have tremendous impact in others' lives and apply specialized skills and training—to provide injured or dying patients, and their comrades, empathic and sensitive care at critical moments. And by utilizing their unique capabilities in these chaotic situations, these psychologists may also find a very deep and personal connection to, and purpose in, their work (Kraft, 2007).

EVEN IF YOU DON'T KICK DOWN DOORS, SOMETIMES IT FEELS LIKE YOU HAVE

Deployed mental health providers are asked to assess and treat a variety of presenting symptoms in their warrior patients. The diagnoses made on combat deployments might include an array of Axis I and Axis II conditions, as well as a host of V codes. However, the symptoms most combat psychologists expect to treat, and will likely see with frequency, are those associated with combat stress injuries, acute stress disorder, and posttraumatic stress disorder.

These service members, who are brave and insightful enough to seek immediate assistance in dealing with a traumatic experience, will present with stories to tell. Those narratives can be frankly dreadful to hear; patients whose thoughts and dreams are haunted by wartime events have often lived through remarkable horror, terror, grief, and inner conflict.

What is the effect on the deployed psychologist who listens to such stories day after day? Compassion fatigue is a concept that has been well defined in the caregiver literature. For years, nurses, counselors, and family caregivers have been warned of this phenomenon, and encouraged to practice self-care to buffer themselves from the detrimental effects of the cost of empathy. In recent years, additional concepts have emerged, potentially even more

critical in understanding the effect of treating trauma on the healer. Secondary trauma, vicarious trauma, and shared trauma are concepts now commonly discussed in the treatment literature (e.g., Bride & Figley, 2009). The idea is that clinicians actually absorb the memories of their patients through their empathy and care, and that they are at risk for developing similar symptoms to those they are treating.

When clinicians learn evidence-based treatment protocols for PTSD, they are educated on the importance of preserving distance between their patients' traumatic memories and memories of their own. Combat psychologists quickly learn that this can be quite difficult to achieve while in the same environment with their patients (and even upon returning home). In fact, some patients' narratives relate such familiar or triggering situations that providers have described blurred capacity to differentiate—in nightmares and intrusive thoughts—that through which they have lived, and that which is voiced by their patients (e.g., Kraft, 2007).

The chronic, exhausting nature of the work of providing mental health treatment on deployment, especially when the provider is treating acute trauma, can take its toll. Without the right safeguards in place (which are addressed in the recommendations to follow), psychologists are at risk of not only decreased effectiveness, but potential personal suffering, as a result of their important work.

SOMETIMES, THE ETHICAL CHOICE IS ELUSIVE AT BEST

For generations, war has been described using words like fog, chaos, and insanity. Combat veterans know the reason why. Some new military psychologists might be tempted to believe it will always be simple, while on a combat deployment, to follow the ethics standards that have always helped them make good decisions in their past clinical practices. The truth is, the chaotic situations of a wartime environment can lead to confusing ethical challenges in mental health care, and even test the psychologist's inner compass—the same one that has typically pointed to what is right.

Johnson (2008) described some of the many ethical dilemmas that face military mental health providers in a variety of operational settings. These include mixed agency and/or mission predicaments, questions of loyalty on the part of the provider, unavoidable personal familiarity with patients, and issues of competence, among many others (Johnson, 2008). The deployed psychologist will certainly face many of these quandaries, as well as some of the unique situations that could only surface in a situation as chaotic and morally confusing as combat.

For instance, military psychologists may be faced with great risk to their personal safety as a result of treating unstable patients who are carrying loaded weapons (Kraft, 2011). They might find themselves needing to explain their recommendations for MEDEVAC to combat commanders, who only know they will be left short-handed for an important upcoming mission. They often develop very close personal relationships with those in their unit, and then might be forced to deal with the intricacies of that necessary shifting of roles—from comrades to mental health providers—when there are no other psychologists to provide needed care for their friends. They may see firsthand the effect of sleep deprivation, exposure, and exceedingly high operational tempo on the medical personnel with whom they work and live—and in their roles of command consultants, may need to point out impending danger for patients in order to affect change. They could come face-to-face with the stark realization that their own vicarious trauma, compassion fatigue, grief, guilt, and exhaustion have severely compromised their abilities to care for their patients and themselves. These are only a few of the unusual, dynamic, and always thought-provoking ethical challenges that might face a combat psychologist (Kraft, 2007).

The military psychologist might be thrust into unfamiliar and perplexing scenarios each day of his or her wartime deployment. A solid background based on the ethical principles that guide our work will be an important start to navigating these moments. And certainly, as with any ethical conundrum, it is vital for the psychologist to know when he or she needs to seek the council of a supervisor, trusted mentor, or peer. Above all, it is important not to try to solve every puzzle alone.

As long wars march on, there is a drastic need for other military psychologists in these unique roles to write about and share their experiences, so that others can continue to learn from them. With each new ethical situation described and explored, the potential for innovative solutions is expanded, allowing us all the opportunity to make better psychologists of ourselves and our students, on the battlefield and at home.

RECOMMENDATIONS FOR MANAGING RISKS AND CHALLENGES IN COMBAT PSYCHOLOGY

Although military psychologists will not be able to avoid some of the risks they will face on combat deployments, there are a number of steps they can take to decrease the possibility that those risks will lead to either short- or long-term dysfunction, professionally or personally. Each of the following recommendations is suggested based on available evidence and proven training protocols, as well as personal anecdotes from those who have survived and thrived in combat mental health roles.

1. **Know what to expect, and know what is expected from you**. Psychologists deployed with combat forces will face risks in many situations. Some may involve very real physical danger, including threat to life itself—providing shocking moments of impact trauma. The preparation of psychologists for deployment in these unfamiliar conditions is paramount. Knowledge and training can act as buffers for frightening, overwhelming scenarios. If at all possible, before deployment, psychologists should know what types of risks might face them there, as well as what actions will be required from them to be part of the solution—for their people, their patients, and themselves. Above all, in any combat situation, psychologists should be quick to trust and follow the combat troops with whom they serve.

As medical personnel have always sworn to care for their marines, sailors, soldiers, and airmen during times of war, they must also realize that feeling is mutual.

2. **Search for meaningful moments as a psychologist in a trauma hospital**. Many military psychologists during wartime will experience significant exposure to combat casualties. Even those with physical trauma experience may find the injuries in a combat hospital triage unit quite shocking. If it is not possible to minimize exposure to the sights and sounds of grievously injured service members, psychologists might be able to reduce the personal effect of these experiences by seeking meaning in their roles within the hospital. For instance, learning protocol and procedure from the trauma and surgery staff might allow psychologists to better offer their services at critical times in patients' hospital experiences, and to feel part of the team. Knowing the flow and roles of personnel during mass casualty situations can provide psychologists with the chance to engage with patients at unique and meaningful times such as before surgery or before death, as well as to keep a watchful eye on the hospital staff. Above all, it is essential to remember that psychologists have unique opportunities to provide significant connection and comfort amid chaos—for frightened and injured patients, as well as for their exhausted healers. Those moments can frame the deployment experience in a positive and life-changing way.

3. **Seek supervision, or consultation, with at least one trusted colleague—and promise to care for each other**. Combat deployments can be isolating for mental health care providers, for whom the exposure to vicarious trauma is often high and ethical conundrums often confusing. Even if a psychologist is the only mental health asset in the area, it is essential to seek out trusted consultation with another medical provider or chaplain. This professional relationship should be one that fosters open and honest discussions about patient care, ethical dilemmas, and personal/professional concerns about competence, loyalty, and multiple relationships. In addition, the

psychologist should ensure that he or she has at least one trusted comrade who is certain to help make sure self-care practices are in place. All medical providers in combat environments need to make time for sleep, exercise, and rejuvenating moments—of connection with home, pleasure, and self-expression. Yet medical people are often the worst at caring for themselves. Thus, the need for a trusted comrade—sometimes the same person who fulfills the role of consultant or supervisor, sometimes someone else—to help ensure the psychologist places his or her own well-being and care as a priority, is paramount.

References

Bride, B. E., & Figley, C. R. (2009). Secondary trauma and military veteran caregivers. *Smith College Studies in Social Work, 79*(3), 314–329.

Grossman, D. A. (1995). *On killing: The psychological cost of learning to kill in war and society*. New York, NY: Little, Brown.

Johnson, W. B. (2008). Top ethical challenges for military clinical psychologists. *Military Psychology, 20*, 49–62.

Johnson, W. B., & Kennedy, C. H. (2010). Preparing psychologists for high-risk jobs: Key ethical considerations for military clinical supervisors. *Professional Psychology: Research and Practice, 41*(4), 298–304.

Kraft, H. S. (2007). *Rule number two: Lessons I learned in a combat hospital*. New York, NY: Little, Brown.

Kraft, H. S. (2011). Psychotic, homicidal, and armed: The delicate balance between personal safety and effectiveness in a combat environment. In W. B. Johnson & G. P. Koocher (Eds.), *Ethical conundrums, quandaries, and predicaments in mental health practice: A casebook from the files of experts*. New York, NY: Oxford University Press.

Moore, B. A., & Reger, G. M. (2006). Clinician to frontline soldier: A look at the roles and challenges of Army clinical psychologists in Iraq. *Journal of Clinical Psychology, 62*(3), 395–403.

Reger, G. M., & Moore, B. A. (2009). Challenges and threats of deployment. In S. Freeman, B. A. Moore, & A. Freeman (Eds.), *Living and surviving in harm's way: A psychological treatment handbook for pre- and post-deployment of military personnel* (pp. 51–65). New York, NY: Routledge.

40 POSTDEPLOYMENT ADJUSTMENT

David S. Riggs

Returning from a combat deployment, though often a happy and joyous occasion, can also be stressful and challenging for military personnel and their families. Change and transition can be difficult regardless of the situation; however, there are certain unique challenges inherent in the transition from combat to garrison or civilian life. Returning from combat and reintegrating with family, friends, and community requires service members to negotiate significant change and to adjust to the differences between their existence in combat settings and at home. The changes that service members must negotiate can be conceptualized in several overlapping ways. This transition is clearly a time when service members may have to come to terms with existential issues raised by their experiences of war. The challenge of reintegration can also be understood as learning that the skills and behaviors that worked effectively in combat may not be functional or adaptive in home settings. Also, some service members will likely experience emotional, cognitive, and behavioral responses to the stress of a combat deployment that will complicate the reintegration process.

EXISTENTIAL CHALLENGES

Chaplain (LTC) John Morris (2008) of the Minnesota National Guard identifies five existential "challenges" for service members transitioning from a wartime deployment to home:

- overcoming the sense of alienation and reconnecting with family, friends, and community
- moving from simplicity to complexity
- replacing war with another form of "high"
- moving beyond war to find meaning in life
- coming to peace with what one has seen, experienced, or done while at war

Difficulties in resolving any of these issues can lead to difficulties for returning service members.

Overcoming Alienation

Service members returning from combat often feel as though they do not fit in with families and communities to which they are returning. This feeling of alienation is based, in part, on the fact that service members, their families, and others have changed during the deployment period. Service members have had life-changing experiences, some positive and others negative. Family change is also expected during a deployment. Children grow and develop, family routines and responsibilities adjust in response to the missing service member, and family dynamics and relationships change. Overcoming this sense of alienation requires time and effort. Service members and those around them must accept changes that have occurred and negotiate additional changes as reintegration proceeds. It is important that service members and families recognize that things will not be exactly the same after the deployment and that they work to find a "new normal."

Simplicity to Complexity

Service members tend to be mission-focused, and this is reinforced during deployments. This focus on mission means that the deployed environment, while challenging, can also be relatively straightforward. With clear goals and a limited number of choices or options available in many situations, service members often report that "life was pretty simple." In contrast, the life that they return to is filled with numerous competing goals and requirements, a myriad of choices and a sense of great complexity. This transition can prove challenging for some service members, as the complexity and choices can seem overwhelming. Furthermore, service members may find that they have more difficulty making decisions and choices. Choices in the deployed setting were often very important, perhaps literally "life-or-death." For some, it is difficult to let go of this and recognize that some decisions are relatively unimportant. Other service members find it difficult to understand why people seem to place so much importance on decisions that they see as falling so far short of the decisions they had to make while deployed.

A New "High"

The experience of combat is incredibly intense and includes frequent periods of emotional and physical arousal. Service members may describe experiencing a rush or sense of high resulting from these experiences that is rarely matched by the day-to-day routine that they find when they return from war. Attempts to replace this sense of intense purpose and physical and emotional arousal can create significant challenges for service members returning

from combat. Some service members will seek out opportunities to try to replicate those feelings that might place them at risk for injury. Others may seek the sense of high by turning to drugs or alcohol, which can create additional problems for individuals and families.

Finding Meaning

The intensity and focus on mission associated with combat deployments provides service members with a great sense of purpose and meaning. It is often difficult to find that same sense of meaning in the day-to-day activities that they must deal with upon returning from deployment. These issues may be particularly salient for individuals who leave military service, either personnel who are permanently separating from the force or members of the Guard and Reserves who demobilize and return to their civilian lives. However, even if service members remain in the military they may find that their duties away from combat fail to provide the same fulfillment as they found during deployment. For some, this search for purpose will lead them back into combat; others will find meaning or purpose with a new mission, often one of service to others; but some service members struggle with this loss of purpose for a significant time.

Coming to Peace

The realities of combat include exposure to events that will challenge one's sense of humanity. Service members may be called on to engage in behaviors (e.g., killing someone) and exposed to events or the aftermath of events (e.g., civilian casualties of bomb attacks) that they would not experience in a noncombat setting. These events can lead some service members to question their sense of self, their sense of humanity, or their religious/spiritual beliefs. These events and the questions they raise may also create significant problems reconnecting with family and friends as these service members may worry that other people will not be able to accept them after what they have done, experienced, or witnessed.

SKILLS AND BEHAVIORS THAT FUNCTION WHILE DEPLOYED

Surviving a combat deployment requires the development and honing of skills and behaviors that work while deployed but that may create difficulties when service members return home. The need to "turn off" these skills upon returning is complicated by the requirement to deploy service members repeatedly into combat. Service members often express a concern that if they turn off these skills while at home they will be unable to turn them back on for the next deployment.

During a combat deployment, service members are focused on identifying potential dangers and maintaining safety. Service members learn to rapidly identify situations and individuals that may represent a risk. Further, if one cannot reasonably assure safety then it is prudent to assume that a risk exists until proven otherwise. This approach leads to a number of potential problems as the service member returns from deployment. Situations and events that were routine prior to the deployment, such as going to the supermarket or driving in traffic, may now be seen as risky. Such situations may illicit attempts to exert control to maintain safety or be avoided completely. Family members and friends who do not perceive these situations as dangerous may find the service member's behaviors difficult to tolerate. Also, the tendency to see individuals as potential threats leads to problems with trust and intimacy that can create significant reintegration difficulties.

As mentioned above, combat deployments reinforce service members' focus on mission and rapid problem-solving. In short, activities that distract from accomplishing the assigned mission are perceived as potentially threatening. As service members reintegrate into their families, competing goals and responsibilities may become frustrating and cause conflicts with others in the family. Also, problems in the deployed setting often require rapid decision-making and action with limited opportunity for discussion or group processing. During reintegration, this approach to problem solving can create difficulties when it conflicts with more cooperative or egalitarian approaches that have typically been used by the family. Conflicts may also arise when family members or colleagues do not respond to the service member's "orders" once decisions are made.

Related to the focus on safety, service members often perceive a need for control over situations. Upon returning from deployment, this can manifest in demands for family members to maintain order around the house or discomfort in situations where the service member has limited control. Control also pertains to information. While deployed, maintaining informational security may make the difference between the success and failure of a mission. At home, a pattern of disclosing information on an "as needed" basis may interfere with communication and be perceived as "keeping secrets" by family and friends. Issues of control may also be seen in service members' attempts to maintain emotional control that may help to maintain mission readiness while deployed. Limited emotional expression when they return home may make them appear cold or disconnected from people in their lives.

COMBAT AND OPERATIONAL STRESS

The stress of combat deployment, both the trauma of combat and the cumulative stress of extended operations, can result in physical, behavioral, and emotional changes in service members (Department of Defense, 2012). These changes are normal, expected, and are typically transient, but may persist over time. Many combat stress reactions may actually serve a useful function during the deployment (e.g., heightened vigilance), but when they persist they can create difficulties when the service member returns from deployment. Further, many of the normative responses to combat stress may complicate the resolution of issues described above. Elevated emotional reactions, for example, may create additional difficulties for service members who are already alert to potential dangers in their environment. Similarly, withdrawal from social or recreational activities can exacerbate the service member's sense of alienation from family and friends.

Combat stress responses have been broadly categorized as "reactions," "injuries," or "illnesses." Combat stress illnesses largely constitute diagnosable disorders, such as PTSD, depression, anxiety disorders, and substance abuse/dependence, that will not be discussed in this chapter. Combat stress reactions and injuries, while less severe than the diagnosable illnesses, may persist and cause difficulties on return from a combat deployment. For the most part, combat stress reactions and injuries constitute similar behavioral and emotional responses that differ primarily in severity and persistence.

Combat Stress Reactions

Combat stress reactions tend to be rather mild and are expected to remit relatively quickly. Often, it is expected that service members will be largely recovered from combat stress reactions prior to returning from deployment. However, it is quite possible that some reactions may continue as the service member returns or that the return from deployment may exacerbate reactions that have been "under control" during the deployment. It is also possible that reactions seen as minimal problems during the deployment (e.g., difficulty relaxing) may be identified as problematic as the service member negotiates the reintegration process.

Combat stress reactions include intense emotions such as fear, anger, sadness, and guilt that are appropriate and expected reactions to situations that arise in combat. If these emotional reactions persist they can create difficulties for the returning service members as they work to reconnect with family and friends. Anger and irritability can cause conflict between the service member and the people around them. Persistent feelings of sadness or guilt can lead service members to withdraw from family and other social relationships. If feelings of fear persist in the form of anxiety and worry, service members may struggle to readjust and fully participate in family and community activities. Problems may be exacerbated when these emotions arise suddenly

and in unexpected situations leaving all to wonder why the service member is upset.

Related to possible concerns about safety and security described above, service members experiencing a combat stress reaction will often experience a sense of increased vigilance and a greater tendency to startle. These responses can be conceptualized as part of the effort to identify potential threats early in order to counter them. Although these reactions to combat and operational stress may be seen as adaptive in the deployed setting, they can create difficulties once the service member returns home. Increased vigilance and elevated startle reactions may cause a service member to find certain activities such as attending a party or a child's soccer game less enjoyable or even unpleasant. This may lead the service member to avoid such situations or to a sense of disappointment or even conflict with family and friends.

Combat stress reactions such as difficulties in concentration and memory can create problems for service members as they work to reintegrate. Such difficulties are common in the immediate aftermath of combat trauma, but may persist through the return home. One can imagine that as service members work to negotiate the transition from the relative simplicity of the deployed setting to the more complicated home setting with its competing priorities and demands, difficulties in memory or concentration may complicate matters.

Service members experiencing combat stress reactions may experience nightmares and difficulties sleeping. Sleep problems and nightmares may be indicators of more severe problems, such as PTSD or depression, that require a thorough assessment. In the absence of more severe problems, sleep difficulties upon returning from a deployment are expected to be transient. Persistent problems may be related to other reintegration issues; for example, some service members may have problems sleeping because they feel insecure without someone standing guard during the night. Even mild disruptions in sleep may contribute to difficulties following a deployment. Loss of sleep can exacerbate memory and concentration problems, leading to greater adjustment difficulties at work and at home. Additionally, nightmares and sleep difficulties may disrupt a spouse's sleep, leading to possible relationship difficulties.

As mentioned above, service members may experience adjustment difficulties in the areas of social and recreational activities. This may result from the discomfort that they experience in these situations or from a sense that the activity is less important or meaningful than their combat duties. It is also possible that they no longer gain the same sense of pleasure from these activities as they did prior to deployment. As with sleep difficulties, the loss of interest or enjoyment in activities may be indicative of PTSD or depression, but it is also commonly seen among service members experiencing a combat stress reaction that falls short of these diagnosable disorders. It appears that these sorts of issues are more likely to be identified as significant problems following the deployment (when they are identified by family or friends) than during the deployment when opportunities for recreational and social activities were more limited. Regardless of the factors leading to this loss of interest or enjoyment, these problems may lead to significant difficulties in the reintegration process. Refusal to attend activities that are valued by other members of the family such as children's school or sporting events and social get-togethers can be emotionally painful to others and can further contribute to the sense of alienation and disconnect among family members.

Combat Stress Injuries

Combat stress injuries are generally seen as more severe and persistent manifestations of the combat stress reactions described above. The presence of one or more of these indicators of combat stress injury at the point that the service member is returning from deployment probably warrants additional assessment to identify or rule out more significant problems such as PTSD, depression, or another diagnosable disorder. Assessments should also attend to the risk that service members with combat stress injuries may experience suicidal or homicidal thoughts.

The emotions of fear and anger experienced as part of a combat stress reaction may manifest as anxiety/panic attacks or angry/violent outbursts in service members with a combat stress injury. These more severe manifestations can significantly complicate adjustment following the deployment. Anxiety or panic attacks can lead to avoidance of situations, making it significantly more difficult for the service member to fully reintegrate with his or her surroundings. Angry outbursts, particularly when accompanied by violence or threats of violence can directly impact family members, friends and coworkers as well as the service member. Other intense emotional reactions, such as uncontrolled crying, may also create difficulties for the service member.

Combat stress injuries also include more severe manifestations of the sleep disturbance and nightmares that may be present in combat stress reactions. In the case of sleep disturbances, service members with a combat stress injury may report significant loss of sleep due to difficulties falling asleep or staying asleep. Nightmares, rather than being infrequent and causing relatively little disruption to sleep, may cause service members to awaken and have difficulty returning to sleep. As with the more intense emotional reactions, as sleep difficulties and nightmares become more severe, they can lead to greater difficulties with reintegration and adjustment following deployment. Insufficient sleep is associated with significant psychological problems, but also can exacerbate the memory and concentration problems described above.

Combat stress injuries may also manifest in ways that do not have parallel presentations in combat stress reactions. Most significantly, service members struggling with combat stress injuries may experience suicidal or homicidal thoughts. These alone represent serious adjustment issues that must be addressed. However, it is also important to recognize how some of the other issues discussed above can interact with and complicate attempts to address issues of suicidal and homicidal thoughts. For example, a sense of alienation from others and withdrawal from social interactions may exacerbate thoughts of suicide. Similarly, difficulties and conflict in familial and other interpersonal relationships may increase homicidal thoughts. Further complicating efforts to address these and other adjustment problems, difficulties with trust and safety concerns may interfere with service members seeking help for the adjustment and reintegration problems.

Managing the transition home from a combat deployment can be challenging for many service members and their families. Reintegration issues including existential issues, behaviors learned in combat, and symptoms developed during deployment can create difficulties. Counselors and therapists have a potentially important role to play in supporting families through the reintegration process,

but it is important to recognize that conflict and difficulties during this phase of the deployment cycle are to be expected and are not necessarily pathological.

References

Department of Defense. (2012). *Military deployment guide: Preparing you and your family for the road ahead.* Available at http://www.militaryhomefront.dod.mil/12038/Project%20Documents/MilitaryHOMEFRONT/Troops%20and%20Families/Deployment%20Connections/Pre-Deployment%20Guide.pdf

Morris, J. (October 2008). *Minnesota Army National Guard Reintegration Initiative.* Seminar presented at the Center for Deployment Psychology, Bethesda, MD.

41 COMBAT AND OPERATIONAL STRESS CONTROL

Kristin N. Williams-Washington and Jared A. Jackson

Military environments can produce stressors that are not normally found in other environments. In combat environments, stressors can be extreme in both intensity and duration. Unlike the vast majority of people who experience traumatic exposures, service members (SM) are not often afforded the opportunity to remove themselves from a stressful event. Combat environments also limit the resources available to adequately and appropriately cope with stressors. The term "combat and operational stress" (COS) has been coined to describe the physiological, emotional, and psychological stressors associated with the demands and dangers of the combat environment.

It is important to note that COS applies to all combat-related activities and operations

and is not restricted to those directly involved in combat. COS can be experienced by all military personnel, spanning all military branches and all types of military operations. COS may also occur in training environments or other operations that produce combat-like conditions or responses, like a simulation, for example. Specifically, COS may be found in:

• training,
• all phases of deployment,
• peacekeeping,
• humanitarian missions,
• stability and reconstruction,
• government support missions, and
• those missions that may include weapons of mass destruction (WMD) and/or chemical, biological, radiological, nuclear, and

explosive (CBRNE) weapons (Department of Defense [DoD], 1999).

As with all stressors, COS can cause impairments in functioning and increase the risk of experiencing debilitating physiological and psychological symptoms. The term "combat operational stress reaction" (COSR) is used to describe the physical, psychological, and emotional symptoms produced by exposure to COS. While COSR may share some of the symptoms found in clinical diagnoses such as posttraumatic stress disorder or major depression, COSR lacks the criteria needed to satisfy a clinical diagnosis and is by nature considered subclinical and generally transient. With the appropriate attention and time, individuals with COSR have a high recovery rate and more often than not will exhibit subclinical symptoms (DoD, 1999). Without appropriate attention and adequate time to recover, COSRs can be exacerbated and lead to significant impairment in functioning. When an SM's symptoms prevent them from performing their duties effectively, or at all, it is deemed a COS casualty, inferring that the SM has been compromised to the point of being unfit to return to duty.

COSC MISSION

In an effort to treat and reduce the effects of COSR and prevent COS casualties, the Department of Defense (DoD) developed the Combat Operational Stress Control (COSC) program. According to DoD regulations, the Combat Stress Control (CSC) program "Ensures appropriate prevention and management of COSR casualties to preserve mission effectiveness and warfighting, and to minimize the short- and long-term adverse effects of combat on the physical, psychological, intellectual, and social health of SMs" (DoD, 1999). In essence, the goal of COSC is twofold: initially, prevent COS casualties; and, if prevention fails, treat COSR and COS casualties. In both the prevention and treatment aspects of COSC, the goal is to improve functioning and resiliency in SMs. There are personnel

specifically assigned to address COS and CSC. COSC personnel often include, but are not limited to: psychologists, psychiatrists, social workers, chaplains and ministry teams, and occupational therapists.

The responsibility of COSC falls on military commanders at all levels. Commanders are ultimately responsible for mission effectiveness and providing SMs with the tools and training needed to effectively complete missions. Thus, commanders are ultimately responsible for helping SMs cope with COS and for reducing COSR. However, a commander's primary concern is often the timely completion of mission objectives. It is the duty of COSC personnel to work with commanders to ensure SMs are getting their behavioral health needs met and are effectively coping with COS while also achieving mission objectives.

Commanders may have personnel on staff specifically assigned to COSC. If a commander does not have sufficient personnel to address the COSC needs of his/her SMs, personnel may be temporarily assigned, or commanders may receive assistance from a COSC detachment team. COSC detachment teams are independent behavioral health units responsible for aiding military units who are not currently equipped with COSC personnel or units with insufficient COSC personnel.

COSC programs are under the purview of DoD Directive 6490.5 to provide treatment to SMs throughout each branch of the military in an effort to "enhance readiness, contribute to combat effectiveness, enhance the physical and mental health of military personnel, and to prevent or minimize adverse effects of Combat Stress Reactions (CSRs)" (DoD, 1999).

COSC programs are designed to bring treatment to the SM, not to bring the SM to treatment. The DoD has directed that all SMs experiencing CSRs shall be treated or managed "within the unit or as close to the operational front or near the comprehensive unit needs assessments service member's unit as possible" (DoD, 1999). The goal of treatment is to keep the SM in the fight, to keep the service member with his unit and within his or her Area of Operation, as this has been found to increase resiliency. Evacuating and/or removing SMs from their unit has been found to disrupt unit cohesion and social support networks. In addition, DoD has found the "evacuation and separation of CSR casualties from his or her military unit greatly increases the risk of subsequent, serious, long-term social and psychiatric complications, and is, therefore, indicated only when absolutely mission essential" (DoD, 1999).

In working with SMs and military units, COSC treatment usually consists of the following:

• enhancing adaptive stress reactions,
• normalizing appropriate CSRs,
• preventing maladaptive stress reactions,
• assisting SMs with controlling COSRs, and
• assisting SMs with behavioral disorders

Treatment occurs in a variety of settings (group, individual, classroom, chapel, and informal/out of office) and may vary with the type of personnel providing the treatment (psychologist, physiatrist, social worker, chaplain, etc.). SMs can often report to a behavioral health clinic for formal treatment. However, COSC personnel try to bring treatment to the SM and often interact with SMs in their place of work or area of operation. This reduces interference with mission operations, gives SMs greater access to treatment and counsel, and helps COSC personnel understand the stressors the SMs are facing. As COSC personnel interact with the SMs in their places of work, they begin to build relationships of trust, know the COSC personnel understand their situation and stressors, and often become more willing to seek additional help when needed.

SMs needing more treatment can follow-up with COSC personnel in a more formal setting. This can be accomplished in a private room close to the SMs' work area or at a designated COSC clinic. Whatever the setting or type of care being provided, the goal of COSC is the same, to maintain/enhance SM functioning and enhance combat effectiveness. COSC casualties and maladaptive CSRs reduce combat effectiveness and compromise mission objectives. Thus, COSC personnel place a high emphasis on preventing maladaptive CSRs and COS casualties.

PREVENTION

The key to providing effective treatment for COS is threefold: establishing a positive behavioral health presence within the unit; utilizing comprehensive unit needs assessments; and appropriate traumatic event management. In order for a COSC provider to establish a positive presence, they must do just that, be present. Engaging with the assigned unit prior to a traumatic event serves to assist in cohesion. Unit cohesion and morale, combined, have been found to be the best predictor of resiliency within a unit and the behavioral health provider should aspire to be a part of the unit with command support. Command support can be secured through command consultation and showing ways in which a COSC provider can be a force multiplier by keeping SMs in the fight through provision of preventative care.

Appropriate and effective coping skills are needed in a deployed environment. Some SMs do not know how to cope with the stressors they are facing and must be taught new strategies. Coping strategies often used in nondeployed settings may not be available in deployed settings. SMs have limited contact with family and friends, have limited entertainment options, are often confined to a small area of operation, and have a small area for personal space when they are provided with "downtime." SMs who coped with stress by spending time with family, hiking, riding bikes, going for a drive, going to a movie, and so forth, must find new methods of coping. COSC personnel increase SM resiliency by teaching them available and effective coping strategies (i.e., building friendships within their unit, long-distance communication with family and friends, relaxation techniques, journaling, physical exercise, etc.). Resiliency is also increased by providing a different perspective, teaching SMs to use each other for support, and normalizing reactions.

Often, several SMs from a unit will display difficulties coping with stressors. A Unit needs assessment (UNA) is a tool that enables the COSC provider to better understand the specific needs of the unit for which he/she is responsible for providing behavioral health services. UNAs are typically conducted via anonymous survey to screen for those individuals currently having difficulty, those with the propensity for difficulty in relationship status due to separation, and those with a predisposition for anxiety and/or depression. This assessment can also identify patterns of difficulties SMs have with workloads, fatigue, interactions with Commanders, confidence in equipment, confidence in Command, and unit cohesion. A UNA provides Commanders and SMs with information about areas that need improvement within the unit. It is an excellent means to inform Command of unit need and to achieve "buy-in" from unit Commanders who may have thought that his/her unit had no concerns. By providing the Commander with aggregate data, possible hidden concerns are brought to light and services can be provided to the entire unit in a preventative manner.

Traumatic event management (TEM) is the somewhat systematic response to potentially traumatizing events. Per Field Manual 4-02.51 ([FM], 2006), the following events can likely be classified as potentially traumatic events for individual SMs and their units:

Heavy or continuous combat operations, death of unit members due to enemy or friendly fire, accidents, serious injury, suicide/homicide, environmental devastation/human suffering, significant homefront issues, and operations resulting in the death of civilians or combatants. (Chapter 6)

Events are considered potentially traumatic if intense feelings of "terror, horror, helplessness, and/or helplessness" are experienced (FM, 2006, p. 6-1). The overall goal of TEM is to address the concern with the expectation of the individual and unit to return to action. For example, a unit returns to their base following a mission in which one soldier was killed in action. The command would alert the COSC provider, who would conduct a briefing on the occurrences and discuss possible reactions the soldiers within the unit will experience. The normalization of possible reactions mitigates the fear of a soldier, who may perceive their response as excessive, and reduces additional stress and/or anxiety by preventing maladaptive stress responses. This briefing also provides a platform for the unit to become more cohesive as they grieve together and permits soldiers to look out for changes in one another.

IDENTIFYING THOSE WHO NEED CARE

The Navy and Marine Corps make use of a Combat and Operational Stress Continuum Model as a means of visually expressing the varying levels of coping and impairment from Ready (Green Zone) to Ill (Red Zone) (Marine Corps Reference Publication [MCRP], 2010). This comprehensive model provides examples of symptoms or features individuals may possess when in each of the four, color-coded stress zones (Figure 41.1).

This model also provides information for whose responsibility it is to maintain the SM in each zone. For example, an SM who is in the Ill or Red Zone may have posttraumatic stress disorder (PTSD) with symptoms that may worsen over time. This SM's care and concern is placed on the caregiver, either a psychologist, psychiatrist, or social worker. However, an individual in the Ready or Green Zone who is optimally effective would be the responsibility of the unit leader. Leaders, in general, must be able to "identify not only the stress reactions, injuries, and illness experienced..., but also the day-to-day stressors they encounter so they can recognize occasions of high risk for stress problems" (MCRP, 2010). Leaders who are adept at mitigating the aforementioned stressors are typically able to maintain their SMs within the Green Zone. The overall goal is to maintain every SM in the Ready or Green Zone. Thus, having SMs who are well trained, prepared, and in control.

COSC requires providers to interact with and treat SMs in unorthodox environments and settings. This is because they are caring for SMs who are faced with uncommon stressors and who have limited resources for coping. COSC personnel do not wait for SMs to become symptomatic and report to their clinic for help. COSC personnel spend much of their time outside of the conventional clinic

READY (Green Zone)	REACTING (Yellow Zone)	INJURED (Orange Zone)	ILL (Red Zone)
Definition - Adaptive coping and mastery - Optimal functioning - Wellness **Features** - Well trained and prepared - Fit and focused - In control - Optimally effective - Behaving ethically - Having fun	**Definition** - Mild and transient distress or loss of optimal functioning - Always goes away - Low risk for illness **Features** - Irritable, angry - Anxious or depressed - Physically too pumped up or tired - Loss of complete self-control - Poor focus - Poor sleep - Not having fun	**Definition** - More severe and persistent distress or loss of function - Leaves a "scar" - Higher risk for illness **Causes** - Life threat - Loss - Inner conflict - Wear and tear **Features** - Panic or rage - Loss of control of body or mind - Can't sleep - Recurrent nightmares or bad memories - Persistent shame, guilt, or blame - Loss of moral values and beliefs	**Definition** - Persistent and disabling distress or loss of function - Clinical mental disorders - Unhealed stress injuries **Types** - PTSD - Depression - Anxiety - Substance abuse **Features** - Symptoms and disability persist over many weeks - Symptoms and disability get worse over time
Unit Responsibility	**Individual, Peer, Family Responsibility**		**Caregiver Responsibility**

FIGURE 41.1 Combat and Operational Stress Continuum Model (Model reproduced from MCWP 6-11C/NTTP 1-15M Combat Stress, December 2010 edition)

interacting with SMs at their places of work and trying to prevent COSR and COS casualties. Prevention is the primary concern, but appropriate care and treatment is provided if preventative efforts fail. COSC personnel work with SMs and their Commands to understand the stressors SMs are experiencing and to understand needs of individual SMs and the unit as a whole. COSC works with Commands to provide accessible care and to ensure all have the needed skills and resources available to preserve and enhance overall well-being. SMs work and train as a team to effectively achieve their mission objectives. COSC personnel also work as a team with Commands and units to keep SMs physically, psychologically, and emotionally strong.

References

Department of Defense (DoD). (1999, February 23). *Combat Stress Control (CSC) Programs*, directive 6490.5. Washington, DC: Author.

Field Manual 4-02.51 (FM). (2006, July 6). *Combat and Operational Stress Control*. Washington, DC: Department of Defense.

Marine Corps Reference Publication 6-11C and Navy Tactics, Techniques, and Procedures 1-15M (MCRP). (2010, December). *Combat and Operational Stress Control*. Washington, DC: Department of Defense.

42 TRAUMA AND POSTTRAUMATIC STRESS DISORDER

Blair E. Wisco, Brian P. Marx, and Terence M. Keane

DEFINITIONS

Posttraumatic stress disorder (PTSD) is an anxiety disorder that can develop after a traumatic event. Although the deleterious effects of war have been recognized for many years, PTSD was not officially recognized as a mental health concern by the medical community until 1980. Exposure to a traumatic event is a necessary criterion for a diagnosis of PTSD. Traumatic events are defined by the *Diagnostic and Statistical Manual* (DSM-IV-TR) as:

direct personal experience of an event that involves actual or threatened death or serious injury, or other threat to one's physical integrity; or witnessing an event that involves death, injury, or a threat to the physical integrity of another person; or learning about unexpected or violent death, serious harm, or threat of death or injury experienced by a family member or other close associate. (Criterion A1; American Psychiatric Association [APA], 2000, p. 463)

Examples of traumatic events include physical or sexual assault, exposure to a natural disaster, being involved in a serious accident, and combat exposure.

In the current version of the DSM, one must also respond to the traumatic event with intense fear, helplessness, or horror to meet diagnostic criteria for PTSD (Criterion A2). The traumatic event must be reexperienced in at least one of the following ways: recurrent intrusive memories of the trauma, recurrent distressing dreams related to the trauma, flashbacks in which one acts or feels as if the traumatic event were recurring, emotional distress in response to trauma reminders, or physiological reactivity to trauma reminders (Criterion B). The individual must demonstrate persistent avoidance and/or emotional numbing as indicated by at least three of the following symptoms: avoidance of thoughts, feelings, or conversations related to the trauma, avoidance of activities, places, or people that provoke memories of the trauma, inability to recall important aspects of the traumatic event, markedly diminished interest or participation in activities, feelings of detachment or estrangement from others, restricted range of affect, or a sense of foreshortened future (Criterion C). Finally, the individual must exhibit at least two persistent hyperarousal symptoms: sleep disturbance, irritability or angry outbursts, concentration difficulty, hypervigilance, or exaggerated startle response (Criterion D).

Additionally, the aforementioned symptoms must last for more than 1 month (Criterion E) and must cause clinically significant distress or functional impairment (Criterion F). Diagnostic specifiers clarify PTSD symptom onset and duration. If symptom duration is less than 3 months, PTSD is specified as "acute"; if longer than 3 months, it is considered "chronic." A specifier of "delayed onset" is added if symptoms developed at least 6 months after the experience of the traumatic event.

The American Psychiatric Association (APA) periodically revises the DSM to reflect the latest research. One proposal for the next edition, DSM-5, is to remove the requirement that individuals react to trauma with fear, helplessness, or horror, given the limited predictive validity of this criterion. PTSD may also be moved from the anxiety disorder category to a new diagnostic category, Trauma and Stressor-Related Disorders, to reflect the central importance of trauma exposure and to reflect the significance of emotions other than fear in the experience of PTSD. At the time of this writing, APA has not yet released DSM-5, and the fate of these and other proposed changes to the diagnosis remains uncertain.

TRAUMA AND THE MILITARY

Although trauma exposure used to be considered a relatively rare occurrence, prevalence estimates from large nationally representative samples suggest that 50–60% of the general population will be exposed to at least one potentially traumatic event in their lifetime (Kessler, Sonnega, Bromet, Hughes, & Nelson, 1995). Most people exposed to potentially traumatic events do not develop PTSD, with lifetime PTSD prevalence estimates of about 8%. Compared with the general population, military service members are at increased risk for both trauma exposure and development of PTSD. Veterans who served in Vietnam have an increased risk of lifetime trauma exposure compared with civilians and up to 30% meet criteria for lifetime PTSD (Kulka et al., 1990). Among military service members serving in more recent conflicts, Operation Enduring Freedom and Operation Iraqi Freedom, PTSD prevalence estimates range from 5 to 20% (Ramchand, Schell, Jaycox, & Tanielian, 2011).

PTSD prevalence differs by gender in the general population, with women twice as likely to develop PTSD as their male counterparts (Kessler et al., 1995). Among military personnel, however, this gender difference is attenuated or even eliminated (Vogt et al., 2011). One possible explanation for the lack of gender differences among military personnel is

that women who choose military service may be more resilient to the effects of trauma than women in the general population. Alternatively, the types of stressor exposures associated with military service may negate the effects of other variables that typically account for gender differences in PTSD prevalence. Consistent with this possibility, past research suggests that gender differences in PTSD are less pronounced for combat trauma than other kinds of trauma exposure (Vogt et al., 2011). Although it is commonly assumed that female service members are protected from the most intense forms of combat exposure due to exclusion policies, in fact women are frequently exposed to combat during the course of their military service (Vogt et al., 2011).

Unfortunately, both male and female service members are also at risk for exposure to a number of other potentially traumatic events, including all of the traumatic events that affect the general population. In recent years, there has been increased recognition of military sexual trauma (MST), defined as sexual assault or severe sexual harassment that occurs during Active Duty service or training. Female veterans are at increased risk for MST, with approximately 1% of male veterans and 20% of female veterans reporting exposure. Given the high rates of PTSD and other mental health concerns associated with MST, awareness and sensitive assessment of MST is essential for military psychologists.

ASSESSMENT

Several screening instruments, self-report questionnaires, and semistructured clinical interviews for PTSD possess excellent psychometric properties when examined in military samples. Screening tools, such as the four-item Primary Care PTSD screen (PC-PTSD), are used to identify individuals with likely PTSD diagnoses for more extensive assessment (National Center for PTSD [NCPTSD], 2012). Longer self-report measures can provide information about likely PTSD diagnoses and symptom severity. The PTSD Checklist (PCL) is a 17-item questionnaire with excellent psychometric properties for military and veteran samples (NCPTSD, 2012). The items of the PCL correspond directly to DSM-IV-TR symptoms, and the total score gives an index of symptom severity. Scores above recommended cutoffs suggest a probable PTSD diagnosis, with lower cutoffs suggested for service members of OEF/OIF than veterans of the Vietnam War. Other self-report measures of PTSD include the PK scale of the Minnesota Multiphasic Personality Inventory-2, the Mississippi Scale for Combat-Related PTSD, the Impact of Events Scale, and the Posttraumatic Diagnostic Scale (NCPTSD, 2012).

Although self-report measures have the advantage of being relatively quick to administer, they can be influenced by response biases and idiosyncratic interpretations of test questions. Semistructured clinical interviews may be less influenced by these factors and are considered the gold standard for assessing PTSD. The Clinician-Administered PTSD Scale (CAPS) is a standardized diagnostic interview, designed specifically for the assessment of PTSD, that is well validated for use with military and veteran populations (NCPTSD, 2012). The CAPS includes questions about both the frequency and intensity of symptoms, and the total score serves as an index of PTSD symptom severity. The Structured Clinical Interview for the DSM-IV (SCID) is another standardized diagnostic interview that assesses PTSD along with other Axis I disorders. The SCID can establish whether or not a PTSD diagnosis is present, but provides little information about PTSD symptom severity (NCPTSD, 2012).

A complete PTSD assessment would also include evaluation of lifetime trauma exposure, current occupational and social functioning, and comorbid psychiatric and medical complaints. Other conditions commonly comorbid with PTSD in military service members include traumatic brain injury, chronic pain, major depression, and substance use disorders.

TREATMENT

Evidence-based treatments (EBTs) for military-related PTSD include psychotherapy and pharmacotherapy options (Department of Veterans Affairs & Department of Defense [VA/DOD], 2010). Two individual psychotherapies with strong research support are prolonged exposure (PE) and cognitive processing therapy (CPT). As part of a national mandate that all veterans have access to EBTs, the Department of Veterans Affairs (VA) has disseminated PE and CPT throughout the VA health care system.

PE is similar to other exposure-based treatments that were previously developed for military veterans and shown to have efficacy for combat-related PTSD (Keane, Fairbank, Caddell, & Zimering, 1989; Keane & Kaloupek, 1982). The PE treatment protocol was initially developed and tested with civilians, particularly women with PTSD related to sexual assault (VA/DOD, 2010). PE includes psychoeducation about reactions to trauma, breathing exercises, and two types of exposure. In vivo exposure is planned exposure to places, people, or activities that the patient is currently avoiding due to trauma-relevant fears. Imaginal exposure entails describing the worst traumatic event out loud for a prolonged period of time during session and listening to a recording of the session at home. PE also includes processing thoughts and feelings aroused by the exposure exercises. Preliminary evidence suggests that PE is efficacious in military samples.

Like PE, CPT was initially developed and tested with civilians (VA/DOD, 2010), but has more recently been tested and found to be effective with military veterans (Monson et al., 2006). CPT includes cognitive restructuring and exposure elements. The cognitive restructuring component involves teaching patients how to identify and challenge maladaptive thinking patterns ("stuck points") related to the trauma. Patients are encouraged to examine self-blame related to their trauma experience and their beliefs about themselves and about others in five specific content areas (safety, trust, power/control, esteem, and intimacy). The exposure component of CPT involves the writing of a narrative, or "written account," describing the traumatic event and the patient's emotions related to it. In order to encourage emotional engagement with the exercise, the patient writes two versions of the account and reads them at home between sessions. CPT has been directly compared with PE and was found to be equally efficacious (VA/DOD, 2010).

Another individual psychotherapy for PTSD that has received considerable attention in the research literature is eye-movement desensitization and reprocessing (EMDR). EMDR includes many components found in other EBTs, including exposure to trauma memories and identifying negative and positive cognitions associated with the trauma. A core component of EMDR is desensitization and reprocessing, which involves recalling the trauma memory while making alternating eye movements. Although eye movements are conceptualized as an integral component of this treatment, EMDR is equally effective with or without eye movements (VA/DOD, 2010). EMDR has been shown to be comparable in efficacy with other EBTs for PTSD (VA/DOD, 2010).

Anxiety management is another individual therapy option that focuses on techniques for reducing symptoms of anxious arousal. Anxiety management strategies include relaxation training, or teaching patients to systematically relax major muscle groups, and breathing retraining, or instruction in slow, deep breathing to promote relaxation. Anxiety management can also include components such as psychoeducation about the relaxation response, positive self-talk, and assertiveness training. Some versions of anxiety management therapies may include in vivo or imaginal exposures, but exposure is not necessarily a component of the treatment. Stress Inoculation Training is one anxiety management protocol that has demonstrated efficacy in treating PTSD (VA/DOD, 2010).

Given the rising demands for mental health care, group psychotherapy is becoming increasingly popular. Group therapy for PTSD has been shown to possess efficacy, but group treatments specifically designed to target PTSD symptoms have not tended to outperform nonspecific treatment approaches (Sloan et al., 2011; VA/DOD, 2010). Continued research on group PTSD treatments, particularly among military service members, is an important topic of future research.

Evidence-based pharmacological treatments are also effective for treating military-related PTSD and can be used in conjunction with psychotherapy or as stand-alone treatments. Selective serotonin reuptake inhibitors (SSRIs) have strong empirical support and are considered a first-line PTSD treatment option according to current practice guidelines (VA/DOD, 2010). Tricyclic antidepressants and mirtazapine also have some empirical support but are considered second-line treatment options for PTSD (VA/DOD, 2010). Prazosin is also recommended as an adjunctive treatment for targeted treatment of nightmares and sleep disturbance, but is not recommended as a stand-alone treatment (VA/DOD, 2010). Current guidelines do not recommend the use of benzodiazepines or atypical antipsychotics (e.g., risperidone) in the treatment of PTSD (VA/DOD, 2010). Benzodiazepines have the potential to be addictive and do not treat the core symptoms of PTSD. Atypical antipsychotics have significant side effects and have limited evidence supporting their efficacy in the treatment of PTSD.

In summary, several effective psycho- and pharmacotherapy options exist for treatment of PTSD. CPT, PE, and EMDR are all effective in treating the core symptoms of PTSD, with similar effect sizes seen for each of these therapies. The relative efficacy of psychotherapy and medication is less clear, because no large controlled trials have directly compared psychotherapy with medication in the treatment of PTSD. Although several effective treatment options exist, little is known about which treatments work best for which patients. Identifying factors that moderate treatment outcome that can be used to match patients with treatment is an important topic of future research.

References

American Psychiatric Association. (2000). *Diagnostic and statistical manual of mental disorders* (4th ed., text revision). Washington, DC: Author.

Department of Veterans Affairs and Department of Defense (VA/DoD). (2010). *VA/DoD clinical practice guideline for management of posttraumatic stress*. Washington, DC: Author. Retrieved from http://www.healthquality.va.gov/Post_Traumatic_Stress_Disorder_PTSD.asp

Keane, T. M., Fairbank, J. A., Caddell, J. M., & Zimering, R. T. (1989). Implosive (flooding) therapy reduces symptoms of PTSD in Vietnam combat veterans. *Behavior Therapy, 20*, 245–260.

Keane, T. M., & Kaloupek, D. G. (1982). Imaginal flooding in the treatment of posttraumatic stress disorder. *Journal of Consulting and Clinical Psychology, 50*, 138–140.

Kessler, R. C., Sonnega, A., Bromet, E., Hughes, M., & Nelson, C. B. (1995). Posttraumatic stress disorder in the National Comorbidity Study. *Archives of General Psychiatry, 52*, 1048–1060.

Kulka, R. A., Schlenger, W. E., Fairbank, J. A., Hough, R. L., Jordan, B. K., Marmar, C. R., & Weiss, D. S. (1990). *Trauma and the Vietnam War generation: Report of findings from the National Vietnam Veterans Readjustment Study*. New York, NY: Brunner/Mazel.

Monson, C. M., Schnurr, P. A., Resick, P. A., Friedman, M. J., Young-Xu, Y., & Stevens, S. P. (2006). Cognitive processing therapy for veterans with military-related posttraumatic stress disorder. *Journal of Consulting and Clinical Psychology, 74*, 898–907.

National Center for PTSD. (2012). *Professional section: Assessment*. Retrieved from http://www. ptsd.va.gov/professional/pages/assessments/assessment.asp

Ramchand, R., Schell, T. L., Jaycox, L. H., & Tanielian, T. (2011). Epidemiology of trauma events and mental health outcomes among service members deployed to Iraq and Afghanistan. In J. I. Ruzek, P. P. Schnurr, J. J. Vasterling, & M. J. Friedman (Eds.), *Caring for veterans with deployment-related stress disorders* (pp. 13–34). Washington, DC: American Psychological Association.

Sloan, D. M., Feinstein, B. A., Gallagher, M. W., Beck, J. G., & Keane, T. M. (2011). Efficacy of group treatment for Posttraumatic Stress Disorder symptoms: A meta-analysis. *Psychological Trauma: Theory, Research, Practice, Policy*. Advance online publication. doi:10.1037/a0026291

Vogt, D., Vaughn, R., Glickman, M. E., Schultz, M., Drainoni, M., Elwy, R., & Eisen, S. (2011). Gender differences in combat-related stressors and their association with postdeployment mental health in a nationally representative sample of U.S. OEF/OIF veterans. *Journal of Abnormal Psychology, 120*, 797–806. doi:10.1037/a002345.

43 ANXIETY DISORDERS AND DEPRESSION IN MILITARY PERSONNEL

Nathan A. Kimbrel and Eric C. Meyer

INTRODUCTION

Military psychologists are likely to routinely work with personnel experiencing anxiety disorders and depression, as these conditions are common in both the general population (Kessler, Chiu, Demler, & Walters, 2005) and among military personnel (Fiedler et al., 2006; Sareen et al., 2007). The aim of this chapter is to provide an overview of the prevalence, assessment, and treatment of these disorders and to discuss their potential impact on military personnel. Notably, posttraumatic stress disorder (PTSD) is not included here, as it is reviewed elsewhere in this book (see Chapter 42 for a discussion of PTSD).

PREVALENCE

Anxiety disorders are the most common class of mental health disorders in the US general population. The 12-month prevalence rate for any anxiety disorder was 18.1% in the National Comorbidity Survey—Replication study (NCS-R; Kessler et al., 2005). Specific phobia (8.7% 12-month prevalence rate) was the most common anxiety disorder, followed by social anxiety disorder (6.8%), generalized anxiety disorder (GAD; 3.1%), panic disorder (2.7%), and obsessive-compulsive disorder (OCD; 1.0%). Mood disorders were the second most common class (9.5% 12-month prevalence rate), with major depressive disorder (6.7% 12-month prevalence rate) accounting for most of these cases. There is substantial co-occurrence among these disorders, and co-occurence is associated with greater symptom severity, greater functional impairment, and worse treatment response (Meyer, Kimbrel, Tull, & Morissette, 2011).

Estimating the prevalence of anxiety disorders and depression among Active Duty military personnel and veterans is challenging for two reasons. First, considerable evidence suggests that while military personnel are trained for combat and peacekeeping operations, deployment as part of such operations is associated with increased risk for mental health problems, including anxiety disorders and depression (e.g., Fiedler et al., 2006; Sareen et al., 2007). Second, the majority of prevalence data regarding mental health diagnoses in military personnel come from epidemiological studies employing brief, self-report measures. Only a few large-scale epidemiological studies of active duty military personnel and/or veterans have used a structured diagnostic interview. We briefly review two of these studies here.

Sareen and colleagues (2007) assessed anxiety disorders and depression in a large, representative sample of Canadian military personnel. They reported the 12-month prevalence of major depression (6.9%) to be quite similar to the rates reported for the general US population in the NCS-R (6.7%). In contrast, they reported lower 12-month prevalence rates among military personnel for social anxiety disorder (3.2% vs. 6.8%), GAD (1.7% vs. 3.1%), and panic disorder (1.8% vs. 2.7%) compared to rates reported in the NCS-R. They further reported that military personnel who had witnessed atrocities or massacres during deployment were at increased risk for the development of anxiety disorders and depression. Fiedler et al. (2006) assessed US veterans deployed to the 1991 Gulf

War and nondeployed veterans of the same era 10 years after their deployments. They reported higher 12-month prevalence rates for depression (15.1% vs. 7.8%), GAD (6.0% vs. 2.7%), social anxiety disorder (3.6% vs. 1.7%), OCD (2.8% vs. 1.1%), and panic attacks (1.6% vs. 0.5%) among deployed veterans compared to nondeployed veterans. Taken together, these findings suggest that rates of anxiety disorders and depression among military personnel are likely to vary as a function of deployment-related experiences, with significantly higher rates occurring among personnel that have been deployed to a warzone.

ASSESSMENT

Accurate assessment and diagnosis are critical for informing treatment selection and for monitoring treatment progress. Thus, a careful history of the presenting problem should be taken at the outset of the assessment. A functional analysis should also be conducted to clarify factors that may be maintaining the disorder(s). It is further recommended that, when feasible, psychologists working with military personnel include a semistructured clinical interview in their assessment batteries, as such interviews include detailed prompts to help interviewers establish whether clients meet criteria for specific diagnoses. For example, the Structured Clinical Interview for DSM-IV (SCID-IV; First, Spitzer, Gibbon, & Williams, 1996) is one of the most commonly used general diagnostic interviews and is routinely used to assess anxiety disorders, depression, and other relevant Axis I conditions (e.g., substance use disorders, psychosis). We also recommend the use of the Anxiety Disorders Interview Schedule for DSM-IV (ADIS-IV; Brown, Di Nardo, & Barlow, 1994), as the ADIS-IV was designed to aid in the differential diagnosis of anxiety disorders and depression. Another advantage of this interview is that it enables clinicians to make clinical severity ratings for both symptoms and diagnoses.

A final recommendation is to have clients complete self-report measures of symptom severity throughout treatment in order to monitor treatment progress (Barlow, 2008). A multitude of reliable and validated self-report measures of anxiety and depression are available for this purpose (e.g., Beck Depression Inventory—II [BDI-II; Beck, Steer, & Brown, 1996]; Depression Anxiety Stress Scales [DASS; Lovibond & Lovibond, 1995]; Social Phobia Scale [SPS; Mattick & Clark, 1998]). In addition, one particularly efficient self-report measure for military psychologists to consider using is the Psychiatric Diagnostic Screening Questionnaire (PDSQ; Zimmerman, 2002). The PDSQ was designed for use in primary care settings and takes about 15–20 minutes to complete. A unique advantage of this measure is that it provides symptom severity scores as well as clinical cutoffs for 13 different Axis I disorders, including social anxiety disorder, panic disorder, agoraphobia, GAD, OCD, and depression. Thus, in addition to assessing symptom severity, the PDSQ can also facilitate diagnostic efficiency.

TREATMENT

A brief overview of some of the available evidence-based treatments for anxiety disorders and depression is provided here. For more comprehensive information on evidence-based treatments for these conditions, including step-by-step treatment protocols and illustrative case examples, the interested reader is directed to Barlow (2008) or other similar treatment manuals. The Society of Clinical Psychology (Division 12 of the American Psychological Association) also provides information regarding evidence-based treatments at: http://www.apa.org/divisions/div12/cppi.html.

Specific Phobia

While specific phobia is the most common anxiety disorder (and also one of the most treatable), it is relatively rare for people to seek treatment for specific phobias (Barlow, 2002). Nonetheless, when patients do seek treatment for a specific phobia, there is clear consensus that treatment should involve exposure to the feared objects and situations. Indeed, exposure appears to be both necessary and sufficient

for successfully treating the vast majority of people with specific phobias (Barlow, 2002). In vivo exposure to actual feared objects and situations appears to be more effective and more efficient than imaginal exposure. Despite the considerable variability in exposure-based treatments for specific phobia (e.g., duration of treatment, group versus individual treatment, level of therapist involvement), the effect of exposure as a treatment modality appears to be quite robust (Barlow, 2002).

Social Anxiety Disorder

Cognitive-behavioral therapy (CBT) combines key elements from traditional cognitive and behavioral approaches and is the most well supported intervention for treating social anxiety disorder (Barlow, 2008). CBT for social anxiety disorder involves cognitive restructuring aimed at helping clients view social situations in a more realistic and less threatening way. It also includes a significant exposure component whereby clients remain in feared social situations for prolonged periods of time, which typically leads to reductions in anxiety. Assessment, psychoeducation, role plays, construction of a fear hierarchy, and homework assignments are additional components of this approach. CBT for social anxiety disorder is ideally suited for group therapy because of the ample opportunities for exposure and social support. Thus, most of the research on CBT for social anxiety disorder has used a group modality. However, when group treatment is not an option (e.g., when there are not enough clients available to constitute a group), the available data suggest that individual treatment is also effective (Barlow, 2008).

Generalized Anxiety Disorder

Evidence suggests that CBT is also an efficacious treatment for GAD and co-occurring symptoms of depression (Barlow, 2002). Although elements of CBT for GAD vary, common components include the following: (1) psychoeducation, (2) worry monitoring and early identification of situational triggers for worrying, (3) applied relaxation, (4) cognitive restructuring targeting maladaptive appraisals of future outcomes and information-processing biases, and (5) imaginal and in vivo rehearsal of acquired coping skills (e.g., through behavioral experiments or self-control desensitization). Most studies have examined individual CBT for GAD, although some studies have examined group CBT for GAD. More recent developments in the treatment of GAD are based on evidence suggesting that worry represents unsuccessful attempts to reduce distress and that worry interferes with present-moment focus, emotional processing, and interpersonal relationships. Accordingly, treatments that emphasize awareness and acceptance of present-moment experience, in contrast to future-oriented worry, have increasingly been used in treating GAD, as well as other anxiety disorders and depression. These mindfulness and acceptance-based approaches (e.g., acceptance and commitment therapy; ACT; Hayes, Strosahl, & Wilson, 2012) remain rooted in behavior therapy by encouraging mindful action in the service of valued life directions (e.g., being more present in one's relationships).

Panic Disorder and Agoraphobia

CBT is a highly effective and well-established treatment for panic disorder and agoraphobia (Barlow, 2008). The key components of CBT for panic disorder and agoraphobia are interoceptive exposure, which involves repeated exposure to feared bodily sensations (e.g., shortness of breath, increased heart rate, light-headedness, and so forth) and graded in vivo exposure to feared situations. The latter involves construction of an agoraphobia hierarchy that ranges from the least feared to most feared situation. In addition, cognitive restructuring is used to teach clients to identify and monitor catastrophic interpretations of bodily sensations and to evaluate these thoughts more objectively. Assessment, self-monitoring (e.g., keeping a panic attack record), psychoeducation, homework assignments, breathing retraining, and applied relaxation are other typical components of CBT for panic disorder and agoraphobia. This type of therapy may be applied in either individual or group formats,

typically in weekly sessions for 10–20 weeks, although briefer treatments may be effective as well (Barlow, 2008).

Obsessive-Compulsive Disorder

The most well supported psychosocial intervention for the treatment of OCD is exposure with response prevention (EX/RP). After an OCD diagnosis has been established through a thorough diagnostic assessment, it is recommended that psychotherapists spend an additional 4–6 hours with clients developing rapport and gathering information necessary for the development of a treatment plan. In particular, psychotherapists should provide education about the intervention, collect information about the nature and course of the disorder (including a history of prior treatment), teach the client ritual monitoring skills, identify the threat cues that cause the client distress, and identify the client's beliefs about the perceived consequences of refraining from compulsive or ritualistic behavior. The EX/RP treatment plan should then be developed and implemented. During treatment, clients engage in repeated, graduated exposure (imaginal and in vivo) designed to bring about obsessional distress while voluntarily refraining from ritualizing and avoidance. Home visits are explicitly encouraged in order to facilitate generalization to the home environment. Overall, EX/RP has been shown to be effective for the treatment of OCD; however, there has been some debate as to whether intensive treatment EX/RP programs (e.g., daily sessions for 3 to 4 weeks) are more effective than less frequent treatment programs. Current recommendations are that, if feasible, clients with very severe OCD symptoms receive intensive EX/RP (Barlow, 2008).

Depression

There are several well-established treatments for depression, including cognitive therapy, behavior therapy, and interpersonal therapy. Cognitive therapy for depression is a structured therapeutic approach that emphasizes the role of negative information processing biases in depression (Barlow, 2008). It is based on the premise that depressed clients often have negative automatic thoughts about themselves, their environment, and the future. The cognitive theory of depression further posits that depressed clients tend to distort their interpretations of events, which serves to maintain and strengthen their negative views. Examples of cognitive distortions include overgeneralization, all-or-nothing thinking, and emotional reasoning. The goal of cognitive therapy is to help clients identify, evaluate, and change their dysfunctional belief patterns. It is assumed that decreasing maladaptive thinking patterns will ultimately lead to decreases in depressive symptoms. To change clients' maladaptive thinking patterns, psychotherapists and clients work collaboratively to evaluate the evidence for and against the client's depressive thoughts. Cognitive therapists often utilize Socratic questioning as part of this process. Behavioral experiments are also commonly assigned to help clients test the validity of their negative expectations about future events (Barlow, 2008).

Behavior therapy for depression is based on learning theory and the premise that depressed individuals' current behavioral patterns do not lead to sufficient amounts of positive reinforcement (Barlow, 2008). Therefore, the goal of behavior therapy for depression is to help clients increase the frequency and quality of pleasant activities that they experience. Behavior therapy for depression can take place in individual or group settings. Typical treatment components include conducting a functional analysis of the maintaining conditions, orienting the client to the behavioral model, self-monitoring, activity scheduling, graded task assignment, social-skills training, relaxation training, and problem solving (Barlow, 2008).

Interpersonal psychotherapy (IPT) for depression is based on interpersonal and attachment theories as well as empirical findings indicating that depression is associated with interpersonal deficits, recent stressors, complicated bereavement, role transitions, and role disputes (Barlow, 2008). In IPT, the depressive episode is conceptualized as stemming from recent life events, and the primary goals are to (1) address an important interpersonal problem (e.g., complicated bereavement, role transition, role dispute) and (2) bolster interpersonal skills. The initial phase of IPT involves assessment, identification of the focal problem, development of a therapeutic alliance, and goal setting. In the middle phase, the therapist helps the client to resolve the focal problem by linking the client's mood to specific events, the use of communication analysis, exploration of the client's wishes and options, decision analysis, role plays, and other interpersonal techniques. During the termination phase, progress is reviewed regarding symptom reduction and resolution of the focal interpersonal problem. Psychotherapists' and clients' feelings about termination are also processed during the termination phase.

Future Directions for Treatment

With respect to future directions, there is a growing movement toward more broad-based interventions for anxiety disorders and depression. For example, Barlow (2008) describes a unified protocol (UP) for emotional disorders (i.e., anxiety disorders and depression) that contains therapeutic elements common to many of the CBT-based approaches described above (e.g., psychoeducation, cognitive restructuring, exposure, prevention of emotional avoidance). While the UP awaits empirical validation, the core components of this approach are already well supported. One potential advantage of this and other transdiagnostic approaches (e.g., ACT; Hayes et al., 2012) is that they may facilitate training and dissemination efforts by providing a single set of therapeutic principles to be learned rather than many disorder-specific protocols. Moreover, in theory, integrative, transdiagnostic approaches should be better able to address the high rates of co-occurrence among the anxiety disorders and depression (Meyer et al., 2011).

RELEVANCE AND IMPACT OF ANXIETY DISORDERS AND DEPRESSION ON THE MILITARY

The assessment and treatment of anxiety disorders and depression is of high relevance to military psychologists for several reasons.

First and foremost, these disorders are associated with significant distress, reduced quality of life, decreased productivity, higher rates of absenteeism, and high utilization of health care resources (e.g., Barlow, 2008; Meyer et al., 2011). Anxiety disorders and depression also directly impact military retention. Creamer et al. (2006) conducted a 10-year retrospective study of Australian Navy personnel and found that 8% of the sample developed an anxiety disorder (excluding PTSD) and 17% developed a mood disorder (including, but not limited to depression) during that 10-year period. Both anxiety and mood disorders were associated with greater risk of early separation from military service. Even more important was the finding that the greatest risk of early separation occurred in the year immediately following the onset of the disorder, when risk of early separation was nearly twice as high for personnel with mood and anxiety disorders. Importantly, personnel with mood and anxiety disorders who remained in the military past the first year of onset were at no greater risk for separation during subsequent years than personnel without these disorders (Creamer et al., 2006). Taken together, these findings suggest that early identification and intervention is critical and may ultimately reduce suffering, improve quality of life, increase productivity, enhance morale, and improve retention rates (Barlow, 2008; Creamer et al., 2006; Meyer et al., 2011).

References

Barlow, D. H. (2002). *Anxiety and its disorders* (2nd ed.). New York, NY: Guilford.

Barlow, D. H. (2008). *Clinical handbook of psychological disorders: A step-by-step treatment manual.* New York, NY: Guilford.

Beck, A. T., Steer, R. A., & Brown, G. K. (1996). *Manual for the Beck Depression Inventory* (2nd ed.). San Antonio, TX: Psychological Corporation.

Brown, T. A., Di Nardo, P. A., & Barlow, D. H. (1994). *Anxiety Disorders Interview Schedule for DSM-IV (ADIS-IV).* San Antonio, TX: Psychological Corporation.

Creamer, M., Carboon, I., Forbes, A. B., McKenzie, D. P., McFarlane, A. C., Kelsall, H. L., & Sim, M. R. (2006). Psychiatric disorder and separation from military services: A 10-year retrospective study. *American Journal of Psychiatry, 163,* 733–734.

Fiedler, N., Ozakinci, G., Hallman, W., Wartenberg, D., Brewer, N. T., Barrett, D. H., & Kipen, H. M. (2006). Military deployment to the Gulf War as a risk factor for psychiatric illness among US troops. *British Journal of Psychiatry, 188,* 453–459.

First, M., Spitzer, R., Gibbon, M., & Williams, J. (1996). *Structured Clinical Interview for DSM-IV.* Washington, DC: American Psychiatric.

Hayes, S. C., Strosahl, K. D., & Wilson, K. G. (2012). *Acceptance and commitment therapy: The process and practice of mindful change* (2nd ed.). New York, NY: Guilford.

Kessler, R. C., Chiu, W. T., Demler, O., & Walters, E. E. (2005). Prevalence, severity, and comorbidity of 12-month DSM-IV disorders in the National Comorbidity Survey Replication. *Archives of General Psychiatry, 62,* 617–627.

Lovibond, S. H., & Lovibond, P. F. (1995). *Manual for the Depression Anxiety Stress Scales* (2nd ed.). Sydney, Australia: Psychology Foundation.

Mattick, R. P., & Clarke, J. C. (1998). Development and validation of measures of social phobia scrutiny fear and social interaction anxiety. *Behaviour Research and Therapy, 36,* 455–470.

Meyer, E. C., Kimbrel, N. A., Tull, M. T., & Morissette, S. B. (2011). PTSD and co-occurring affective and anxiety disorders. In B. Moore & W. E. Penk (Eds.), *Handbook for the treatment of PTSD in military personnel.* New York, NY: Guilford.

Sareen, J., Cox, B. J., Afifi, T. O., Stein, M. B., Belik, S.-L., Meadows, G., & Asmundson, G. J. G. (2007). Combat and peacekeeping operations in relation to prevalence of mental disorders and perceived need for mental health care. *Archives of General Psychiatry, 64,* 843–852.

Zimmerman, M. (2002). *Manual for the Psychiatric Diagnostic Screening Questionnaire.* Los Angeles, CA: Western Psychological Services.

44 SERIOUS MENTAL ILLNESS IN THE MILITARY SETTING

David F. Tharp and Eric C. Meyer

PREVALENCE

Serious mental illnesses (SMIs) include schizophrenia, schizoaffective disorder, other psychotic disorders, and bipolar disorder. While SMI represents a small proportion of mental health disorders, the associated functional impairment and costs of treatment are disproportionately high. The 12-month prevalence of any SMI in the general population is 6.3% (Kessler et al., 2005). Prevalence data on SMI in military personnel are limited. According to the Defense Medical Epidemiology Database (DMED), from 2000–2009, 1,976 military personnel had an initial hospitalization for schizophrenia, which reflects an incidence rate of 0.14 individuals per every 1,000 per year. The DMED also indicated there were 3,317 initial hospitalizations for bipolar I disorder from 1997 to 2006, reflecting an incidence of 0.24 individuals per every 1,000 per year.

SMI AS A DISQUALIFYING CONDITION FOR MILITARY SERVICE

SMI is a disqualifying condition for military service (Department of Defense Instruction, Number 6130.03, Sept 13, 2011). However, the peak age of risk for schizophrenia ranges from approximately 18–30 years of age, which includes the age at which most military Active Duty (AD) personnel begin their military careers. Therefore, it is often the case that SMI does not emerge until after a service member joins the military. Once a service member is diagnosed, the median time between illness onset and initiation of medical discharge is 1.6 and 1.1 years for schizophrenia and bipolar disorder, respectively (Millikan et al., 2007). Because it is well known that SMI is a disqualifying condition, military personnel experiencing symptoms of SMI may be hesitant to report these experiences or to request treatment. Moreover, concerns over perceived stigma in the military related to mental disorders have been well documented.

ASSESSMENT

Because SMI is a disqualifying condition, military psychologists are likely to encounter personnel with SMIs during screening for enlistment, during the prodromal phase, or when SMI has not yet been formally diagnosed. This underscores the importance of early identification and accurate assessment, as well as general familiarity with signs and symptoms of SMI. Some of the more common symptoms of SMI include recent withdrawal or loss of interest in others; mood swings; decrease in functioning; memory or concentration problems; illogical thought and speech that are difficult to explain; feeling disconnected from others or one's surroundings; sense of unreality; unusual or exaggerated sense of personal powers to understand meanings or influence events; illogical or "magical" thinking; fear or suspiciousness of others; deterioration in personal hygiene; and dramatic changes in sleep and appetite.

More accurate methods for identifying individuals at clinical high risk (i.e., putatively prodromal) for SMI based on exhibiting attenuated symptoms have recently been developed (e.g., Loewy, Pearson, Vinogradov, Bearden, & Cannon, 2011). A brief self-report measure (Prodromal Questionnaire—Brief Version) was used to identify young adult and adolescent civilians at clinical high risk for SMI with 88% sensitivity and 68% specificity (Loewy et al., 2011). These authors suggested a two-stage evaluation process in which people who screen positive on the self-report measure are interviewed using the Structured Interview for Prodromal Syndromes (Loewy et al., 2011). Several prospective, longitudinal studies have found that a sizable proportion of people identified as being at clinical high risk will go on to develop SMI in the near future.

In cases involving an established service member, accurate assessment and diagnosis are extremely important. However, resources for conducting assessments are typically dependent on duty location and context. Whenever feasible, it is recommended that a semistructured clinical interview such as the Structured Clinical Interview for DSM-IV (SCID-IV; First, Spitzer, Gibbon, & Williams, 1996) be used. When logistical challenges preclude such structured assessment, a thorough clinical interview that includes a detailed history is a minimum standard for diagnosing SMI. Based on the first author's experience as the Medical Advisor in Kandahar, Afghanistan, the amount of time to assess SMI in the field is often quite limited. In such cases, clinical judgment and objective measures available in one's toolkit, usually via deployed laptop computers, must be used to decide whether to transfer a service member with SMI to the closest Medical Treatment Facility (MTF).

Deployed mental health professionals often have a laptop computer equipped with the Automated Neurological Assessment Measure and self-report measures that may aid in assessing SMI (e.g., Minnesota Multiphasic Personality Inventory, Personality Assessment Inventory, and so forth). Documenting relevant history and contextual information that may impact the assessment, such as time in theater and level of exposure to stressors, is critical. Recording objective and subjective data in the Armed Forces Health Longitudinal Technology Application (AHLTA) maximizes continuity of care and the ability of providers at the MTF to conduct more accurate and thorough follow-up assessment. In certain situations in which the clinical picture is ambiguous, a mission-essential service member may be exhibiting signs of a potential SMI. In such instances, it is recommended that ongoing assessment occur using regular symptom monitoring via self-report measures, along with objective performance data and clinical judgment.

CLINICAL MANAGEMENT AND TREATMENT

Recommended guidelines for comprehensive rehabilitation for SMI include a combination of evidence-based pharmacological and psychosocial treatment (Mueser, Torrey, Lynde, Singer, & Drake, 2003; Vieta et al., 2009). Guidelines for evaluation, management, and treatment of schizophrenia are provided by the US Department of Health and Human Services (http://guidelines.gov/content.aspx?id=5217). In the absence of more comprehensive intervention, acute management of SMI almost invariably involves the use of medications, typically including one or more of the following: antipsychotics, mood stabilizers, antidepressants, and anxiolytics. This is particularly true in the field environment, prior to the service member being medically evacuated to a higher echelon of care, where the focus is likely to be on short-term management of acute symptoms and where intensive psychosocial intervention may not be possible or relevant. For this reason, it is imperative that military psychologists work collaboratively with medication providers. Of note, treatments for schizophrenia and bipolar disorder are typically applied to the treatment of schizoaffective disorder and less common psychotic disorders.

Psychopharmacological Treatment

Antipsychotic medications are first-line treatments for SMI. Some of the more common first-generation medications include fluphenazine, trifluoperazine, perphenazine, chlorpromazine, haloperidol, and thiothixene. More recently developed second- and third-generation agents available in the United States include clozapine, aripiprazole, risperidone, olanzapine, ziprasidone, and quetiapine. It was hoped that these newer medications would be superior in terms of treating negative symptoms and reducing extrapyramidal side effects.

A large meta-analysis of randomized controlled trials compared all first- and second-generation antipsychotic medications for schizophrenia (Leucht et al., 2008). Three of the six second-generation medications approved in the United States (clozapine, olanzapine, and risperidone) were found to be superior to first-generation medications in reducing both positive and negative symptoms. The other second-generation medications were no more efficacious than first-generation drugs, even when looking only at negative symptoms. Second-generation medications were generally associated with fewer extrapyramidal side effects but greater weight gain and sedation, with some exceptions. While demonstrating the general efficacy of antipsychotic medications, these findings, combined with studies that have found high rates of treatment discontinuation due to side effects for all antipsychotic medications, highlight the difficulty of selecting the most effective medication for long-term treatment (Leucht et al., 2008).

Treatment guidelines for bipolar disorder have changed in recent years. Recent meta-analyses have concluded that antipsychotic medications are both more acceptable and more efficacious than mood stabilizers (e.g., lithium) in treating the manic phase of bipolar disorder (e.g., Cipriani et al., 2011). In particular, risperidone, olanzapine, and haloperidol were recommended as being among the best options for treating mania (Cipriani et al., 2011). Quetiapine has the most empirical support for treating bipolar depression (De Fruyt et al., 2012), although smaller meta-analyses also support the use of divalproex sodium and lamotrigine.

Psychosocial Treatment

In combination with collaborative psychopharmacology, comprehensive intervention for SMI includes the following evidence-based psychosocial treatment components, as indicated: recovery-based psychoeducation, cognitive-behavioral or interpersonal psychotherapy, social skills training, rehabilitation aimed at improving neurocognition and social cognition, family psychoeducation, assertive community treatment, supported employment, and integrated treatment for individuals with co-occurring substance use disorders (e.g., Mueser et al., 2003; Pilling et al., 2002; Vieta et al., 2009; Wykes, Huddy, Cellard, McGurk, & Czobor, 2011). An overview of evidence-based psychosocial treatment components most relevant for military psychologists has been highlighted. Additional information regarding evidence-based psychosocial treatments is available from the American Psychological Association (http://www.apa.org/divisions/div12/cppi.html).

There is empirical support for psychoeducation as a stand-alone psychosocial intervention for SMI (Mueser et al., 2003; Vieta et al., 2009). Moreover, psychoeducation forms the basis for more comprehensive psychosocial intervention (Mueser et al., 2003; Vieta et al., 2009). Psychoeducation provides people with SMI a conceptual and practical approach to understanding and managing their illness. It extends beyond providing information about symptoms and treatment options to teaching coping skills and preventative health behaviors (e.g., sleep hygiene) and targeting barriers to treatment compliance. Psychoeducation programs may be provided individually or in a group format. One widely used program is Illness Management and Recovery (Mueser et al., 2003).

Cognitive or cognitive-behavioral psychotherapy is a widely used, empirically supported treatment for SMI (Beck, Rector,

Solar, & Grant, 2008; Pilling et al., 2002; Vieta et al., 2009). Cognitive therapy targets delusional beliefs and catastrophic appraisals of hallucinations. The building of rapport and trust is essential in cognitive therapy, as psychotherapist and patient work collaboratively to identify, explore, and replace maladaptive thinking that leads to and maintains negative self-evaluation, distress, and isolation. In contrast to pharmacological interventions for SMI, treatment discontinuation rates for cognitive therapy are low (Pilling et al., 2002).

For bipolar disorder, findings suggest that intensive cognitive-behavioral therapy may be more effective than short-term therapy (Vieta et al., 2009). Another form of empirically supported psychotherapy used primarily in treating bipolar disorder is interpersonal social rhythm therapy, which is an extension of interpersonal therapy, a widely used, empirically supported treatment for depression. This approach emphasizes the importance of interpersonal relationships in maintaining healthy daily patterns (e.g., sleep schedule, activity level) and preventing subsequent mood episodes.

Cognitive rehabilitation programs have gained empirical support in the treatment of schizophrenia in several meta-analyses, particularly when combined with other psychosocial treatments (e.g., Wykes et al., 2011). Cognitive impairment is a core feature of SMI that persists regardless of phase of illness (i.e., level of remission of psychotic or mood symptoms) and is a strong predictor of level of functional impairment. Cognitive rehabilitation programs typically involve computerized modules or a combination of computerized modules and live therapy. Computerized modules are aimed at preserving or enhancing neurocognition (e.g., memory, attention, executive functioning) and processing of social information (i.e., social cognition). Cognitive enhancement therapy is one form of cognitive rehabilitation found to slow or prevent the progression of schizophrenia.

SMI typically affects families and caregivers, and stress within the support system is associated with risk of relapse or symptom exacerbation for the individual. Accordingly, there is substantial empirical support for family-based interventions for SMI, including improved medication compliance and reduced relapse and hospital readmission rates (Meuser et al., 2003; Pilling et al., 2002; Vieta et al., 2009). Treatment with individual families or multi-family groups typically involves psychoeducation, communication training, goal setting, and problem solving.

IMPACT OF SMI ON THE MILITARY

SMI is an excluded condition for military service due to the adverse impact of SMI on a person's ability to function effectively and efficiently, particularly when exposed to stressful situations. Use of recently developed assessment measures may promote more accurate screening and early identification of individuals at risk of developing SMI. More effective screening and early identification would promote considerable savings in terms of resources associated with enlistment, training, later identification, treatment, and discharge of military personnel with SMI. Next, military personnel with SMI (i.e., those in the prodromal phase or who have not been formally diagnosed) may be able to perform at an acceptable level until confronted with an unknown amount of stress. Clinical decompensation can occur rapidly, and referral for Command Directed Evaluation and possible removal from duty may be required under such circumstances. This, in turn, increases stress on the unit as a whole. In summary, early identification, accurate assessment, and empirically supported treatment are essential for minimizing the impact of SMI on the military and for promoting the health and welfare of the individual service member with SMI and their unit.

References

Beck, A. T., Rector, N. A., Solar, N., & Grant, P. (2008). *Schizophrenia: Cognitive theory, research, and therapy.* New York, NY: Guilford.

Cipriani, A., Barbui, C., Salanti, G., Rendell, J., Brown, R., Stockton, S.,... Geddes, J. (2011). Comparative efficacy and acceptability of antimanic drugs in acute mania: A multiple-treatments meta-analysis. *Lancet, 378,* 1306–1315.

De Fruyt, J., Deschepper, E., Audenaert, K., Constant, E., Floris, M., Pitchot, W.,…Claes, S. (2012). Second generation antipsychotics in the treatment of bipolar depression: A systematic review and meta-analysis. *Journal of Psychopharmacology, 26*, 603–617.

First, M., Spitzer, R., Gibbon, M., & Williams, J. (1996). *Structured Clinical Interview for DSM-IV*. Washington, DC: American Psychiatric.

Kessler, R. C., Demler, O., Frank, R. G., Olfson, M., Pincus, H. A., Walters, E. E.,…Zaslavsky, A. M. (2005). Prevalence and treatment of mental disorders 1990 to 2003. *New England Journal of Medicine, 352*, 2515–2523.

Leucht, S., Corves, C., Arbter, D., Engel, R., Li, C., & Davis, J. (2009). Second-generation versus first-generation antipsychotic drugs for schizophrenia: A meta-analysis. *Lancet, 373*, 31–41.

Loewy, R., Pearson, R., Vinogradov, S., Bearden, C., & Cannon, T. (2011). Psychosis risk screening with the Prodromal Questionnaire—Brief version (PQ-B). *Schizophrenia Research, 129*, 42–46.

Millikan, A., Weber, N., Niebuhr, D., Torrey, E., Cowan, D., Li, Y., & Kaminski, B. (2007). Evaluation of data obtained from military disability medical administrative databases for service members with schizophrenia or bipolar disorder. *Military Medicine, 172*, 1032–1038.

Mueser, K. T., Torrey, W. C., Lynde, D., Singer, P., & Drake, R. E. (2003). Implementing evidence-based practices for people with severe mental illness. *Behavior Modification, 27*, 387–411.

Pilling, S., Bebbington, P., Kuipers, E., Garety, P., Geddes, J., & Orbach, G. (2002). Psychological treatments in schizophrenia: I. Meta-analysis of family intervention and cognitive behaviour therapy. *Psychological Medicine, 32*, 763–782.

Vieta, E., Pacchiarotti, I., Valentí, M., Berk, M., Scott, J., & Colom, F. (2009). A critical update on psychological interventions for bipolar disorders. *Current Psychiatry Reports, 11*, 494–502.

Wykes, T., Huddy, V., Cellard, C., McGurk, S. R., & Czobor, P. (2011). A meta-analysis of cognitive remediation for schizophrenia: Methodology and effect sizes. *American Journal of Psychiatry, 168*, 472–485.

45 SUBSTANCE USE IN THE US ACTIVE DUTY MILITARY

Robert M. Bray

Substance use and abuse (illicit drug use, the misuse of prescription drugs, excessive alcohol use, and tobacco use) have been long-standing concerns in the military because of their negative impact on military readiness and performance and their association with disease. Over the years the Department of Defense (DoD) has developed programs and policies to help prevent, deter, and decrease substance abuse using education, training, urinalysis and breathalyzer testing, treatment and/or rehabilitation, periodic assessment of the nature and extent of drug and alcohol abuse, and a focus on health promotion.

HEALTH BEHAVIOR SURVEYS

To better understand and monitor substance use in the active duty military, DoD initiated a series of comprehensive surveys that included personnel in the Army, Navy, Marine Corps, and Air Force. Ten DoD Surveys of Health Related Behaviors (HRB) among Active Duty Military Personnel were conducted from 1980 to 2008 (Bray et al., 2009; Bray et al., 2010); an 11th survey has recently been completed in 2011, but results have not yet been released. These HRB surveys are cross-sectional studies repeated every 3 to 4 years with a different sample of participants in each survey. They provide the most definitive data on substance abuse in the military and are particularly valuable in that they are population-based surveys with large sample sizes (ranging from 12,000 to nearly 25,000 participants) designed to represent the DoD Active Duty population. Respondents were asked to answer all questions anonymously to encourage them to provide honest answers on sensitive questions.

SOCIODEMOGRAPHIC CHARACTERISTICS OF ACTIVE DUTY PERSONNEL

It is important to understand the characteristics of the military population to gain an appropriate context for substance use and abuse in the military. Table 45.1 presents estimates of the distribution of sociodemographic characteristics among military personnel in four survey years: 1980, 1988, 1998, and 2008. As shown, the military population is predominantly male, white, has some college or a college degree, is relatively young with just under half aged 25 or younger, married, and concentrated in enlisted pay grades E4 to E6 and E1 to E3. There have been some shifts in the distribution of characteristics across the 28-year period. In particular, the proportion of women, Hispanic and other racial/ethnic groups, college-educated personnel, and personnel aged 35 years or older increased significantly

TABLE 45.1. Estimated Sociodemographic Characteristics of Active Duty Military Personnel in Selected Survey Years

Sociodemographic Characteristic	Survey Year			
	1980 (N = 15,268)	1988 (N = 18,673)	1998 (N = 17,264)	2008 (N = 24,690)
Gender				
Male	91.2 (0.7)	88.8 (1.0)	86.3 (0.7)	85.7 (0.8)
Female	8.8 (0.7)	11.2 (1.0)	13.7 (0.7)	14.3 (0.8)
Race/ethnicity				
White, non-Hispanic	70.7 (1.4)	69.4 (0.9)	64.5 (0.9)	64.0 (1.0)
African American, non-Hispanic	18.8 (1.3)	18.5 (0.8)	17.6 (0.8)	16.7 (0.8)
Hispanic	4.6 (0.4)	8.0 (0.6)	10.8 (0.5)	10.4 (0.4)
Other	5.8 (0.4)	4.1 (0.3)	7.1 (0.4)	8.9 (0.5)
Education				
High school diploma or less	53.9 (1.6)	42.9 (1.5)	31.3 (1.2)	32.8 (1.4)
Some college	30.4 (1.2)	34.7 (0.9)	46.3 (1.0)	45.0 (0.8)
College degree or more	15.7 (1.2)	19.4 (1.4)	22.4 (1.4)	22.3 (1.6)
Age				
20 or younger	21.3 (1.4)	13.8 (1.1)	10.2 (0.6)	14.7 (1.0)
21–25	35.2 (1.1)	30.4 (1.2)	28.4 (0.9)	32.2 (1.4)
26–34	27.8 (1.1)	34.4 (1.0)	34.4 (0.7)	29.3 (0.7)
35 or older	15.6 (1.1)	21.4 (1.4)	27.0 (1.0)	23.8 (1.4)
Family Status				
Not married	47.1 (1.4)	39.3 (1.9)	39.9 (0.7)	45.7 (1.1)
Married	52.9 (1.4)	60.7 (1.9)	60.1 (0.7)	54.3 (1.1)
Rank				
E1–E3	27.2 (1.5)	21.0 (1.4)	18.9 (0.9)	21.0 (1.3)
E4–E6	50.2 (1.0)	51.9 (1.0)	52.5 (1.2)	51.7 (2.4)
E7–E9	8.2 (0.6)	10.4 (0.6)	10.8 (0.4)	10.2 (0.5)
W1–W5	1.1 (0.2)	1.0 (0.1)	1.2 (0.1)	1.4 (0.7)
O1–O3	8.3 (0.6)	9.6 (0.7)	9.5 (0.8)	9.3 (0.7)
O4–O10	5.0 (0.7)	6.1 (0.7)	7.2 (0.7)	6.4 (0.8)

Note: Table entries are column percentages (with standard errors in parentheses).
Source: DoD Surveys of Health Related Behaviors among Active Duty Personnel: 1980, 1988, 1998, 2008.

between 1980 and 2008, and for some of these characteristics they have nearly doubled.

TRENDS IN SUBSTANCE USE

Figure 45.1 presents the trends from 1980 to 2008 indicating the percentage of Active Duty military personnel who engaged in heavy alcohol use, illicit drug use, and cigarette use during the 30 days prior to the survey. As shown, the percentage of military personnel who smoked cigarettes in the past 30 days declined dramatically from 1980 to 1998. It increased significantly from 1998 to 2002 and continued a gradual decline such that the rate in 2008 was similar to the rate in 1998. Of interest, 15% of personnel indicated that they started smoking cigarettes after joining the military, and among current cigarette smokers this rate was 30%.

Additional information on other forms of tobacco use indicates that smokeless tobacco use rates have been around 12% from 1995 to 2002, then increased to 15% in 2005 and remained at about this level (14%) in 2008. Men were far more likely to use smokeless tobacco than women, with the highest rates among men aged 18 to 24. The rates of use

varied from 17% to 22% between 1995 and 2008 with the most recent rate in 2008 at 19%. In 2008, 24% of service members reported smoking cigars during the past year, and 4% reported smoking a pipe.

Returning to Figure 45.1, heavy alcohol use (defined as five or more drinks on the same occasion at least once a week in the past 30 days) decreased between 1980 and 1988, showed some fluctuations between 1988 and 1998, increased significantly from 1998 to 2002, and continued to increase gradually in 2005 and 2008. The heavy drinking rate for 2008 was not significantly different from when the survey series began in 1980, although heavy alcohol use showed a gradual and significant increase during the decade from 1998 to 2008 (from 15% to 20%). Similar to the increase in heavy drinking, rates of binge drinking (five or more drinks/occasion for men, four or more for women, at least once in the past month) also showed an increase from 35% to 47% during the 10-year period from 1998 to 2008. Indeed, excessive alcohol use has become part of military tradition and culture. Further, the stresses of war and exposure to combat are associated with increases in binge and heavy drinking (Bray et al., 2009; Jacobson et al., 2008).

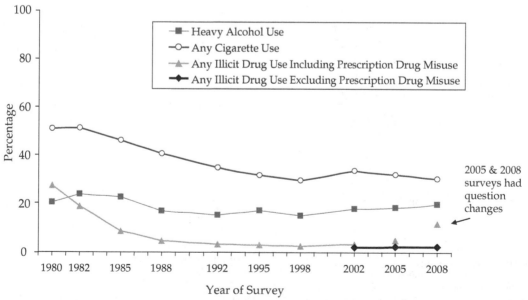

FIGURE 45.1 Substance Use Trends for Active Duty Military Personnel, Past 30 Days, 1980–2008 (*Source:* DoD Surveys of Health Related Behaviors, 1980–2008.)

Figure 45.1 also shows that the prevalence of illicit drug use (including prescription drug misuse) declined from 28% in 1980 to 3% in 2002. In 2005 the prevalence was 5%, and in 2008 it was 12%. These higher rates in the latter two surveys are largely a function of increases in misuse of prescription pain medications but may also partially be due to improved question wording in 2005 and 2008. Because of wording changes, data from 2005 and 2008 are not entirely comparable to prior surveys and consequently are not included as part of the trend line. An additional line from 2002 to 2008 shows estimates of illicit drug use excluding prescription drug misuse. These latter rates were very low (2% in 2008) and did not change across these three iterations of the survey. Any illicit drug use including prescription drug misuse was defined as use of marijuana, cocaine (including crack), hallucinogens (PCP/LSD/MDMA), heroin, methamphetamine, inhalants, or GHB/GBL or nonmedical use of prescription-type amphetamines/stimulants, tranquilizers/muscle relaxers, barbiturates/sedatives, or pain relievers. Any illicit drug use excluding prescription drug misuse was defined as use of marijuana, cocaine (including crack), hallucinogens (PCP/LSD/MDMA), heroin, inhalants, or GHB/GBL.

Figure 45.2 provides additional information about the use of specific drugs or drug classes during the past 30 days before the survey for 2002, 2005, and 2008. As shown, during this period rates for illicit drugs were all quite low in the 1% to 3% range. In contrast, pain relievers showed very large increases across the years and in 2008 were the most commonly misused drugs in the past 30 days at 10%, followed by tranquilizers and muscle relaxers at 3%. This indicates that prescription drug misuse, which was at 11% in 2008, is largely accounted for by the misuse of prescription pain medications. The relatively high rate of prescription pain medication misuse is of considerable concern, but

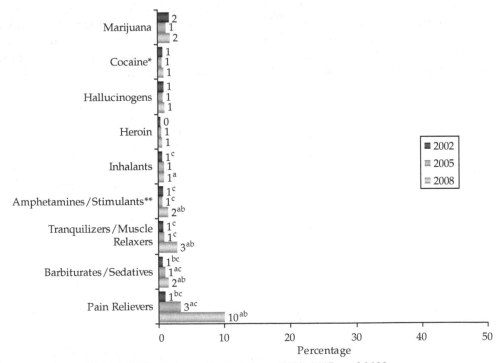

FIGURE 45.2 Use of Selected Illicit Drugs, Past 30 Days, 2002, 2005, and 2008

[a] Estimate is significantly different from the 2002 estimate at .05 level.
[b] Estimate is significantly different from the 2005 estimate at .05 level.
[c] Estimate is significantly different from the 2008 estimate at .05 level.
* Includes crack.
** Methamphetamine included in this estimate.

perhaps should not be surprising in view of the high rates of injuries among service members associated with deployments to Afghanistan and Iraq and the corresponding increases in prescriptions for pain medications among military members. Receiving pain medications from a doctor for legitimate use can sometimes lead to misuse, perhaps in part because of the ease of obtaining the medications. The Army's *Health Promotion, Risk Reduction, and Suicide Prevention* report issued in 2010 (US Army, 2010) estimated that over one-third of US Active Duty military service members were taking some form of prescription medications and that 14% of the force was taking some form of opioid medication.

Two relatively new types of drugs, spice and bath salts, have recently been gaining in popularity among civilians, but the extent of their use among service members is not well documented. Spice is a synthetic cannabinoid that has been detected in herbal smoking mixtures and produces effects similar to THC, the active ingredient in marijuana. Intoxication, withdrawal, psychosis, and death have been reported after consumption. Bath salts, known by such street names as "Ivory Wave," "Purple Wave," "Vanilla Sky," and "Bliss," is a new drug in the form of synthetic powder that can be used to get high and is usually taken orally, inhaled, or injected. Bath salts, which can be obtained legally in mini-marts, smoke shops, or over the Internet, contain amphetamine-like chemicals that that can trigger intense cravings and pose a high risk for overdose. Viewed as a cocaine substitute, bath salts can result in chest pains, increased blood pressure, increased heart rate, agitation, hallucinations, extreme paranoia, and delusions. There are no data at present on the use of bath salts in the military.

SOCIODEMOGRAPHIC CHARACTERISTICS OF SUBSTANCE USERS

Multivariate analyses of HRB data help provide a better understanding of the sociodemographic characteristics of heavy alcohol users, illicit drug users, and cigarette users. These analyses showed that higher rates of heavy alcohol use occurred among persons who were serving in the Marine Corps or Army, were men, were white or Hispanic, had less than a college degree, were single or married but unaccompanied by their spouse, and were of any rank (pay grade) except senior officers (O4–O10). Drug users were more likely to be serving in the Army, Navy, or Marine Corps relative to the Air Force, were more likely to be men, and to be single or married but unaccompanied by their spouse. Cigarette smokers were more likely to be serving in the Army, Navy, or Marine Corps, were more likely to be men, to be white non-Hispanic, to have less than a college degree, to be single, to be enlisted (especially pay grades E1–E6), and to be stationed outside the continental United States. Of note, these user characteristics were highly similar for heavy drinkers and for cigarette smokers.

FACTORS INFLUENCING MILITARY SUBSTANCE USE AND ABUSE

A variety of complex factors contribute to substance use and misuse in the active duty military that span individual, social, cultural, and environmental influences. Individual factors include demographic, genetic, and psychological characteristics and are possible risk factors for substance use. For example, young adults and males are more likely to engage in substance use than their older or female counterparts. Genetic markers are also associated with addiction, and drugs such as opiates and nicotine are well known for their addictive properties. Psychological components include beliefs, attitudes, intentions, and values, some of which may be associated with higher risk for substance use. Further, civilian personnel with a history of heavy drinking, smoking, or perhaps prior drug use may disproportionately select to join the Armed Forces, bringing their substance use problems into the military. The military screens for illicit drug use prior to entry but is more accepting of tobacco and alcohol use.

Social factors include family, friends, and norms about desired behavior. Peer pressure to "fit in" with friends may lead service members to engage in heavy drinking or drug use or to initiate smoking. Research indicates that service members are at increased risk of becoming smokers if they have friends who smoke and view smoking positively. Social factors identified with initiation of smoking after joining the military are curiosity, friends smoking, and wanting to be "cool." Socialization about the regulations and norms can help buffer substance use.

Cultural factors include perceptions about traditions and acceptable practices, and acceptance, support, and tolerance for use. Over the years a culture and resulting stereotype has developed that the military is composed of heavy-smoking, hard-drinking service members. Indeed, there is evidence that military culture may encourage excessive drinking and tobacco use. Many positive steps have been taken to modify this stereotype and attempt to change the culture (e.g., the ban on tobacco use during basic training, smoke-free federal buildings, smoking only in outdoor designated areas, campaigns to reduce DWIs, efforts to encourage and promote responsible alcohol use such as the culture of responsible choices program, substance abuse treatment programs). Unfortunately, despite these efforts a culture persists that still encourages excessive alcohol and tobacco use.

Environmental factors include external features of the immediate environment such as availability and easy access to substances, advertising that promotes use, and poor enforcement of policies designed to control and deter use. Unfortunately, a number of these environmental factors encourage use. For example, alcohol and tobacco are readily available to military personnel at reduced prices. Tobacco and alcohol advertisements in military news publications (e.g., *Army* and *Navy Times*) encourage use. Drug testing appears to decrease illicit drugs, but users still have internal networks that allow them to gain access to drugs if they are determined to use them. These factors may contradict and interfere with some policies and programmatic efforts to reduce substance use in the military. However, drug use policy appears as a positive example of what can be achieved with a rigorous and clear protocol that is strongly encouraged and closely monitored at all DoD levels.

The military has made notable progress in combating illicit drug and cigarette use. Illicit drug use has shown dramatic declines, but the military is facing new challenges with rising rates of prescription drug misuse. Although cigarette use has shown impressive reductions, about one-third of personnel are still smokers. There has been little progress in reducing binge and heavy drinking, and their rates appear to be rising. Despite commendable progress, much more remains to be done. Further reductions will need to take into account individual, social, cultural, and environmental factors within the Armed Forces.

References

Bray, R. M., Pemberton, M. R., Hourani, L. L., Witt, M., Olmsted, K. L., Brown, J. M.,...Scheffler, S. (2009). *2008 Department of Defense survey of health related behaviors among active duty military personnel*. Research Triangle Park, NC: RTI International. Retrieved from http://www.tricare.mil/2008HealthBehaviors.pdf

Bray, R.M., Pemberton, M.R., Lane, M.E., Hourani, L.L., Mattiko, M. J., & Babeu, L. A. (2010). Substance use and mental health trends among U.S. military active duty personnel: Key findings from the 2008 DoD Health Behavior Survey. *Military Medicine, 175*(6), 390–399.

Jacobson, I. G., Ryan, M. A. K., Hooper, T. I., Smith, T. C., Amoroso, P. J., Boyko, E. J.,...Bell, N. S. (2008). Alcohol use and alcohol-related problems before and after military combat deployment. *Journal of the American Medical Association, 300*(6), 663–675. doi:10.1001/jama.300.6.663

US Army. (2010). *Army health promotion risk reduction suicide prevention report 2010*. Washington, DC: Department of the Army.

46 SUBSTANCE USE DISORDERS AMONG MILITARY PERSONNEL

Joseph Westermeyer and Nathan A. Kimbrel

INTRODUCTION

Alcohol and drug problems (i.e., substance use disorders, or SUDs) have occurred among military personnel from time immemorial, and heavy alcohol use (i.e., 14 or more standard drinks per week for men, 7 or more for women), binge drinking (i.e., 5 or more standard drinks per day in men, 4 or more in women), and prescription drug abuse remain significant problems in today's military (Bray et al., 2009; Bray et al., 2010). The aim of the present chapter is to summarize the prevalence, psychopharmacology, assessment, and treatment of SUDs as well as the potential impact of these disorders on military personnel.

PREVALENCE OF SUBSTANCE USE DISORDERS IN THE MILITARY

The prevalence of SUDs in the military has changed over time. Both cigarette smoking and illicit drug use have sharply declined since 1980 (Bray et al., 2009; Bray et al., 2010). Cigarette smoking during the past 30 days has dropped from a prevalence rate of 51% in 1980 to a rate of 31% in 2008. Similarly, whereas 28% of US military personnel reported some form of drug misuse (including prescription drug abuse) during the past 30 days in 1980, that number had decreased to 3% by 2002. Since 2002, illicit drug abuse (excluding prescription

drug abuse) has been stable. In fact, only 2% reported using illicit drugs during the past 30 days in 2002, 2005, and 2008. Unfortunately, prescription drug abuse during the past 30 days doubled from 2002 (1.8%) to 2005 (3.8%) and nearly tripled from 2005 (3.8%) to 2008 (11.1%), representing a sixfold increase across a span of 6 years (Bray et al., 2010). Thus, there is a clear need for increased efforts aimed at reducing prescription drug abuse among US military personnel.

In contrast with the rapidly changing patterns of drug abuse, heavy alcohol use (as defined above) among US military personnel has shown a fairly consistent and high prevalence rate (approximately 20% during the past 30 days) since 1980 (Bray et al., 2010). Indeed, heavy drinking occurs almost twice as often among men in the military as among civilian men. The prevalence of heavy drinking among military men is also about four times higher than among military women, although the prevalence of heavy drinking among military women is about 50% higher than among civilian women (Bray et al., 2002). Branches of the service also show differing rates, with the highest prevalence in the Marines (35%), the lowest prevalence in the Air Force (20%), and intermediate rates in the Army and Navy (28% and 26% respectively), even after rates have been standardized for gender, age, education, race-ethnicity, and marital status (Bray et al., 2002). Deployment to a combat zone is also associated with heavy alcohol use (Bray et al., 2009).

NEUROBIOLOGY OF SUBSTANCE USE DISORDERS

Substances of abuse can mimic, precipitate, or contribute to a variety of psychiatric conditions (Cavacuiti, 2011). For example, hallucinogens, stimulants, cocaine, and cannabis can precipitate psychosis, especially in vulnerable people. Withdrawal from addictive levels of alcohol or sedatives (such as benzodiazepines and sleeping medications) can produce hallucinations soon after withdrawal as well as seizures (delirium tremens) later (about 72 hours after alcohol discontinuation; weeks or even months after discontinuation of some sedatives). Chronic use of alcohol, sedatives, and opiates can produce depressive symptoms, which may or may not resolve within a few weeks of establishing abstinence. Discontinuation of cocaine and stimulants (e.g., amphetamine, methylphenidate) can also precipitate depression, and cocaine and stimulants can mimic hypomania and mania in the midst of ongoing use.

Alcohol and drugs exert their effects on behavior, emotions, and cognition by acting on neurotransmitters. Thus, understanding the actions of different substances requires knowledge of neurotransmitters, their actions, and localizations (Cavacuiti, 2011). For example, stimulants mimic the impacts of adrenalin. Opiates impersonate the effects of endorphins. Hallucinogens affect dopamine transmitter sites. Alcohol, with physiological cross-effects highly similar to sedatives, affects the same neurotransmitters and even simple amino acids (such as glutamate) that produce sedation. Alcohol and drugs typically do not affect a single neurotransmitter system, however. This multineurotransmitter effect can complicate clinical pictures.

Some substances are more apt to produce frequent use. For example, nicotine and opioids can produce addictive use in as many as half of people repeatedly exposed to them. In contrast, many cannabis users voluntarily abandon use over time. Alcohol can be highly addictive in some cultures, yet produces virtually no addicted persons in others. In addition, some psychoactive substances produce disability more rapidly than others. For example, those addicted to heroin or cocaine tend to seek treatment about 3 years after starting use. Those addicted to alcohol, cannabis, or opium tend not to seek treatment until a decade or longer of use.

Finally, psychoactive substances can and frequently do affect physiological symptoms outside of the nervous system (Cavacuiti, 2011). For example, alcohol alone can adversely affect the endocrine system (central obesity, hypertension, hypogonadism, thyroid abnormalities), cardiovascular system (hypertension, coronary atherosclerosis), and the gastrointestinal system (esophagitis, gastritis, peptic ulcer, hepatitis, hepatic cirrhosis). Thus, a physical examination is recommended as a standard part of most SUD assessments.

ASSESSMENT OF SUBSTANCE USE DISORDERS

In order to properly diagnose a SUD, a considerable amount of diverse information is needed. Ideally, the clinician will obtain a careful history, including family history, type of substance(s) used, frequency of use, typical dose, pattern of use (e.g., daily, weekends, nights), duration and change of patterns over time, previous attempts to quit or cut-down, and any negative consequences related to the substance use (e.g., unemployment, DUIs, divorce, financial problems). It is further recommended that, when feasible, military psychologists conduct a semistructured diagnostic interview, such as the Structured Clinical Interview for DSM-IV (SCID-IV; First, Spitzer, Gibbon, & Williams, 1996). The advantage of the SCID and other similar diagnostic interviews is that they contain detailed prompts and assess a wide range of clinical disorders (e.g., depression, anxiety disorders) that may be highly relevant to the patient's SUD (e.g., if the patient is "self-medicating"). The structured questions and prompts of diagnostic interviews also help to ensure that clients meet full diagnostic criteria prior to being assigned a clinical diagnosis. However, obtaining all of the necessary information from patients with serious SUDs can be challenging at times, and, in some cases, it may take several days or weeks to obtain all of the relevant information. Collateral sources of information can be valuable in these situations, especially when minimization, denial, or amnesia obstruct accurate data collection. Identification of biomedical problems, via medical histories, physical examination, and laboratory tests can further aid in this process. With longer sobriety and greater understanding, most patients provide increasingly reliable and relevant information. Lastly, it should be noted that launching directly into the core information for an SUD diagnosis can sometimes be off-putting for patients with SUDs. As a result, it is sometimes more efficient over the long run to begin with related topics other than usage. For example, one might ask about parental use of substances while the patient was growing up, as clinical experience suggests that this line of questioning often favors nondefensive responses.

CLINICAL INTERVENTIONS FOR SUBSTANCE USE DISORDERS

As in the rest of society, resources available to the military psychologist can differ greatly depending on the size of the military post as well as access to nearby military, VA, or civilian resources. Moreover, available resources during deployments can differ substantially from those that are available in nondeployment situations (see Adler, Bliese, and Castro [2010] for deployment-specific strategies). Regardless of post size, creating a multidisciplinary team can be important in addressing a biopsychosociocultural problem like an SUD that cuts across disciplines. Chaplains, nurses, physicians, and social workers can and should collaborate with psychologists (and vice versa) in providing relevant services.

Given the challenging nature of SUD treatment, evidence-based interventions should be individualized (when appropriate) to maximize their potential impact. For example, when working with patients in the early stages of addiction (e.g., heavy use with few consequences), the military psychologist might manage the case entirely alone using a motivational

interviewing (MI) or a cognitive-behavioral approach. In contrast, in cases of advanced SUDs, a team approach and a phased recovery may be more appropriate. Unlike some other disorders with clear empirical rationales for one treatment approach, numerous evidence-based treatments exist for SUD. The latter include cognitive behavioral/relapse prevention (CB/RP), MI, 12-step facilitation, behavioral couples therapy, contingency management, cue exposure treatment, and the community reinforcement approach (CRA; Higgins, Sigmon, & Heil, 2008; McCrady, 2008). In practice, it is common for clinicians to incorporate elements of different modalities in order to individualize and maximize clients' treatment programs. As McCrady (2008) points out: "A key therapist responsibility is to help a client find a treatment approach that is effective for him or her, rather than slavishly adhering to a particular treatment model or setting" (p. 495).

Some of the many domains that should be considered during treatment planning include the severity of the problem, current stage of change, patients' expectations and preferences, treatment setting, prior treatment, available social support, life stressors, and variables currently maintaining the addiction. A potentially helpful way of planning SUD treatment within a military setting is to apply the military principles of strategy and tactics to the stages of change model (Prochaska, DiClemente, & Norcross, 1992). The stages of change model proposes that changes in addictive behaviors involve a progression through five stages of change: precontemplation, contemplation, preparation, action, and maintenance. Thus, the latter four stages of change can be conceived as goals to be addressed strategically; however, these goals can be addressed through numerous tactics (i.e., evidence-based treatments), depending on the particulars of the case (e.g., severity of the problem, patient preferences). Each of the stages is described below.

1. The *Precontemplation Stage* is one in which patients have no immediate plans to change their addictive behavior. Many patients in this stage are not aware of the severity of their addiction and are likely to minimize the impact of their SUD on themselves and others. Precontemplators are likely to have been coerced into treatment by commanding officers, concerned family members, or others (e.g., legal concerns). It is important to recognize that precontemplators are likely to return to their old ways once the pressure for them to change decreases (Prochaska et al., 1992), as they likely lack the internal motivation necessary to achieve long-term sobriety. Thus, a primary goal when working with precontemplators should be to increase their awareness of their problem in order to facilitate their movement into the contemplation stage. Motivational approaches, such as MI, are often better choices at this stage than more action-oriented approaches (e.g., stimulus control and other behavioral techniques).

2. Patients in the *Contemplation Stage* have developed an awareness that an SUD exists and are giving serious consideration to attempting to overcome it during the next 6 months; however, they have not yet committed to action (Prochaska et al., 1992). As a group, contemplators are most open to consciousness-raising techniques. They are also likely to reevaluate the effects that their addictive behaviors have had on their lives and on those closest to them. They are also likely to be actively weighing the pros and cons of their addictive behaviors (Prochaska et al., 1992). Again, MI and other similar approaches are likely to be good therapeutic options for patients at this stage as opposed to more action-oriented approaches.

3. Patients in the *Preparation Stage* plan to take action in the next 30 days and have often already begun to make small behavioral changes (e.g., reducing their normal drinking amount by several drinks); however, they have not yet reached abstinence. Individuals in the preparation stage often begin to use behavioral techniques such as stimulus control to help them begin to reduce their substance use (Prochaska et al., 1992). Thus, MI, CB/RP, 12-step facilitation, behavioral couples therapy, contingency management, and other evidence-based techniques aimed at facilitating abstinence are all likely to be beneficial during this stage.

4. Patients in the *Action Stage* are actively attempting to overcome their SUD and have successfully modified their addictive behavior for at least 1 day, but not longer than 6 months (Prochaska et al., 1992). This stage is difficult and requires a great deal of time and effort on the part of the patient. Moreover, in the case of the person with dependence and frequent or daily use, the action stage may require medication, especially for severe alcohol-sedative or opiate withdrawal. Hospitalization may be required for withdrawal seizures or delirium tremens. Some individuals may prefer admission to a residential recovery program or an intensive day program in order to maximize their chances of successfully quitting by having additional support available to them during the acute phase of quitting. The latter approach can be particularly helpful when cravings are likely to be quite strong. Patients in the action stage are likely to benefit from a variety of evidence-based interventions, including CB/RP, MI, 12-step facilitation, behavioral couples therapy, contingency management, cue exposure treatment, and CRA. Again, choice of treatment should be based on a variety of concerns (e.g., patient preferences) as outlined above. However, regardless of treatment modality, a primary goal should be to help the patient to reestablish work, residence, and a social network while remaining abstinent. Some slips and lapses are expected to occur during this period and should be framed as temporary and understandable lapses (as opposed to "failures" or "relapses").

5. Patients in the *Maintenance Stage* have been free of their addictive behaviors for at least 6 months and are continuing to work at preventing relapse (Prochaska et al., 1992). Like the action stage, some slips and lapses are expected to occur during this period as well. New couples, marital, or family issues might also surface. Comorbid psychiatric disorders that have not previously been addressed may become more apparent during this phase. New recreational, spiritual, and avocational activities are also likely to begin in this phase. Some members of self-help groups may choose to become sponsors during this time. At this point, patients no longer rely so strongly on their clinicians for guidance of their recovery. Among those requiring ongoing care for comorbid conditions, the therapeutic issues are more typical of nonaddicted patients. The recovering patient is more in charge of his/her own recovery, with the clinician acting more like a consultant for special problems that may arise on occasion. In short, the optimal outcome for patients with SUDs is to progress to—and remain in—the maintenance stage for the remainder of their lives.

RELEVANCE AND IMPACT OF SUBSTANCE-USE DISORDERS ON THE MILITARY

As noted above, heavy drinking occurs at much higher rates among men and women in the military relative to their civilian counterparts. This issue has high relevance to the military because heavy drinking may cause or worsen biomedical, psychological, and interpersonal processes. Even "light" drinking can result in blood levels of alcohol that are illegal for certain activities (such as driving or flying), and blood levels from a single drink can impair tasks involving judgment, rapidity, and coordination. Thus, even moderate alcohol use has the potential to negatively impact military readiness. Similarly, illicit drug use of any kind poses special liabilities in the military. Cannabis (marijuana) impedes the ability to discern speed-by-space relationships when motion is involved. Sedatives or opioids can reduce the anxiety associated with high-risk situations, but then impede the ability to cope with risky eventualities that arise. Stimulants can facilitate remaining awake and alert, while undermining the ability to fully assess a situation and decide how best to proceed. Given the complexities of modern warfare, even small amounts of psychoactive substances can compromise neural functions critical to combat performance (e.g., perception, interpretation, recall, judgment, hand-eye coordination, energy level, split-second problem-solving, determining priorities in complex situations). In sum, in addition to the devastating direct consequences that SUDs have on military personnel and their families, these disorders also have a high potential to negatively impact military readiness. Military psychologists should be vigilant to the signs and symptoms of SUDs and be prepared to assess and treat these challenging disorders whenever they may arise.

References

Adler, A. B., Bliese, P. D., & Castro, C. A. (2010). *Deployment psychology: Evidence-based strategies to promote mental health in the military.* Washington, DC: American Psychological Association.

Barlow, D. H. (2008). *Clinical handbook of psychological disorders: A step-by-step treatment manual.* New York, NY: Guilford.

Bray, R. M., Hourani, L.L., Rae, K. L., et al. (2002). Department of Defense Survey of Health-Related Behaviors Among Military Personnel. Research Triangle Park, NC: RTI International, 2003.

Bray, R. M., Pemberton, M. R., Hourani, L. L., Witt, M., Rae Olmsted, K. L., Brown, J. M.,…Bradshaw, M. R. (2009). 2008 Department of Defense survey of health related behaviors among active duty military personnel. Report prepared for TRICARE Management Activity, Office of the Assistant Secretary of Defense (Health Affairs) and U.S. Coast Guard. Available online at http://www.tricare.mil/2008HealthBehaviors.pdf

Bray, R. M., Pemberton, M. R., Lane, M. E., Hourani, L. L., Mattiko, M. J., & Babeu, L. A. (2010). Substance use and mental health trends among U.S. military active duty personnel: Key findings from the 2008 DoD Health Behavior Survey. *Military Medicine, 175,* 390–399.

Cavacuiti, C. A. (2011). *Principles of addiction medicine.* Philadelphia, PA: Lippincott, Williams, and Wilkins.

First, M., Spitzer, R., Gibbon, M., & Williams, J. (1996). *Structured clinical interview for DSM-IV.* Washington, DC: American Psychiatric Press.

Higgins, S. T., Sigmon, S. C., & Heil, S. H. (2008). Drug abuse and dependence. In D. H. Barlow (Ed.), *Clinical handbook of psychological disorders: A step-by-step treatment manual.* New York, NY: Guilford.

McCrady, B. S. (2008). Alcohol use disorders. In D. H. Barlow (Ed.), *Clinical handbook of psychological disorders: A step-by-step treatment manual.* New York, NY: Guilford.

Prochaska, J. O., DiClemente, C. C., & Norcross, J. C. (1992). In search of how people change: Applications to addictive behaviors. *American Psychologist, 47,* 1102–1114.

47 TRAUMATIC BRAIN INJURY

Melissa M. Amick, Beeta Homaifar, and
Jennifer J. Vasterling

Traumatic brain injury (TBI) has been one of the most common injuries sustained during Operation Enduring Freedom (OEF) and Operation Iraqi Freedom (OIF). Encounters with improvised explosive devices have been the source of many OEF/OIF TBIs and are thought to possibly lead to blast injury (i.e., propagation of a blast wave through the cranium, compressing the brain), in addition to possible blunt and penetrating brain injuries. As compared to previous wars, in which blunt or penetrating injuries predominated, modern war-zone-acquired TBI is more likely to include a blast component in addition to other injury components. Military TBI is not limited to warfare, however, and may result from training accidents and other sources of injury (e.g., motor vehicle accidents or falls).

TBIs acquired during OEF/OIF are more likely to be mild in severity rather than moderate to severe (Department of Defense [DoD], 2009). Although neuroimaging and medical records can help document moderate to severe TBI, diagnosis of mild TBI (mTBI) mainly relies on assessing the patient's subjective recall and review of the medical record for altered mental status or associated TBI symptoms immediately following the injury. Further, brain injuries sustained during military service may be influenced by other co-occurring clinical conditions that may overlap with TBI in terms of symptom presentation. The following sections have been written to aid the military health

care provider in the difficult task of assessment and treatment of TBI.

DIAGNOSTIC CRITERIA FOR TBI

Departments of Defense (DoD) and Veterans Affairs (VA) consensus criteria define a TBI as "a traumatically induced structural injury and/or physiological disruption of brain functioning as a result of external force that is indicated by new onset or worsening of at least one of the following clinical signs immediately following the event" (DOD, 2009, p. 19). Clinical signs include one or more of the following: (1) decreased/loss of consciousness; (2) posttraumatic amnesia (PTA) defined by DoD/VA as a gap in memory for events occurring immediately before or after the injury; (3) alteration of mental state (AMS) at time of injury evidenced by confusion, disorientation, or slowed thinking; (4) transient or persistent neurological deficits (e.g., weakness, disruption of balance, loss of smell); or (5) intracranial brain lesion (DOD, 2009). It should be noted that other classification systems are also widely used (e.g., American Congress of Rehabilitation Medicine).

TBI severity is categorized according to the duration of loss of consciousness (LOC), altered mental state (AMS), and posttraumatic amnesia (PTA), as well as the severity of neuroimaging findings or other behavioral

TABLE 47.1. DOD and VA Consensus Criteria for TBI severity (DoD, 2009)

Criteria	Mild	Moderate	Severe
Structural imaging	Normal	Normal or abnormal	Normal or abnormal
Loss of consciousness (LOC)	0–30 min	> 30 min and < 24 hrs	> 24 hrs
Alteration of mental state (AMS)	A moment up to 24 hrs	> 24 hours. Severity based on other criteria	
Posttraumatic amnesia (PTA)	0–1 day	> 1 and < 7 days	> 7 days
Glasgow Coma Scale (best available score in first 24 hours)	13–15	9–12	< 9

symptoms (Table 47.1). Although current TBI classification schemes use seemingly discrete severity labels, it is important to recognize that TBI severity actually reflects a continuum in which physiological processes are transient at the mildest end but increasingly involve more significant axonal damage (and related behavioral manifestations) as the severity increases (Bigler & Maxwell, 2012). Because mTBI, sometimes also referred to as "concussion," composes the majority of injuries sustained in military settings (Defense and Veterans Brain Injury Center [DVBIC], 2012) some of the unique challenges in the assessment and treatment of mTBI are highlighted.

POSTCONCUSSIVE SYMPTOMS

It is important to distinguish the TBI (i.e., the actual physical injury) from its possible sequelae, often referred to as postconcussive symptoms (PCS). PCS include a range of physical/somatic (e.g., dizziness, headaches, photophobia, fatigue), cognitive (e.g., poor concentration, decreased memory), and emotional (e.g., depression, anxiety, irritability) symptoms. The symptoms are also likely to change over time, with more severe injuries typically associated with more prolonged symptoms, as compared to milder injuries. Despite onset after a blow to the head, symptoms characteristic of PCS are not typically indicative of a TBI if they occur in the absence of AMS, PTA, or LOC, or if they develop or worsen significantly after the injury. There are, however, rare exceptions (e.g., slowly developing subdural hematomas) in which symptom onset may be delayed. Although most studies of PCS report that symptom

resolution occurs within a few days to weeks among the majority of mTBI cases, emerging findings suggest that some symptoms may last up to a year in some people. PCS are nonspecific and overlap with a number of psychological and minor medical conditions and occur fairly frequently among healthy individuals. The etiology of prolonged PCS in mTBI is unknown but may be influenced by a combination of biological and psychosocial factors, including injury characteristics not detectable by conventional imaging, psychological factors (e.g., depression, anxiety, posttraumatic stress), attribution error, preinjury variables, (e.g., history of prior TBI, genetics, psychological diagnoses, and social support), and contextual and motivational factors (e.g., clinical compared to litigation contexts) (Iverson, 2012).

TBI ASSESSMENT

The multiple factors that may contribute to the onset and maintenance of PCS in mTBI underscore the importance of thorough clinical assessment for both historical TBI attributes and symptoms and the presence of comorbid psychological disorders and preinjury variables.

Clinical Interview

For moderate to severe brain injuries, TBI diagnosis is made on the basis of witnessed disruption of mental status (LOC, PTA, AMS), neuroimaging, and/or Glasgow Coma Scale (see Table 47.1), as individuals with these more severe injuries are likely to seek immediate medical attention. For milder injuries, however, the current gold standard for TBI diagnosis is the clinical interview, which is often based on retrospective recall to determine whether or not the individual experienced AMS, LOC, or PTA; medical record review; and (if possible) witness observations.

It is important to recognize that TBI screening and TBI diagnosis differ. The former uses a few brief questions to identify individuals who may have sustained a mTBI, whereas the latter is achieved through a comprehensive evaluation. Increased awareness of the high prevalence of TBI has led to greater efforts to assess for TBI in the battlefield. Consequently, DOD providers may have access to medical records generated close in time to the injury. By contrast, for the majority of military TBI cases seen within VA facilities, diagnoses are retrospective and based on recall of events that occurred months to years earlier.

AMS is typically assessed by asking individuals if they felt "dazed" or "confused" after an injury. Simple yes/no questions, however, are not likely to reliably distinguish biomechanically induced AMS from AMS due to psychological factors, as fear or stress at the time of the injury can also result in an altered mental state (Ruff et al., 2009) and/or actual stress-related physiological responses. It is recommended that providers start by asking the individual in an open-ended manner to provide as many details as possible about the events leading up to, during, and following the injury. Narratives that describe disorientation to time and/or place, inability to do simple math or other basic tasks, or being incoherent are suggestive of AMS due to a biomechanical brain injury and can be used to establish the duration of these symptoms. Contrasting the AMS experienced consequent to the TBI event with other psychologically stressful events that did not involve a blow to the head may also be helpful to parse out a psychological or psychobiological, as opposed to biomechanical, cause of the AMS.

Although establishing LOC is somewhat more straightforward than determining AMS, if the patient lost consciousness, they may have difficulty estimating the duration. For all disturbances of mental state/consciousness, determining whether others witnessed the injury or were present at any point immediately post injury avoids relying solely on the individual's recall. However, even in the relatively likely event that the provider does not have direct access to a witness, it may help the patient reconstruct an accurate account of the event by prompting them to recall what witnesses may have described to them about their behavior. For LOC, for example, providers are encouraged to ask, "Has anyone told you that you were unconscious?" rather than asking if an individual lost consciousness as a result of their injury.

Although DoD/VA criteria define PTA as memory gaps both immediately before and after the event, frequently, PTA is defined more narrowly as a gap in memory after injury (see McCrea et al., 2008). To evaluate PTA as defined by DoD/VA criteria, providers can ask, "What was the last thing you remember before the injury?" and "What was the first thing you remember after the injury?" It may be difficult to separate a patient's direct autobiographical recall of the period following their injury, after having regained consciousness, versus their secondary semantic memory of what they have been told by other eyewitnesses. Distinguishing between PTA and LOC in the absence of a witness may be particularly challenging. For example, an individual may perceive the gap in their memory as LOC, when in fact they were responsive but simply do not recall the interaction (i.e., PTA) (Ruff et al., 2009). Establishing a timeline between these preinjury and postinjury memories can help providers identify whether a gap in memory occurred and help to approximate the duration of PTA. It is important to consider that other factors such as intoxication or sedation immediately post injury may complicate the determination of the presence and duration of AMS, LOC, or PTA (Ruff et al., 2009).

Neuroimaging

The use of conventional neuroimaging approaches may be helpful in diagnosing TBI, though in mTBI cases, the chance of detecting a significant finding is low. Therefore, conventional imaging may be best suited for moderate to severe injuries (Bigler & Maxwell, 2012). If imaging is obtained in the acute stage post-TBI, computed tomography (CT) can be helpful in determining the presence, extent, and location of large vascular lesions, whereas magnetic resonance imaging may help detect smaller hemorrhagic changes not easily identified with a CT scan. In both the acute and chronic phases, positron emission tomography can be used to identify metabolic changes. Although conventional imaging is available in larger combat hospitals and within some theaters of operation (e.g., Kuwait and Afghanistan), immediate imaging following battlefield injuries may not always be possible, particularly for less severe injuries. Additionally, abnormalities on neuroimaging do not always translate into cognitive and/or functional impairment (Bigler & Maxwell, 2012).

Neuropsychological Assessment

Neuropsychological testing may be helpful in determining the presence and clinical significance of acute and, particularly for moderate to severe injuries, persisting symptoms. The Military Acute Concussion Evaluation (MACE) is a brief cognitive screening tool widely used in Iraq and Afghanistan to evaluate possible symptoms of a TBI (French, McCrea, & Baggett, 2008). The MACE includes both details of the injury event as well as objective assessment of cognitive functioning, questions that have been derived from a civilian acute mental status screening assessment. The MACE may have less utility when administered more than 12 hours post injury. Detailed neuropsychological assessment may not be feasible in the acute phase due to logistical barriers inherent to the combat setting (McCrea et al., 2008). At more intensive levels of in theater care, however, standardized neuropsychological assessments extending beyond the MACE may be conducted.

Repeat assessments over time can be helpful in the detection of persisting cognitive symptoms. One way to track subtle changes and/or improvement over time is to use a reliable change score, which entails determining if the difference between an initial and follow-up score is greater than a certain level. This score can be useful in identifying significant clinical change associated with possible functional consequences. It is important to note that factors other than TBI may also influence neuropsychological performance (e.g., depression, stress, fatigue, medication, substance abuse, motivation).

TREATMENT FOLLOWING TBI

Once the presence of a TBI has been established, providers must determine what recommendations to make with regard to treatment and/or rehabilitation. In the acute stages of post-TBI recovery, it is helpful to emphasize the value of rest (i.e., ample sleep and rest throughout the day), reduced activity (i.e., avoidance of activities that are physically demanding or require a great deal of concentration), avoidance of activities that could slow recovery or increase risk of a future TBI (i.e., activities with increased risk for blows to the head or alcohol use), and gradual return to regular routine. For mTBI, education and reassurance about the high likelihood of complete recovery is often beneficial.

For persistent cognitive deficits, cognitive rehabilitation therapy (i.e., an umbrella term used to describe a range of goal-oriented interventions used to compensate for, or overcome, neurocognitive deficits) is sometimes implemented. According to a recent Institute of Medicine report (Koehler et al., 2011), several cognitive rehabilitation interventions show promise, but their evidence base is not yet fully developed. Examples include the treatment of memory impairments via the use of external memory aids (e.g., notebooks, alerting devices) and retraining of language and social communication skills.

CHALLENGES AND FUTURE DIRECTIONS

Throughout this chapter, the particular challenges of assessing and diagnosing TBI, particularly mTBI, acquired in the military setting have been highlighted. There remain, however, several significant issues that will need to be addressed as military care for TBI continues to evolve. Summarized below are a few of the critical topics military providers are encouraged to consider in their work with injured service members.

- The criteria required for diagnosing the severity of TBI has not received unanimous consensus. There are multiple diagnostic classification systems for TBI, which differ in their criteria, including the definition of PTA. In the case of mTBI, there is disagreement as to which symptoms indicate altered mental status, and it remains to be determined if there is a different trajectory of recovery for those with brief AMS compared to LOC. More refined classification systems and increased consensus among classification systems may improve diagnosis and prognosis and, as a result, help direct treatment.
- Diagnosis and treatment of TBI and common comorbidities (e.g., PTSD, Depression, chronic pain) present a unique challenge for military clinicians in terms of diagnostic assessment and case conceptualization. Providers, however, are cautioned against dismissing all symptoms as psychological in nature, or conversely, assuming that all symptoms are a direct result of the brain injury, given the high stakes of repeated TBI and potential enhanced recovery from TBI if managed appropriately.
- An optimal service delivery model for mTBI has yet to be identified. Brenner and Colleagues (2009) have proposed a stepwise model that focuses on education about possible TBI symptoms and addressing psychiatric comorbidities before treating somatic complaints or providing other interventions for cognitive symptoms. This model is one of several emerging approaches to service delivery, but at this time all models lack empirical support.
- Educational and supportive programs for families of individuals with TBI and common comorbidities are needed. These programs are likely to foster a more supportive environment for the service member and lead to the most successful community reintegration.

References

Brenner, L. A., Vanderploeg, R. D., & Terrio, H. (2009). Assessment and diagnosis of mild traumatic brain injury, posttraumatic stress disorder, and other polytrauma conditions: Burden of adversity hypothesis. *Rehabilitation Psychology, 54*(3), 239–246.

Bigler, E. D., & Maxwell, W. L. (2012). Understanding mild traumatic brain injury: Neuropathology and neuroimaging of mTBI. In J. J. Vasterling, R. A. Bryant, & T. M. Keane (Eds.), *PTSD and mild traumatic brain injury* (pp. 15–36). New York, NY: Guilford.

Defense and Veterans Brain Injury Center. (2012). DoD worldwide numbers for TBI. Retrieved from http://www.dvbic.org/dod-worldwide-numbers-tbi

Department of Veterans Affairs and Department of Defense. (2009). VA/DOD clinical practice guideline for the management of concussion/mild traumatic brain injury. Retrieved from http://www.healthquality.va.gov/mtbi/concussion_mtbi_full_1_0.pdf

French, L., McCrea, M., & Baggett, M. (2008). The Military Acute Concussion Evaluation (MACE). *Journal of Special Operations Medicine, 8*(1), 68–77.

Iverson, G. L. (2012). A biopsychosocial conceptualization of poor outcome from mild traumatic brain injury. In J. J. Vasterling, R. Bryant, & T. M. Keane (Eds.), *PTSD and mild traumatic brain injury* (pp. 37–60). New York, NY: Guilford.

Koehler, R., Wilhelm, E., & Shoulson, I. (Eds.). (2011). *Cognitive rehabilitation therapy for traumatic brain injury: Evaluating the evidence*. Washington, DC: The National Academies Press.

McCrea, M., Pliskin, N., Barth, J., Cox, D., Fink, J., French, L., . . . Yoash-Gantz, R. (2008). Official position of the military TBI task force on the role of neuropsychology and rehabilitation psychology in the evaluation, management, and research of military veterans with traumatic brain injury. *Clinical Neuropsychologist, 22*, 10–26.

Ruff, R. M., Iverson, G. L., Barthe, J. T., Shane, S., Bush, S. S., Brosheke, D. K., & the NAN Policy and Planning Committee. (2009). Recommendations for diagnosing a mild traumatic brain injury: A National Academy of Neuropsychology Education Paper. *Archives of Clinical Neuropsychology, 24*, 3–10.

48 AGGRESSION AND VIOLENCE

Eric B. Elbogen and Connor Sullivan

When military veterans or service members become violent, the costs to individuals, their families, and their communities are great. Many service members or veterans suffer from posttraumatic stress disorder (PTSD), traumatic brain injury (TBI), or substance abuse, each of which has been found to be associated with higher violence rates among service members from previous conflicts (Marshall, Panuzio, & Taft, 2005; Taft et al., 2007). Significant prevalence rates of various types of violence in military populations have been documented. For example, rates of domestic violence vary from 13.5% to 58% based on the type of measure used, time period, and collateral reports (Marshall et al., 2005). Empirical studies have found that aggression toward others is a problem in up to one-third of military service members and veterans of the Iraq and Afghanistan Wars (Elbogen et al., 2012). Although it is unknown whether military and civilian populations differ in terms of violence prevalence, the notable incidence of violence by service members and veterans highlights the need to better understand causes of violence in military populations, determine which individuals are most at risk, and develop methods for decreasing such risk.

ASSESSMENT OF VIOLENCE RISK IN MILITARY POPULATIONS

To date, military psychologists have received relatively little guidance on how to systematically assess the risk of service members engaging in violence and aggression. Research shows that without structured assessments, clinicians perform only modestly better than chance when assessing risk of violence; in particular, clinicians are prone to decision-making errors (Elbogen et al., 2010), including:

- discounting risk factors that are less accessible but have empirical support
- rating categories of risk factors as relevant because they are readily available
- utilizing salient variables with no empirically demonstrated link to violence
- underrating situational factors and overrating individual-level factors.

To reduce such errors, violence risk assessment would ideally be grounded in an evidence-based framework and be informed by empirically validated risk factors (Elbogen et al., 2010). A number of excellent risk assessment tools have been developed in civilian populations and it is likely that the methods used to construct these instruments could be used to develop effective violence risk assessments for military populations. But at present, instruments to effectively assess violence risk in service members and veterans are lacking, especially when assessing variables related to military experience (e.g., war-zone exposure, weapons training, combat-related PTSD) and their outcomes on violence.

Despite this, military psychologists can be guided by the conceptual framework underlying effective risk assessment expounded on in the existing literature. Specifically, even without risk assessment tools that have been validated, military psychologists can structure decisions based on empirically validated risk factors to improve accuracy of violence risk assessment. Empirical research has uncovered a number of violence risk factors among military service members and veterans (Elbogen et al., 2010; Marshall et al., 2005).

A consistent association between PTSD and elevated aggression has been shown in military populations, which is sustained even when accounting for predeployment adjustment and combat experience (Marshall et al., 2005; Taft et al., 2007). It is important to note the link between PTSD and anger/aggression may be due to the association of PTSD with other factors (Taft et al., 2007), such as depression, lack of communication, poor marital adjustment, and heightened anger reactivity. Moreover, certain PTSD symptoms have been shown to be stronger predictors of violent behavior than others. Hyperarousal symptoms of PTSD have specifically been related to increased aggression in several analyses (Elbogen et al., 2010; Taft et al., 2007). While anger itself is a hyperarousal symptom of PTSD, meta-analyses suggest that among those who experienced traumatic events, effect sizes between PTSD and anger are even larger among military compared to civilian populations.

Alcohol and drug abuse has also been linked to violent and aggressive behavior in Active Duty service members and veterans. The recent multidisciplinary Epidemiologic Consultation (EPICON) found that Active Duty service members allegedly involved in crimes related to homicide were at risk for engaging in violent behavior if they had prior psychopathology including alcohol/drug disorders, mood disorders, and anxiety disorders (Millikan et al., 2012). In this sample, over 80% of those with alcohol/drug abuse problems were charged for criminal activity or misconduct while in the military *before* the alleged homicide and were at particularly high risk for continued criminal and/or violent behavior.

Veterans with TBI have been shown to be more likely to be aggressive, especially if lesions were in the mediofrontal or orbitofrontal regions of the brain (Grafman et al., 1996). The combination of TBI and PTSD—not uncommon among veterans who served in recent conflicts—has been linked to increased risk of violence in military veterans (Elbogen et al., 2010). In particular, cognitive and affective sequelae associated with TBI may be compounded by PTSD symptoms, potentially increasing risk of executive dysfunction and impulsive behavior.

Military factors have been investigated with respect to violence and aggression including firing a weapon and high combat exposure. Nonmilitary risk factors, however, also provide insight into predicting elevated violence rates. Just as in civilian populations, history of violence or a history of criminal arrest has been linked to violent behavior in service members and veterans (Elbogen et al., 2012; Millikan et al., 2012). Other significant factors related to aggression and violence in military populations include: younger age, depression, childhood abuse, witnessing family violence, and financial and employment instability (Elbogen et al., 2010; Marshall et al., 2005; Millikan et al., 2012; Taft et al., 2007).

MANAGING VIOLENCE RISK IN MILITARY POPULATIONS

There are very few empirical studies examining effects of specific interventions on managing or reducing violence and aggression in military populations. One randomized trial of cognitive-behavioral therapy (CBT) specifically aimed at reducing anger among veterans with PTSD found increased ability to control anger in the experimental group (Chemtob, Novaco, Hamada, & Gross, 1997). Another study examining CBT showed that among those in a trauma-focused group, veterans with better pretreatment relationships reported reduced posttreatment violence (Monson, Rodriguez, & Warner, 2005). CBT included components such as:

• setting treatment goals and exploring motivation,

- psychoeducation on anger
- self-monitoring and stress management
- assertiveness and communication skills training

To our knowledge, there have been no studies in military populations examining interventions specifically designed to reduce aggression. Also, it is largely unknown how (or whether) health services utilization reduces violent behavior. Treatment addressing aggression in the context of TBI, depression, or substance abuse has yet to be empirically evaluated.

Regardless, given that PTSD, depression, and substance abuse relate to violence and are modifiable (as opposed to violence history, which is static), psychotherapy and psychopharmacology would be important to consider in any risk management plan for a veteran or military service member. TBI is often comorbid with PTSD or depression; thus mental health treatment would be a possible option for reducing aggressive behavior in TBI, as well.

Bolstering protective factors is another approach to violence risk management, which can be conceptualized from the framework of psychosocial rehabilitation (Elbogen et al., 2012). Psychosocial rehabilitation encourages interventions to focus on both treatable symptoms and competence in various domains of basic functioning and psychosocial and physical well-being. The central tenets of this framework involve empowering individuals to set their own recovery goals and promoting active collaboration between individuals and clinicians. Interventions involve teaching skills to improve functioning at work, home, or social environments.

Recent research has confirmed that Iraq and Afghanistan War veterans with more psychosocial protective factors showed greatly reduced rates of severe violence and other physical aggression (Elbogen et al., 2012). Specifically, it was found that lower violence was linked to a veteran's overall psychosocial health, defined as his or her level of basic functioning (e.g., living, work, and financial stability) and psychological well-being (e.g., social support, spirituality, and resilience). Veterans with high scores on measures of psychosocial health were over 90% less likely to report severe violence compared to veterans with low scores.

Many aspects of psychosocial health (living stability, employment, social support, self-direction, meeting basic needs) are present when deployed service members live on a military base but may not be present when service members return home. Developing psychosocial health in the community can be seen as a necessary part of postdeployment adjustment. Current VA and DOD programs helping veterans find employment, maintain stable living, manage finances, strengthen resilience, and build a social support network may therefore prevent aggression after veterans return home.

CHALLENGES AND RECOMMENDATIONS

One of the biggest challenges currently facing military psychologists assessing or managing violence risk is how little we know about what factors might increase the chances that a service member or veteran acts violently and what approaches might be used to reduce or prevent violence. Most of the research conducted in this area enrolled male veterans from previous conflicts, whereas women now make up over 15% of the military. To our knowledge there are very few published studies of aggression and violence in military populations with a prospective design. Despite this, there are some steps that military psychologists can take to improve their practice of violence risk assessment and management:

Conduct a Structured Assessment of Risk Factors

To minimize decision-making pitfalls, it is recommended that military psychologists follow some general principles from scholarship on risk assessment in the civilian literature including:

- investigating risk factors shown to have a scientific association with violence
- considering individual traits as well as situational variables related to violence

- recognizing that more risk factors generally implies greater risk
- ensuring that the assessment is conducted in a structured and consistent way

Clinicians could consider the use of risk assessment tools validated with civilians such as the Classification of Violence Risk (COVR; Monahan et al., 2005) or Historical-Clinical-Risk Management-20 (HCR-20; Douglas, Ogloff, Nicholls, & Grant, 1999), with the caveat that these have not yet been validated for military populations and may not include relevant risk and protective factors (Elbogen et al., 2010). Clinicians should review risk factors with empirical support systematically in each individual case and examine how these factors link to, or increase or decrease risk of, violent behavior.

Increase Treatment Engagement

Given that PTSD and substance abuse often play a role in violence risk among veterans and military service members, traditional mental health and substance abuse treatment should be considered. CBT especially appears to show some promise (Chemtob et al., 1997; Monson et al., 2005). Also, like civilians, military service members and veterans may not think they need mental health care or may not want mental health care. Research in the last decade has identified that military service members and veterans may be resistant to getting treatment for these problems. Anger and irritability symptoms are robust predictors of PTSD treatment dropout. Whether it is stigma admitting one has a psychiatric disorder or lacking transportation to get to a mental health center, risk management of violence should address barriers to care and consider issues of treatment adherence and engagement.

Enhance Psychological Well-Being

Rehabilitation to improve resilience, coping, and social support may protect veterans and service members against engaging in violent behavior. After return home, service members may become isolated and need assistance adjusting back to family and social life. The military's current effort to have soldiers complete Master Resiliency Training may be effective in improving coping skills and thereby reducing aggression. Related, interventions that help veterans or service members manage stress and interpersonal conflict would appear to also have promise to reduce violence. Mindfulness or dialectical behavioral therapy modalities, though not tested among veterans with respect to aggression, may be especially helpful for those struggling with hyperarousal symptoms of PTSD related to violence.

Address Financial Literacy, Homelessness, and Employment

Veterans make up a disproportionate percentage of the homelessness population, and empirical data show living stability is linked to lower violence (Elbogen et al., 2012). Relatedly, improving service member and veterans' financial well-being would be important for reducing stress and strain that might contribute to conflict and aggression. Social support and stable employment are also key components to lowering risk of violence in the community. Vocational rehabilitation, VA homelessness programs, and money management interventions should be considered in violence risk management plans if pertinent.

In sum, few approaches have been developed to systematically guide risk assessment despite the pressure for providers to evaluate violence accurately and the strong need to keep veterans, their families, and the public safe. To improve clinical practice, it is recommended that military psychologists: (1) structure risk assessments based on factors already shown to empirically relate to violence and aggression, and (2) develop safety plans that address veterans' mental health, substance abuse, psychological well-being, and social environment.

References

Chemtob, C. M., Novaco, R. W., Hamada, R. S., & Gross, D. M. (1997). Cognitive-behavioral treatment for severe anger in posttraumatic stress disorder. *Journal of Consulting and Clinical Psychology, 65*(1), 184–189.

Douglas, K. S., Ogloff, J. R., Nicholls, T. L., & Grant, I. (1999). Assessing risk for violence among psychiatric patients: The HCR-20 violence risk assessment scheme and the Psychopathy Checklist: Screening Version. *Journal of Consulting and Clinical Psychology, 67*(6), 917–930.

Elbogen, E. B., Fuller, S., Johnson, S. C., Brooks, S., Kinneer, P., Calhoun, P. S., & Beckham, J. C. (2010). Improving risk assessment of violence among military veterans: An evidence-based approach for clinical decision-making. *Clinical Psychology Review, 30*, 595–607.

Elbogen, E. B., Johnson, S. C., Wagner, H. R., Newton, V. M., Timko, C., Vasterling, J. J., & Beckham, J. C. (2012). Protective factors and risk modification of violence in Iraq and Afghanistan war veterans. *Journal of Clinical Psychiatry, 73*(6), e767–773.

Grafman, J., Schwab, K., Warden, D., Pridgen, A., Brown, H. R., & Salazar, A. M. (1996). Frontal lobe injuries, violence, and aggression: A report of the Vietnam head injury study. *Neurology, 46*(5), 1231–1238.

Marshall, A. D., Panuzio, J., & Taft, C. T. (2005). Intimate partner violence among military veterans and active duty servicemen. *Clinical Psychology Review, 25*(7), 862–876.

Millikan, A. M., Bell, M. R., Gallaway, M. S., Lagana, M. T., Cox, A. L., & Sweda, M. G. (2012). An epidemiologic investigation of homicides at Fort Carson, Colorado: Summary of findings. *Military Medicine, 177*(4), 404–411.

Monahan, J., Steadman, H. J., Robbins, P. C., Appelbaum, P., Banks, S., Grisso, T., ... Silver, E. (2005). An actuarial model of violence risk assessment for persons with mental disorders. *Psychiatric Services, 56*(7), 810–815.

Monson, C. M., Rodriguez, B. F., & Warner, R. (2005). Cognitive-behavioral therapy for PTSD in the real world: Do interpersonal relationships make a real difference? *Journal of Clinical Psychology, 61*(6), 751–761.

Taft, C. T., Kaloupek, D. G., Schumm, J. A., Marshall, A. D., Panuzio, J., King, D. W., & Keane, T. M. (2007). Posttraumatic stress disorder symptoms, physiological reactivity, alcohol problems, and aggression among military veterans. *Journal of Abnormal Psychology, 116*(3), 498–507.

49 SLEEP LOSS AND PERFORMANCE

William D. S. Killgore

Sleep loss is a fact of life for most military personnel. In garrison or training environments, soldiers, sailors, airmen, and marines are expected to get up early, put in long hours, and often work late into the evening studying, training, maintaining equipment, or performing additional duties. This problem is often compounded during deployments, where soldiers may need to change time zones, operate for extended periods of time, and obtain sleep in dangerous or other inhospitable environments.

While society as a whole has been struggling with the problem of reduced sleep, service members face a number of challenges to their sleep that typically exceed those of the civilian community. Perhaps the most pervasive problem has been the long-held misperception among military personnel and their leadership that needing sleep is just a sign of laziness or weakness. As will be discussed in this chapter, nothing is further from the truth. Scientific evidence suggests that sleep is a vital contributor to warfighter performance. With the current complexities induced by asymmetric warfare, close contact urban environments, and network-centric operations, lack of sleep can degrade combat effectiveness as much or more than nearly any other element of resupply (Wesensten & Balkin, 2010). Rather than being a sign of weakness, obtaining adequate sleep can actually be a force multiplier and can enhance military capabilities. The present chapter provides an overview of the effects of sleep loss on performance and some methods for managing sleep and sustaining performance when optimal sleep cannot be obtained.

ALERTNESS AND VIGILANCE

Total Sleep Deprivation

Many studies have looked at the effects of going without sleep for one night, while a handful have studied the effects of prolonged wakefulness up to 3 or 4 days at a time. Interestingly, acute total sleep deprivation up to 88 hours has not been associated with adverse health consequences. It should come as no surprise, however, that the primary effect of prolonged sleep deprivation is a severely degraded capacity to remain alert, focused, and responsive to the environment, and an overpowering propensity to fall asleep despite strong motivation to remain awake. Military sleep-deprivation research has shown that alertness and vigilance performance remains relatively stable for about the first day of continuous wakefulness (about 16 to 18 hours awake). However, performance begins to severely degrade as wakefulness is extended beyond the normal bedtime (i.e., around midnight) and continues to decline throughout the nighttime and early morning hours, hitting a low around 0800. However, even with no sleep, normal circadian body rhythms will automatically help restore some modest level of alertness temporarily during the daylight hours, but reaction time performance will severely degrade again as nighttime approaches if sleep is not obtained (Wesensten, Killgore, & Balkin, 2005).

Not only does sleep deprivation produce sluggishness in general reaction time, it also leads to an increased propensity toward lapses in attention. A lapse is a period of nonresponsiveness to a stimulus, usually lasting half a second or longer. Some lapses may last several seconds, comprising a period when the individual is oblivious to incoming stimuli and lacks immediate situational awareness. These types of lapses are almost nonexistent in well-rested individuals, but become much more prevalent as the duration of sleep deprivation is increased. In fact, during the early morning hours of a second night of sleep deprivation, the probability of experiencing a lapse is about 1,000% greater than immediately after a normal night of sleep (Wesensten & Balkin, 2010). While a brief period of nonresponsiveness lasting a half-second or so may not seem like much, it could be the critical difference in shooting or being shot, deciding between friend or foe, failing to notice an improvised explosive device (IED), or simply having a motor vehicle accident. As a rule of thumb, it has been suggested that general alertness and vigilance performance can be expected to degrade by approximately 25% for each 24-hour period of sustained wakefulness (Belenky et al., 1994).

Partial Sleep Restriction

Continuous or sustained operations often preclude the ability to obtain a full night of sleep due to the exigencies of the mission. Over the past century, military leaders have often espoused the blatantly wrong assertion that soldiers can function effectively on 4 hours of sleep per day. While it is true that military personnel can continue to perform physically at a modest level for days or weeks with only limited regular sleep, scientific data suggest that alertness and vigilance capacities, which are critical for operational success in modern operational environments, are severely degraded by chronic sleep restriction. Figure 49.1 shows the reaction time performance for groups of participants given either 9, 7, 5, or 3 hours of time in bed to sleep each night over a weeklong stay in the laboratory. With each 2-hour reduction in nightly sleep, the ability to sustain mental focus and alertness was additionally degraded. Furthermore, those obtaining less sleep declined at a faster rate than those obtaining more sleep (Belenky et al., 2003).

Figure 49.1 also shows that even when the sleep-restricted participants were allowed three days of recovery sleep at end of the study (8 hours in bed per night), their alertness and vigilance performance never returned to prior baseline levels, suggesting that it may not be possible to fully "sleep it off" on the weekend following a period of chronically restricted sleep. It may take up to a week or more of normal sleep for performance to be restored to baseline levels after extended sleep restriction.

Finally, a recent survey of combat soldiers in Iraq (MHAT V) showed that when the amount of nightly sleep was cut by just 1 hour per night, personnel reported significantly greater difficulties handling the stresses of their jobs. Similarly, compared to soldiers reporting 7 to 8 hours of sleep per night, those getting only 4 hours of sleep per night were 250% more likely to report that they had made a mistake or had an accident that adversely affected the mission.

EMOTIONAL STABILITY

Sleep loss has severe effects on mood and emotional functioning. Military personnel who have been deprived of sleep for even a single night show significantly increased ratings of negative mood state. Furthermore, after two nights of sleep deprivation, soldiers showed significant declines in emotional intelligence and coping capacities, including reduced empathy for others, loss of self-esteem, and degraded appreciation of interpersonal dynamics (Killgore, 2010). Without sleep, soldiers showed poor frustration tolerance and increased symptoms associated with depression, anxiety, paranoia, and somatic complaints.

These impairments in emotional functioning may have a number of consequences in stressful military environments when unit cohesion and cooperation are critical. For example, a recent mental health survey of soldiers in combat (MHAT V) reported that rates of mental health problems nearly doubled for every 2 hours of chronically reduced sleep. Furthermore, sleep deprivation has been shown to reduce team performance in military settings (Baranski et al., 2007). Finally, sleep-deprived soldiers appear to be slower to make difficult, highly emotionally charged moral judgments compared to their rested performances, and were more likely to make judgments that violated their typical moral beliefs once sleep deprived (Killgore, 2010). In sum, sleepy soldiers are

more irritable, more willing to compromise moral positions, less empathic, less cooperative and team-focused, and less able to cope with the stresses of combat than when normally rested.

LEARNING AND MEMORY

The ability to encode and retain new information appears to be highly dependent on sleep. Evidence suggests that sleep is necessary before learning in order to prepare the brain to effectively encode information and is also necessary following learning in order to consolidate and integrate information into existing knowledge structures (Diekelmann & Born, 2010). Lack of sleep can have a modestly impairing effect on the ability to encode new semantic information (e.g., warning orders, commander's intent, critical enemy position information, etc.), but appears to have a particularly impairing effect on temporal memory, or the order and timing for which specific events occurred. Thus, without sleep, military personnel may have some difficulty learning new information during training or assimilating information from briefings. They may also have greater difficulty recalling the order of specific events, which could be critical in tactical situations.

Recent evidence also suggests that sleep loss has a differential effect on various types of memory, with its greatest effects on some aspects of emotional memory. Specifically, sleep deprivation impairs subsequent recall of positive and neutral stimuli, but has no significant effect on recall of negative stimuli. While more research is necessary in this area, such preliminary findings raise the possibility that sleep deprivation during stressful combat settings could bias later recall toward the retention of negative and traumatic experiences over positive ones, potentially exacerbating posttraumatic stress or other adjustment problems.

EXECUTIVE FUNCTIONS

Executive functions include a diverse set of cognitive capacities that are involved in the control and coordination of willful action to achieve future goals. These capacities include the ability to direct attention and cognitive resources, maintain information in immediate working memory, ignore irrelevant information, think flexibly, shift mental focus, form abstract concepts, and plan and sequence multiple steps. Considerable evidence now suggests that sleep deprivation impairs some, but not all, of these various capacities (Killgore, 2010).

Other than the previously described deficits in simple alertness and vigilance, current evidence suggests that sleep loss has only minimal if any significant effects on working memory capacity, logical deductive reasoning, reading comprehension, or nonverbal problem solving. In contrast, sleep deprivation appears to significantly impair divergent and innovative thought processes. Without adequate sleep, military personnel are likely to show deficits in the ability to think flexibly and creatively, plan ahead, prioritize information, detect errors and make appropriate corrections, and update courses of action when new information becomes available (Wesensten & Balkin, 2010).

Other evidence also suggests that sleep-deprived individuals tend to show greater risk-taking on behavioral tasks but deny such riskiness when queried on self-report measures, suggesting that self-awareness and judgment become impaired with sleep loss (Killgore, 2010). Consequently, when sleep deprived, a service member or military leader may begin to take greater risks but be unaware of this change. These skills and judgment capacities are often those that are particularly necessary for military commanders, who must maintain situational awareness, approach problems creatively, and adapt strategies and tactics to outmaneuver and defeat the enemy. Thus, the importance of obtaining adequate sleep for military leaders and decision-makers cannot be overemphasized. A sleep-deprived commander is at risk of mission failure due to an inability to think flexibly and adapt to a rapidly changing set of contingencies.

MAXIMIZING PERFORMANCE

Sleep Management

Individuals differ significantly in their biological need for sleep. Some people can function well on only a few hours of sleep per night, while others are likely to show deficits when obtaining less than 8 or 9 hours. On the whole, however, the majority of people begin to show some performance degradation when sleep is reduced below about 7 hours per 24-hour period. To maximize combat effectiveness, commanders should endeavor to provide 7–8 hours of sleep to their personnel within each 24-hour period (Wesensten & Balkin, 2010). Of course, mission requirements and the exigencies of combat will dictate how closely this suggestion can be followed. However, sleep should be considered on par with other vital components of resupply, such as food, water, fuel, and ammunition (Wesensten & Balkin, 2010). Military leaders need to protect the sleep opportunities of their personnel just as vigorously as they would protect these other critical elements.

Environments for sleep should be protected from noise, commotion, and light, and be kept cool and dry. Sleep facilities should be segregated to maintain a quiet and undisturbed area for those sleeping following their shifts. Finally, sleep is usually most efficient when obtained in a single session and at a time that is consistent with the downswing of the circadian rhythm of alertness (i.e., during nighttime hours—usually between 2300 and 0700). However, as long as 7–8 hours of sleep are obtained per 24-hour period, this sleep time may be broken into shorter periods (e.g., two 4-hour sleep periods). It is also important to be aware that most people experience a period of about 20 minutes of "sleep inertia" (i.e., postsleep mental sluggishness) upon awakening from a sleep episode lasting more than 20–30 minutes.

Light Exposure

Exposure to bright light, particularly sunlight in the morning hours, is critical to entraining the circadian rhythm of alertness and sleep.

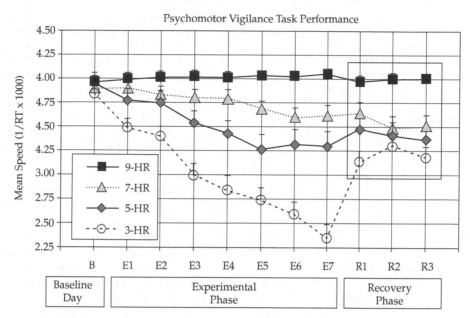

FIGURE 49.1 The effects of 7 days of sleep restriction on psychomotor speed (1/Reaction Time (RT) x 1000). From Belenky et al. (2003). Reprinted with permission.

When traveling across time zones, personnel should attempt to reset their circadian day via exposure to bright light in the morning of the new time zone as soon as possible and limit exposure to light during the evening to maximize the speed of adjustment of the new sleep schedule. When traveling eastward across time zones, most individuals will find that it takes approximately 1 day of adjustment for each time zone crossed before the circadian rhythm of alertness and sleep cycle catch up (e.g., crossing 5 time zones will take approximately 5 days before the sleep rhythm is normalized).

Caffeine

Some individuals are excessively sensitive to caffeine and should avoid this stimulant. However, when used appropriately, caffeine can be an effective temporary countermeasure to sleep loss for most individuals. Caffeine is present in many foods and beverages, but is most commonly found in coffee (approximately 100 mg per 8 oz. cup), tea, soda, and energy drinks. The effects of caffeine on alertness are only noticeable for about 3 to 6 hours after consumption

for most people, but it may cause sleep disruption or sleep onset insomnia for up to about 6 or more hours. Small to moderate repeated doses of caffeine are most effective for sustaining alertness and vigilance during periods of extended wakefulness (Kamimori, Johnson, Thorne, & Belenky, 2005). For overnight operations, 200 mg (i.e., about 1 to 2 cups of coffee) every 2 hours has been shown to be effective at sustaining basic alertness and vigilance. The effectiveness of caffeine can be enhanced by a brief 15-minute nap. For example, if excessively sleepy, a soldier can consume one cup of coffee, set an alarm, and take a brief 15-minute nap. Upon awakening, the combined effects of the nap and caffeine will usually be effective at sustaining alertness for several hours.

Military psychologists must be aware of the effects of sleep loss on cognition, mood, and judgment in order to provide commanders with relevant information and advice on sleep management and in the assessment of unit readiness. In clinical settings, military psychologists need to consider the role that sleep loss may have in psychiatric presentation, including its effects on mood, personality functioning, and judgment.

References

Baranski, J. V., Thompson, M. M., Lichacz, F. M., McCann, C., Gil, V., Pasto, L., & Pigeau, R. A. (2007). Effects of sleep loss on team decision making: Motivational loss or motivational gain? *Human Factors, 49*(4), 646–660.

Belenky, G., Penetar, D., Thorne, D., Popp, K., Leu, J., Thomas, M., ... Redmond, D. (1994). The effects of sleep deprivation on performance during continuous combat operations. In B. M. Marriott (Ed.), *Food components to enhance performance* (pp. 127–135). Washington, DC: National Academy Press.

Belenky, G., Wesensten, N. J., Thorne, D. R., Thomas, M. L., Sing, H. C., Redmond, D. P., ... Balkin, T. J. (2003). Patterns of performance degradation and restoration during sleep restriction and subsequent recovery: A sleep dose-response study. *Journal of Sleep Research, 12*(1), 1–12.

Diekelmann, S., & Born, J. (2010). The memory function of sleep. *National Review of Neuroscience, 11*(2), 114–126.

Kamimori, G. H., Johnson, D., Thorne, D., & Belenky, G. (2005). Multiple caffeine doses maintain vigilance during early morning operations. *Aviation, Space, and Environmental Medicine, 76*(11), 1046–1050.

Killgore, W. D. S. (2010). Effects of sleep deprivation on cognition. *Progress in Brain Research, 185,* 105–129.

Wesensten, N. J., & Balkin, T. J. (2010). Cognitive sequelae of sustained operations. In C. H. Kennedy & J. L. Moore (Eds.), *Military neuropsychology* (pp. 297–320). New York, NY: Springer.

Wesensten, N. J., Killgore, W. D., & Balkin, T. J. (2005). Performance and alertness effects of caffeine, dextroamphetamine, and modafinil during sleep deprivation. *Journal of Sleep Research, 14*(3), 255–266.

50   SLEEP DISORDERS

Vincent F. Capaldi II and Melinda C. Capaldi

Generally sleep complaints may be categorized by the following chief complaints:

- "I cannot fall asleep or stay asleep"
- "I cannot stay awake"
- "I have problems when I am sleeping"

This chapter will address the common sleep disorder diagnoses associated with those patient complaints.

"I CANNOT FALL ASLEEP OR STAY ASLEEP"

Insomnia

Insomnia is the inability to fall asleep or stay asleep resulting in impaired daytime function and subjective patient distress. Generally insomnia is characterized as either transient (lasting less than a month) or chronic (>1 month). The first step in treating insomnia is identification of the underlying cause or causes contributing to the patient's complaint. Insomnia is often a symptom of an underlying psychiatric condition such as depression, posttraumatic stress disorder (PTSD), generalized anxiety disorder (GAD), panic disorder, substance dependence, or bipolar disorder. Psychologists should generally screen for mood, anxiety, substance, and thought disorders prior to referral to a sleep specialist. The causes for patients presenting with transient or short-term insomnia are more easily identified than patients with chronic insomnia. For example, a service member may experience insomnia at the beginning of a deployment because of changes in their sleeping environment, excessive noise, jet lag, shift work, the stress of being separated from their families, unpleasant room temperature, anxiety about death or injury during deployment, and ingestion of stimulants such as caffeine or other medications prescribed for shift work such as modafinil. The majority of psychotropic medications prescribed in combat are hypnotics and antidepressants such as trazodone to target transient insomnia for deployed service members.

Unfortunately, insomnia often does not resolve when the service member leaves a combat zone. Difficulties initiating and maintaining sleep often persist for months after a service member returns home. A psychologist should be aware of the other causes of insomnia including:

- Sleep State Misperception
  - This condition is characterized by the patient perceiving that they are not getting enough sleep without objective evidence (i.e., a normal polysomnography).
- Inadequate Sleep Hygiene
  - Patients engage in non-sleep-promoting activities such as exercise or use of alcohol or stimulants (i.e., caffeinated beverages) prior to bedtime.
- Altitude Insomnia
  - Acute adjustment to high altitude (greater than 4,000 feet above sea level) can contribute to insomnia and increased daytime somnolence due to the stimulation of peripheral chemoreceptors.
- General Medical Disorders
  - Disorders such as congestive heart failure, chronic obstructive pulmonary disease, peptic ulcer disease, pain associated with rheumatic disorders, and gastroesophageal reflux disease are just a few general medical conditions that may contribute to insomnia.
- Restless Leg Syndrome
  - Addressed later in this chapter.
- Periodic Limb Movements in Sleep Disorder
  - Addressed later in this chapter.
- Neurologic Disorders
  - Neurologic conditions such as strokes, traumatic brain injury, headache syndromes (i.e., migraine and cluster headache), trigeminal neuralgia, and neurodegenerative disorders

(i.e., Alzheimer's disease and Parkinson's disease) all can contribute to insomnia and change sleep architecture.

A basic knowledge of how to treat insomnia is an essential tool for the military psychologist. The most effective treatment for chronic insomnia is nonpharmacological. Techniques such as cognitive-behavioral therapy (individual and group) as well as motivational interviewing are often used in the treatment of insomnia. The first step in treating insomnia is the collection of accurate information. Every patient should be asked to complete a sleep diary during treatment. This diary should record when the patient gets into bed, goes to sleep, wakes, and the number of awakenings during the night. Second, patients should be educated about sleep hygiene and stimulus control measures. The following recommendations are useful in improving sleep hygiene and enhancing stimulus control measures.

- Keep a regular sleep-wake schedule (even on weekends). Wake up at the same time every day
- Avoid drinking caffeine after lunch
- Avoid alcoholic beverages near bedtime
- Exercise regularly but not 4–5 hours before bedtime
- Use a worry diary
  - Record worries, anxieties, and stressors to address on the following day before lying down to sleep. This practice decreases the likelihood of rumination contributing to insomnia.
- Go to bed only when sleepy
- Do not use your bed or bedroom for anything except sleep (and sex)
- If unable to fall asleep in 15–20 minutes, leave the bed and go to a room with low light and engage in a nonstimulating activity until sleepy and then return to bed
- Do not take a nap during the daytime. Taking a nap during the daytime will reduce the patient's sleep drive, making it less likely that they will be able to get to sleep at night.
- Avoid using technology with lighted screens while trying to fall asleep (i.e., televisions, computers, phones, iPads)

Chronic insomnia typically abates with improved sleep hygiene and employment of stimulus control measures. These techniques are often explained to the patient in terms of conditioning one's body to associate their bed with sleep. If these initial techniques do not improve symptoms in 2–3 weeks (with diary confirmation), sleep-restriction therapy may be necessary. This technique restricts the patient to less sleep than they are reporting in their sleep diaries (often around 4 hours). As an example, a patient may be told not to go to bed until 0300 and wake at 0700. Gradually the clinician will make the patient's bed time earlier, for example, 0230 every 2–7 days as the patient maintains sleep throughout the night and a set wakeup time. Approximately 25% of patients have a resolution of insomnia with this technique. These techniques may be used in individual or group therapy settings. Some evidence suggests that CBT-oriented group therapy for insomnia utilizing teaching of sleep hygiene and stimulus control is more effective than pharmacotherapy alone.

If nonpharmacologic treatment fails, psychologists should refer the patient to a prescriber for a trial of psychotropic medications. These medications may include benzodiazepine hypnotics such as clonazepam or lorazepam, or nonbenzodiazepine hypnotics such as zolpidem or eszopiclone. Hypnotic medications should only be used for less than 1 month. Sustained intermittent use may be necessary for patients with resistant chronic insomnia. Alternatively, providers may attempt to use antidepressant medications such as trazodone or mirtazapine. When treating male patients, it is important to be aware that in rare cases (approximately 1 in 6,000 cases), the use of trazodone has resulted in the development of priapism (a painful and sustained erection).

"I CANNOT STAY AWAKE"

Obstructive Sleep Apnea

Many patients will complain about not being able to stay awake during the daytime. As mentioned above, the first step is to gather data about the problem by having the patient complete a sleep diary. The diary helps to clarify whether daytime sleepiness is due to lack of sleep or if the symptom is continuing despite a reasonable amount of sleep. Most people require 7–9 hours of sleep per night. In addition to identifying the amount of sleep one is getting, it is also useful to quantify the degree of daytime somnolence by using a rating scale. The Epworth Sleepiness Scale (ESS) is a readily available, robust measure for evaluating the severity of daytime somnolence (Johns, 1993). A score above 10 on this scale is indicative of excessive daytime somnolence (EDS). A reasonable amount of sleep combined with EDS is likely due to obstructive sleep apnea syndrome (OSAS or OSA); in fact one of the most common reasons for EDS is OSA (Lavie, 1983).

In OSA the patient stops breathing multiple times throughout the night due to airway obstruction (usually stemming from increased muscle laxity and increased adipose tissue in the neck). These apneic events result in the patient having microarousals and prevent the patient from having restful and restorative sleep. These short choking episodes throughout the night often produce an increased sympathetic tone in the patient during the daytime making diagnoses such as hypertension, depression, anxiety, and PTSD more difficult to treat.

In addition to excessive daytime somnolence there are a number of predictors of a person having OSA including those outlined in the acronym STOP-BANG (Chung et al., 2008).

- S: Do you **snore** loudly?
- T: Do you feel **tired** or fatigued during the daytime?
- O: Has anyone **observed** you stop breathing while asleep?
- P: Do you have high blood **pressure**?
- B: Is your **B**MI greater than 35?
- A: Is your **age** greater than 50?
- N: Is your **neck** circumference greater than 40 cm?
- G: Is your **gender** male?

One point is given for each positive symptom and a score greater than 3 is highly suggestive of OSA. These patients should be referred to a sleep disorders specialist for overnight polysomnography (PSG). This test is typically administered in a sleep clinic while the patient sleeps overnight. The PSG measures a number of different variables to find out how many times per hour a person stops breathing, has periods of low oxygen (hypopnic events) or has high carbon dioxide (hypercapnic events). The number of times a person becomes apneic or hypopneic is averaged over the number of hours a person sleeps to make the apnea-hypopnea index (AHI) or respiratory distress index (RDI). A score greater than 5–19 indicates mild OSA, moderate is 20–49 and severe is generally >50.

The treatment for this condition typically involves the use of a constant positive airway pressure (CPAP) machine. The patient generally wears a mask over their nose and/or their mouth. This keeps their airways open and reduces the number of apneic events during the course of the night. Most patients have difficulty adjusting to the CPAP mask and require hypnotic medications. This is particularly true for those who experience an extreme suffocating sensation. The military psychologist may aid the patient in acclimating to the CPAP mask using exposure therapy and systematic desensitization in these cases.

Depending on the military service, the diagnosis of OSA may limit the degree of austerity to which a service member can be deployed. As an example, the service member at times may only be deployed to places without electricity to support the use of the CPAP machine. In these cases, it is possible to provide the service member with an oral appliance in lieu of the CPAP machine. Of note, with a diagnosis of OSA, the service member may be entitled to medical benefits upon completing active duty service.

Narcolepsy

Narcolepsy is a rare condition that also causes excessive daytime somnolence. The vast majority of patients with narcolepsy will experience sleep attacks. These sleep attacks generally manifest as overwhelming urges to fall asleep in inappropriate situations. In general, sleep attacks are short, lasting 5–30 minutes during each episode. Patients may also experience cataplexy (a loss of muscle tone). This commonly happens when the person is excited or frightened. Some patients may also describe sleep paralysis as an inability to move just prior to falling asleep or awakening.

Narcoleptic patients also experience hypnogogic and hypnopompic hallucinations at a higher frequency than unaffected individuals. Hypnogogic hallucinations are typically visual hallucinations that occur just prior to falling asleep. Patients that are sleep deprived often endorse this phenomenon even without the diagnosis of narcolepsy. Hypnopompic hallucinations are typically visual hallucinations that occur just prior to a person waking from sleep. These hallucinations are not pathologic in themselves, but rather may be manifestations of profound somnolence.

Given that this is such a rare condition, the military psychologist should always consider the other possible diagnoses that may present with excessive daytime somnolence including: OSA, sleep deprivation, shift work, medications, drugs, alcohol dependence, and neurological conditions (i.e., seizure disorder). Referral to a sleep specialist is recommended. A sleep specialist will likely have the person sleep at the sleep clinic for a night monitored with a PSG. The patient will then undergo a multiple sleep latency test (MSLT) in the morning. The multiple sleep latency test is an objective way to measure the degree of somnolence that a person experiences during the daytime. A person is given 5 opportunities to fall asleep in a quite dark room (after sleeping for an entire night). The diagnosis of narcolepsy or EDS depends on how fast the patient falls asleep and how quickly the person goes into REM sleep.

"I HAVE PROBLEMS WHEN I AM SLEEPING"

Nightmares

Nightmares are often a symptom of an underlying psychiatric condition. The most common condition associated with nightmares is posttraumatic stress disorder (PTSD). Effective treatment of PTSD symptoms using techniques such as prolonged exposure therapy, cognitive processing therapy, or eye movement desensitization and reprocessing (EMDR) may be highly effective in decreasing the frequency and severity of nightmares. Some patients may experience an increase in nightmare symptoms when they first begin these therapies. Therefore, focused treatments of nightmares including nightmare rescripting may be required to treat patients with residual nightmare symptoms after effective treatment of other PTSD symptoms (Davis & Wright, 2006).

Military psychologists should consider alternative reasons for acute onset of recurrent nightmares including the use of certain medications (e.g., antiparkinsonian medications) or withdrawal of medications (e.g., SSRIs). Medications such as hypnotic medications (e.g., zolpidem or eszopiclone) or benzodiazepines may decrease nightmares for some individuals. Psychiatrists may also prescribe medications such as prazosin (an alpha blocker) to decrease the sympathetic response to nightmares during the night. Also, prescribers may use atypical antipsychotics (e.g., quetiapine) to aid with sleep and decrease nightmares for some patients. It is important to remember that all medications carry a degree of risk for the patient and should only be used on a time-limited basis.

Periodic Limb Movements in Sleep and Restless Leg Syndrome

Periodic limb movements of sleep (PLMS) are stereotypic limb movements that recur throughout NREM sleep. A PSG is required for the diagnosis of this condition and prompt referral to a sleep specialist should be considered. Restless leg syndrome (RLS) is a condition in which the patient experiences a sensation of creeping, crawling, tingling, or burning sensations in the calf area that is relieved with movement. This condition is generally a clinical condition that may be treated with dopaminergic medications. Patients with these concerns should also be referred to a sleep specialist. The

medications that these patients are treated with may carry significant psychiatric side effects including hallucinations and increased propensity for pleasure seeking because of increased dopaminergic activation.

Many patients will complain of muscle jerks just prior to falling asleep. Some patients describe it as a sensation of tripping or falling off a cliff just after they fall asleep with a brief period of arousal. These symptoms are consistent with myoclonic jerks or myoclonic twitches. These are normal findings and require no further follow-up.

Sleep Walking: Somnambulism

As with narcolepsy, this condition is quite rare and usually presents during childhood. Sleep walking may be hazardous to the patient and those around them and, depending on their military service, somnambulism may be incompatible with continued military service. As always, military psychologists should screen from other reversible conditions including the use of medications, substances, psychosis, and PTSD. Patients on hypnotic medications are prone to experiencing sleep walking and sleep eating.

References

Chung, F., Yegneswaran, B., Liao, P., Chung, S. A., Vairavanathan, S., Islam, S.,...Shapiro, C. M., (2008). STOP Questionnaire: A tool to screen patients for Obstructive Sleep Apnea. *Anesthesiology, 108*, 812–821.

Davis, J. W., & Wright, D. C. (2006). Exposure, relaxation, and rescripting treatment for trauma related nightmares. *Journal of Trauma and Dissociation, 7*, 5–18.

Johns, M. W. (1993). Daytime sleepiness, snoring, and obstructive sleep apnea: The Epworth Sleepiness Scale. *Chest, 103*, 30–36.

Lavie, P. (1983). Incidence of sleep apnea in a presumably healthy working population: A significant relationship with excessive daytime sleepiness. *Sleep, 6*, 312–318.

51 GRIEF, LOSS, AND WAR

Kent D. Drescher

Death is a universal human experience, and consequently most people at some point in their lives experience the loss, through death, of someone close to them. Hensley and Clayton (2008, p. 650) state, "Bereavement is the reaction to a loss by death. Grief is the emotional and/or psychological reaction to any loss, not limited to death." Death, when it occurs in the military is often unexpected, and possibly violent. It likely involves a young to middle-aged adult and thus is perceived as premature. Military death may also have circumstances that add to the pain of the loss, such as occurring violently and unexpectedly far away, after a long period of separation from family and friends, and bodily remains that may not be intact or viewable. For unit members present at the time of death, exposure to the violence of battle, the horrific devastation of modern weapons, and observing or participating in the struggle to save their friend's life can add additional distress. Military culture and the warrior ethos that includes values of selfless-sacrifice, hero as protector, and stoicism may shape short- and long-term grief responses.

Military grief can have a profound impact on service members, units, parents, spouses, children, and even care-providers. Small unit cohesion is an operational priority that is carefully nurtured during training. The result is that unit members frequently experience extremely close relationships that generally promote resilience in the face of intense stressors. When death occurs within a unit, the pain associated with the loss can be intense. During deployment, while brief memorial ceremonies will likely be provided, frequently the unit will continue to function at full operational status, and unit members will have little time to process the loss. The high operational tempo of military service, particularly during wartime, can significantly interfere with the normal process of grieving and mourning following death. The US Navy and Marine Corps doctrine now identifies grief and loss as one of four causes of combat operational stress injury (US Marine Corps & US Navy, 2010).

In addition to death, many service members experience devastating wounds, resulting in permanent loss of physical and mental functions. There may be loss of relationships with close friends as units return home and some members leave service, and sometimes lost relationships extend to intimate relationships with spouses, partners, children, parents, and others, as the veteran returns "different" from the person who first deployed. Military members may experience loss of self as they find themselves changed and unable to experience the world in the same way as they did prior to deployment. This may include loss of innocence that comes about through exposure to the horrors and carnage of war. Along with each of these different types of loss comes a range of emotional experiences—grief, sadness, rage, bitterness, confusion, and disappointment.

NORMAL VERSUS COMPLICATED GRIEF

While for many years it was thought that "grief work" was required for healthy adaptation to loss, current clinical consensus is that therapeutic intervention for normal grieving is neither necessary nor particularly useful. This however raises the question as to what constitutes normal grieving. The intensity of acute grief and the time period over which it occurs vary and is affected by factors including closeness of the relationship, circumstances of the loss, age of the deceased, expectedness of the death, and the amount of trauma/violence involved. One thing that is clear is that there is a wide variation in grief practices and expectations about appropriate emotional experience and expression across cultures, ethnicities, religions, and families. In accord with ethical principles, and because of the variation in normal grief practices, cultural and religious factors need to be taken into consideration when assessing grief, and in interacting with individuals experiencing acute grief, as they can play a major role in determining the parameters of normal grief for a particular individual.

For example, members of particular religious groups may have strong feelings about who handles a body, how and what procedures are performed, and the time these procedures take that may be very important in the acute aftermath of a death. Members of some ethnic minority groups may express grief differently (e.g., appear more withdrawn or stoic) in the presence of providers with different ethnicity due to issues of trust. Members of certain cultures may have a strong need for the presence of extended family and expectations about which family member(s) will take the lead in guiding important life decisions. As a provider it is important to be alert to the possible impact of religious and cultural differences, and aware of how one's own culture and religion may impact on the clinical relationship, without rushing to judgment or stereotyping.

Grief can produce functional disruption in four main life areas: (1) cognitive organization, (2) mood, (3) physical health and self-care, and (4) social and occupational functioning. Generally, these disruptions are seen most strongly in the immediate aftermath of the death and subside gradually during the first year following the death. Symptoms reported by the bereaved most frequently in the first month following the death include: crying, sleep problems, loss of appetite, fatigue, loss of interest, poor memory, restlessness, and low mood. Maladaptive coping strategies to manage emotional pain such as substance use may also be present (Hensley & Clayton, 2008).

Grief among children (who lose a parent or sibling) may manifest itself somewhat differently. Developmentally, children are generally less able to verbalize their thoughts and emotions and may be less able to tolerate strong emotions for long periods. They may avoid talking about the death, or only engage those feelings intermittently. They may take on mannerisms of the deceased or attach themselves to objects connected to the deceased. In younger children regressive behaviors such as tantrums, bedwetting, and separation anxiety may occur. In older children, school problems, anger outbursts, withdrawal from adults, and even risk-taking behaviors may be seen. It would be normal to expect some behavioral regression and acting-out behavior (Hensley & Clayton, 2008).

Most US studies indicate that for most individuals grief intensity is much reduced after about 6 months. It is also becoming clear that roughly 10% of individuals (those who have lost a loved one to a violent death or suicide and those who have suffered the loss of a child) continue to have strong distress and functional impairment long after the death of someone they cared about (Shear et al., 2011). Current clinical nomenclature uses the term "complicated grief" to describe these extended and problematic grief reactions. Grief may also co-occur with other psychological disorders such as PTSD and depression. A clinician evaluating a bereaved person has the challenge to avoid pathologizing a normal condition while at the same time identifying and treating grief when there is clear and sustained distress and functional impairment.

Complicated grief has been proposed for inclusion as a discrete psychiatric disorder in the next *Diagnostic and Statistical Manual*

(DSM-5). Shear and colleagues (2011) argue that evidence is strong that complicated grief can be distinguished from other disorders, and can benefit from treatment. They also suggest that evidence indicates that complicated grief is an aberrant response to loss that occurs in a minority of individuals. They note several well-validated assessment instruments, the most commonly used being the 19-item Inventory of Complicated Grief (ICG) (Prigerson et al., 1995). There is some symptom overlap with CG and depression (sadness, crying, sleep disturbance, suicidal thoughts), and with PTSD (intrusive images/memories, avoidance behavior, social estrangement, sleep/concentration problems). However, careful assessment of these symptoms will reveal different mechanisms at work. For example, PTSD is characterized by intrusive, painful recollections of a traumatic event. Similarly, CG frequently involves preoccupying ruminative thoughts about the person that died. In both cases recurrent memories are involved, however, in PTSD the memories are aversive and avoided, while in CG the memories of the deceased may be treasured and welcomed. A complicating factor is that the disorders are not mutually exclusive and service members can develop both. One might imagine that, when the disorders are comorbid, the presence of both types of recurrent memories could actually delay recovery and complicate treatment for both disorders.

SUICIDE

Suicide is a mode of death that creates serious emotional challenges for family, friends, unit members, and care providers that are left behind. Suicide "survivors" is a term frequently used to describe individuals in close relationship or contact with a person who dies by suicide. Suicide survivors are those individuals who experience the emotional impact, and have to attempt to make sense of a death that they frequently do not understand. There are a number of problematic issues and questions that arise for suicide survivors. Questions of "why" and feelings of guilt and responsibility

are common. Individuals may overestimate their role in contributing to or failing to prevent the suicide. Families and close friends of the deceased may experience shame, stigma, and a desire to keep the facts about the death secret from others, resulting in social isolation.

The Military Funeral Support instruction provides guidance for military honors and memorial services for service members and Veterans regardless of manner of death. The extent to which those honors would be rendered to a person who committed suicide will vary greatly depending on the request of the family and by the personal beliefs of military commanders. Military unit memorial services, especially when deployed, are greatly influenced by a leader's beliefs and attitudes about honor and the impact of the death on morale and unit cohesion. There have been significant changes supporting the honoring of suicide deaths since 2008. One high visibility policy change includes the 2011 presidential policy change to include writing bereavement letters to survivors of military suicide. Differences in how suicide deaths and deaths from other causes are viewed and dealt with by the military may worsen stigma for survivors.

Social support networks may be at a loss as to how best to relate to grieving suicide survivors, and as a consequence fail to provide needed support. Because military command staff need to understand the reasons for the death, in order to prevent future deaths, as well as to provide information to families and friends, military psychologists may be called on to participate with other staff members in conducting a "psychological autopsy," a formal administrative process designed to identify personal and environmental risk factors that led to the suicide. The postsuicide investigation may be perceived as intrusive by some survivors or intensify guilt as they reflect with hindsight on missed opportunities that may have represented signs of impending suicide. Because the causal factors of suicide are frequently unclear, suicide survivors may experience a range of emotional distress that includes intense anger at the deceased for their "selfish" action. Data also indicate that a suicide can increase the suicide risk of close friends and family who are left behind. These individuals are also at high risk for development of complicated grief (Jordan, 2008).

COMPLICATED GRIEF TREATMENTS

There are currently two evidence-supported treatments that address the challenges of CG. Only one of these treatments, adaptive disclosure (AD) (Gray et al., 2011) has been developed for and evaluated within a military environment, though results from the initial trial must be considered preliminary, and additional successful trials are necessary to confirm its utility. Adaptive Disclosure is a six-session psychological intervention developed for military service members with combat stress issues related to fear and life threat, traumatic grief and loss, and shame/guilt and inner/moral conflict. The treatment utilizes traditional cognitive-behavioral therapeutic techniques such as exposure and combines them with elements of other therapeutic approaches. When significant grief is present, AD utilizes an empty chair technique derived from Gestalt therapy, to allow the client to engage in a guided conversation with the imagined deceased individual or with a moral authority that allows for expression and exploration of previously unexpressed thoughts and emotions about the death.

AD has a number of goals, including changing service member expectations about disclosure of traumatic reactions. It seeks to make thoughts and beliefs about loss and trauma explicit so that they can be examined and reevaluated. It also seeks to increase client self-efficacy about managing recurrent painful memories of trauma and loss. AD is designed to be respectful of military values and culture. The initial open trial of the intervention found large treatment effects for both PTSD and depression, and a comparable attrition rate to other longer evidence-based PTSD treatments. A small to medium treatment effect for posttraumatic growth was also noted. Rates of service member satisfaction with this treatment were very high.

Complicated grief treatment (CGT) (Shear, Frank, Houck, & Reynolds, 2005) is a 16-session psychotherapy process that utilizes elements derived from both cognitive-behavioral and interpersonal treatment models. A successful randomized control trial found CGT superior to interpersonal treatment for depression with a sample of bereaved individuals from the general population. No studies specific to military grief have been performed. Recurrent painful memories of the deceased, and particularly traumatic elements of the death are engaged using "revisiting," an imaginal exposure procedure where the client recounts (and records) the story of the death, and then listens to the audio recording of the revisiting exercise at home between sessions. Additional "in vivo" exposure techniques are utilized to enable people to reengage with previously avoided activities. Another treatment element is the identification of personal life goals and assisting clients in reengagement in these life goals and in significant and meaningful relationships. Treatment goals include reducing the emotional intensity of grief, helping clients reconnect with positive memories of the deceased, and supporting clients in reengaging with life activities and relationships.

SPIRITUALITY AND MEANING

Religion and spirituality, including beliefs, rituals, and practices have been a resource utilized in most cultures throughout recorded human history to assist individuals coping with grief and loss. Religion and spirituality are multidimensional constructs, and so no single statement can fully describe the role that religion and spirituality play in grief recovery. Religious or spiritual involvement may provide benefit for the grieving in several ways. Healthy lifestyles promoted by religious communities may reduce the likelihood of unhelpful coping such as substance use. Spiritual practices such as meditation, prayer, and reading of sacred texts may provide anxiety reduction and helpful emotional comfort, and spiritual beliefs may provide a framework for meaning-making to begin. Involvement in a spiritual community may provide helpful social support that reduces isolation and loneliness. Wortman and

Park (2008) review the complex literature on the relationship of spirituality and bereavement. They note that several studies suggest that meaning-making following a death is one mechanism through which spirituality is related to bereavement. Davis and colleagues (1998) noted that the bereaved frequently attempt to use two mechanisms to find meaning in death. One is "finding benefit," and the other is "making sense." For example, a bereaved person might note as a benefit that they "have a new appreciation for life following the loss." Examples of making sense of loss could include viewing the loss as "God's will," or attributing the death to the deceased's "lifestyle." While these mechanisms are common with nontraumatic deaths, they can be challenging for people to utilize when the death was traumatic or violent. In such situations finding meaning can be a serious challenge.

Chaplains have played an important role assisting grieving military service members in this country since before the Revolutionary War. Chaplains have a unique role in the US military in that they are the only care provider with full confidentiality. Every chaplain is endorsed by a particular faith group and provides religious ministry services to members of their own faith group, yet every chaplain is tasked with providing care, support, and counseling services to all service members, of all ranks, regardless of their faith perspective or lack thereof. For some service members, speaking with a chaplain about grief and loss issues will be extremely helpful in the meaning-making process.

RECOMMENDATIONS

1. Assess for grief and loss issues in initial clinical interviews with military personnel. If significant grief-related impairment is present more than 6 months following a death, assess for complicated grief.
2. Assess how strongly the patient identifies with the military culture and how that influences the grief experience and ascribed meaning.

3. Consider the role grief may play underlying other mental health issues/problems (i.e., PTSD, depression, substance use).
4. Become familiar with novel treatment elements in current complicated grief treatments.
5. Establish relationships with chaplains to allow for effective collaboration and bidirectional referrals around grief and loss.

References

Gray, M. J., Schorr, Y., Nash, W., Lebowitz, L., Amidon, A., Lansing, A.,…Litz, B. (2011). Adaptive disclosure: An open trial of a novel exposure-based intervention for service members with combat-related psychological stress injuries. *Behavior Therapy, 43*(2), 1–9. doi:10.1016/j.beth.2011.09.001

Davis, C. G., Nolen-Hoeksema, S., & Larson, J. (1998). Making sense of loss and benefiting from the experience: Two construals of meaning. *Journal of Personality and Social Psychology, 75*(2), 561–574.

Hensley, P. L., & Clayton, P. J. (2008). Bereavement: Signs, symptoms, and course. *Psychiatric Annals, 38*(10), 649.

Jordan, J. R. (2008). Bereavement after suicide. *Psychiatric Annals, 38*(10), 679.

Prigerson, H. G., Maciejewski, P. K., Reynolds, C. F., Bierhals, A. J., Newsom, J. T., Fasiczkaa, A.,…Miller, M. (1995). Inventory of Complicated Grief: A scale to measure maladaptive symptoms of loss. *Psychiatry Research, 59,* 65–79.

Shear, K., Frank, E., Houck, P. R., & Reynolds, C. F. (2005). Treatment of complicated grief. *JAMA: Journal of the American Medical Association, 293*(21), 2601.

Shear, K. M., Simon, N., Wall, M., Zisook, S., Neimeyer, R., Duan, N.,…Keshaviah, A. (2011). Complicated grief and related bereavement issues for DSM-5. *Depression and Anxiety, 28,* 103–117. doi:10.1002/da.20780

US Marine Corps & US Navy. (2010). *Combat and operational stress control, MCRP 6-11C/NTTP 1-15M.* Quantico, VA: Marine Corps Combat Development Command.

Wortman, J. H., & Park, C. L. (2008). Religion and spirituality in adjustment following bereavement: An integrative review. *Death Studies, 32,* 703–736. doi:10.1080/07481180802289507

52 EARLY INTERVENTIONS WITH MILITARY PERSONNEL

Maria M. Steenkamp and Brett T. Litz

Early interventions are delivered before or soon after the development of clinically significant distress or impaired functional capacities. In intervening early, the aim is to prevent mental disorders or chronic and entrenched problems from forming, to keep the duration of suffering and functional impairments to a minimum, and to prevent secondary problems (for example, aggressive acting out, substance abuse/dependence) from arising. Early interventions target "preclinical" states because they focus on chiefly asymptomatic or subclinically impaired individuals (e.g., those with subsyndromal or partial posttraumatic stress disorder [PTSD]). In the military, early interventions are most commonly designed to redress PTSD, given that the disorder develops after a discernible traumatic event that putatively demarcates the onset of difficulties. PTSD is also the signature mental health problem among the military. In practice, early interventions are delivered much more proximal to exposure to trauma or loss relative to formal mental health *treatment*, which typically is provided after symptoms have persisted for a notable period of time (months and years). Since the start of the wars in Iraq and Afghanistan, the military has implemented several large early intervention programs that span the prevention continuum.

TYPES OF PREVENTION

Historically, there were three types of mental health prevention, namely *primary*, *secondary*, and *tertiary*. Generally, primary prevention entailed broad public health programs to prevent new onset of mental health problems. Secondary prevention entailed strategies to assist individuals with preclinical symptoms early in the development of their disease, to prevent full disorder. Tertiary prevention (treatment) entailed providing care to those with mental health disorders in the hopes of cure, relapse prevention, or rehabilitation. In the trauma context, each type of prevention in this scheme was in part defined based on when it was provided, relative to the trauma. Primary prevention would entail providing intervention before trauma exposure, secondary prevention occurred after trauma exposure (e.g., in emergency rooms), and tertiary prevention was after a trauma-linked mental disorder was present.

The revised and expanded Institute of Medicine (IoM) prevention scheme (Munoz et al., 1996) is especially pertinent to understanding and distinguishing the spectrum of prevention and care resources germane to the challenges and responsibilities of the military. In the IoM framework, mental health prevention entails a continuum of strategies and ways of conceptualizing the needs of individuals at risk, from resilience promotion to after-care and rehabilitation for chronic conditions. In this context, formal prevention interventions are based on *whom they target*: (1) universal prevention targets a whole population; (2) selective prevention targets all members of subgroups at presumed equal increased risk; and (3) indicated prevention targets at-risk individuals with preclinical symptoms and impairments in functioning.

Ideally, all three types of prevention initiatives need to be evidence-based, first drawing on theory and empirical research in their design and implementation, and then examined using outcome research to evaluate program efficacy and effectiveness. In the context of trauma and PTSD, prevention approaches need to draw on research on risk and protective factors to provide an evidence base for their content, targets, procedures, and evaluation. Also ideally, universal, selective, and indicated programs need to work together synergistically within a larger prevention framework, to ensure a continuum of prevention that provides systematically higher levels of care as risk and impairment increase. It is currently unclear to what extent these ideals are being realized. For example, the DoD recently commissioned a study to catalog the various psychological health programs funded or sponsored by the DoD. Over 200 different programs were identified that putatively address psychological health and traumatic brain injury (Weinick et al., 2011). The report found that these programs tended to be decentralized and developed in isolation, and fewer than a third reported having had an outcome evaluation of their services in the past year, making it difficult to draw conclusions about their effectiveness.

Using the IoM framework, we discuss each of these prevention types below, focusing on their use with PTSD and the challenges involved in implementing each type of prevention.

UNIVERSAL PREVENTION IN THE MILITARY

Universal prevention strategies equip individuals with the knowledge and coping skills necessary to foster mental health and encourage bounce-back from psychological challenges that may otherwise result in pathology. They can be aimed at general wellness promotion and/or actual disorder prevention, two constructs that are conceptually distinct. The most notable example of a military universal prevention intervention is the Army's Comprehensive Soldier Fitness (CSF) program, which is aimed at both wellness promotion and disorder prevention (see Seligman & Matthews, 2011). Initiated in 2009, CSF is designed to enhance mental "fitness" and resilience using a positive psychology framework. The multicomponent program includes computerized learning modules on emotional, social, family, and spiritual "fitness," computerized assessment of these domains through self-report measures, and in-person training in advanced resilience skills for noncommissioned officers (NCOs), who apply and disseminate the materials to fellow troops.

An advantage of universal prevention is its breadth, since entire populations of individuals can be targeted. The rationale is that even small changes at the population level can translate into a lower incidence of psychological difficulties and pathology. That is, small effect sizes obtained in an entire population can be substantive and beneficial to public health. However, while the conceptual appeal of universal prevention is clear, its successful implementation in practice is more difficult and, unfortunately, few examples of successful universal preventions exist in either the military or civilian context. There are no rigorous studies testing the effects of universal prevention initiatives for PTSD in either military or civilian settings and, as a result, whether it is even possible to prevent PTSD using a universal prevention framework remains an open question.

Two difficulties related to universal prevention are particularly noteworthy. First, designing universal prevention programs is complicated by our inability to accurately predict who will develop a mental disorder after exposure to putative trauma. To design effective universal prevention programs, developers need to know the causes and mechanisms of susceptibility. However, research on risk and resilience factors tend to lack specificity; those who exhibit or possess a certain risk factor may not necessarily go on to develop a disorder, either because that risk factor is by itself insufficient for the development of the disorder, or because various resilience or protective factors cancel out the effects of the risk factor. This is also true in the case of PTSD, with known predictors of PTSD accounting for less than 20% of the variance in the disorder (Ozer et al., 2003).

Second, service members have historically been provided extensive and varied forms of *informal* universal prevention (and behavioral health promotions) to prepare for military stressors. Any formal universal prevention programs need to demonstrate incremental validity over and above these natural universal prevention factors. The programs that build operational competence and competence on the individual and group level, such as tough realistic training, effective leadership, and unit cohesion-building are not formally designed to prevent mental health disorders in the face of military stress, but these are effectively universal mental health prevention programs in the military. For example, they are delivered to the entire population regardless of risk status, and they improve the overall well-being and resilience of the population while arguably contributing to the goal of universal mental health prevention, namely to positively shift the population distribution of the incidence of mental disorders by addressing their underlying causes. Moreover, the military is a selective institution, requiring that physical and psychological performance standards be met prior to enlistment, while processes such as basic training serve a gate-keeping function for excluding individuals unable to meet physical and psychological demands.

SELECTIVE PREVENTION IN THE MILITARY

In the military, all members of units exposed to shared military trauma and loss are often provided selective prevention strategies, regardless of impairment. For example, the Army's Battlemind Psychological Debriefing is an in-theater group-based method that is either delivered at set intervals during deployment (to address the cumulative impact of deployment stressors) or in response to a traumatic event. Debriefings aim to normalize reactions, promote sharing and unburdening of emotions and depictions of events and experiences (although traumatic events are not recounted in detail); they also attempt to create shared accurate, and in theory more helpful, appraisals of what happened and its meaning moving forward for the group or unit, and to prepare soldiers mentally to return to duty. A separate Battlemind debriefing is also provided once troops return from deployment, focusing on the psychological challenges of the transition from combat to home. There is initial evidence for the effectiveness of these interventions (see Adler et al., 2009).

Similar to universal prevention, one challenge of selective prevention in the military involves demonstrating incremental benefit over natural recovery processes. The vast majority of service members, including those who have significant trauma exposure during deployment, will recover naturally on their own, based on their own resources—such as leader, peer, and family support and meaning-making—and their resourcefulness (e.g., to find respite) without preventive (or therapeutic) interventions. This is true even for service members who may initially exhibit marked emotional distress following a trauma. Service members with PTSD remain a minority—as many as 80% of all service members will not experience clinically significant mental health difficulties related to their deployment. As such, any selective prevention program must guard against disrupting such natural recovery processes.

A related challenge is knowing to whom to provide selective prevention: the vast majority of service members will experience at least one (and typically multiple) traumatic or stressful events while deployed, technically making most service members "at risk" and eligible for selective intervention. It remains unclear how best to distinguish different gradations of risk to identify those service members most in need of formal interventions.

INDICATED PREVENTION IN THE MILITARY

Indicated prevention entails curtailing the duration and exacerbation of subclinical symptoms that have already developed, that is, targeting service members with subthreshold PTSD symptoms so that they do not develop clinically diagnosable PTSD and related disabilities. Historically, symptomatic service members were treated using the forward psychiatry principles of proximity, immediacy, expectancy, and simplicity (PIES), which emphasized providing prompt care within close proximity to the unit, and with expectation of return to the unit. Care consisted of encouraging restoration of function through simple and practical tools such as sleep, nutrition, and hygiene.

An example of an indicated prevention is the Navy and Marine Corps combat and operational stress control doctrine, which includes formalized *psychological first aid* principles for indicated prevention (Nash, Krantz, Stein, Westphal, & Litz, 2011). The model, called combat and operational stress control first aid (COSFA), is based on the stress injury continuum model, which hypothesizes a range of stress reactions, varying in severity from adaptive coping ("ready"), to mild and transient distress and dysfunction ("reacting"), to more severe and persistent— but subclinical—distress and dysfunction ("injured"), to clinically diagnosable conditions ("ill"). It includes both short-term responses to acute symptoms and difficulties, and longer-term interventions that promote healing and recovery of social connectedness, personal and collective competence, and self-confidence.

As is the case with indicated prevention, a notable challenge is knowing when and to whom to provide indicated interventions. There is considerable stigma surrounding mental health difficulties in the military, meaning that programs that rely on service members' ability and willingness to present themselves for formal services may fail to capture a significant portion of those suffering. To this end, the COFSA model explicitly encourages leaders, support personnel, and Marines and sailors themselves to identify individuals in need of indicated prevention through ongoing assessment and, in turn, individuals in need of mental health treatment.

RECOMMENDATIONS

Because most of the extant programs are in development, have little civilian or military precedent to guide them, and have not been well tested, recommendations for early intervention in the military unfortunately cannot be based on solid outcome data at this time. However, given the state of the evidence, the following recommendations are made:

- Prevention should be conceptualized in the context of the continuum of care articulated in the IoM scheme.
- Specialized psychological interventions need not be provided to all service members exposed to high magnitude combat or operational experiences, such as a loss in a unit. In those contexts, good leadership, respite, training, and peer support will be sufficient for most service members.
- However, if service members are exposed to high magnitude combat or operational experiences and are showing signs of performance decrements, preclinical distress/subsyndromal PTSD, or withdrawal or disengagement, we recommend that they be formally assessed and provided a set of CBT-based procedures and skills focusing on ensuring safety, reducing arousal, increasing self-care, engagement with others, and tasks that rebuild confidence and competence.
- The psychological first aid procedures and model is a good starting place for stepping to actual CBT approaches provided by caregivers, if necessary.
- Finally, if mental health professionals are engaged, they have to be familiar with the unit and culture, and they need to coordinate care across disciplines and especially with leaders.

References

Adler, A. B., Bliese, P. D., McGurk, D., Hoge, C. W., & Castro, C. A. (2009). Battlemind debriefing and Battlemind training as early interventions with soldiers returning from Iraq: Randomization by platoon. *Journal of Clinical and Consulting Psychology, 77,* 928–940. doi:10.1037/a0016877

Munoz, R. F., Mrazek, P. J., & Haggerty, R. J. (1996). Institute of Medicine report on the prevention of mental disorders. *American Psychologist, 51*, 1116–1122.

Nash, W. P., Krantz, L., Stein, N., Westphal, R. J., & Litz, B. T. (2011). Comprehensive soldier fitness, Battlemind, and the stress continuum model: Military organizational approaches to prevention. In J. Ruzek, J. Vasterling, P. Schnurr, & M. Friedman (Eds.), *Caring for the veterans with deployment-related stress disorders: Iraq, Afghanistan, and beyond* (pp. 193–214). Washington, DC: American Psychological Association.

Ozer, E. J., Best, S. R., Lipsey, T. L., & Weiss, D. S. (2003). Predictors of posttraumatic stress disorder and symptoms in adults: A meta-analysis. *Psychological Bulletin, 129*, 52–73.

Seligman, M. E. P., & Matthews, M. D. (Eds.). (2011). Comprehensive soldier fitness [Special Issue]. *American Psychologist, 66*(1).

Weinick, R. M., Beckjord, E. B., Farmer, C. M., Martin, L. T., Gillen, E. M., Acosta, J. D.,… Scharf, D. M. (2011). *Programs addressing psychological health and traumatic brain injury among U.S. military servicemembers and their families.* Santa Monica, CA: RAND Corporation.

53 THE PSYCHOSOCIAL ASPECTS AND NATURE OF KILLING

Richard J. Hughbank and Dave Grossman

THE PHENOMENON OF KILLING

There are many reasons why people kill. Whether a person's personality is innate at birth (nature) or developed over time based on societal influences (nurture), the idea of killing and death varies from person to person. One view may see some killing as immoral (e.g., premeditated mass and serial killings) and some killing as necessary (self-defense or in the defense of others). Combat-related killing lies within a gray area and may be seen as either premeditated murder or self-defense, depending on one's viewpoint.

Multiple psychosocial factors influence the act of killing, and psychological motivations for killing have proven a viable subject of research. Some killers have been made to feel like outsiders and are driven by rage, as in the cases of Dylan Klebold and Eric Harris of the Columbine High School massacre, Seung-Hui Cho of the Virginia Tech University massacre, and Steven Kazmierczak of the Northern Illinois University. Other killers' motivations might be rooted in a psychological disorder such as antisocial personality disorder or conduct disorder, as with the serial killers Ted Bundy, Richard Ramirez, Jeffrey Dahmer, and John Wayne Gacy. They killed out of desire, need, fantasy, or social disconnect, all of which are considered unacceptable reasons for killing in most societies.

Others kill based on their perceived need for survival in cases such as self-defense or defense of others. As it relates to self-defense, killing is a tool for survival. When faced with a deadly adversary, the fight or flight response mechanism is triggered, forcing us to make a difficult decision—a response generally identified as the archetypal human response to a stressful situation (Taylor, Klein, Lewis, & Gruenewald, 2000). This immediate decision—whether it is cognitive or reflective—reflects an inherent safety concern; in essence, survival instincts overwhelm us all. As Grossman (1995) noted,

The fight or flight dichotomy is the appropriate set of choices for any creature faced with danger other than that which comes from its own species. When we examine the responses of creatures confronted with aggression from their own species, the set of options expands to include posturing and submission. (p. 5)

Ultimately, survivability depends on one's ability to exhibit an authoritative response to a given threat. The fight or flight mechanism is a critical reactionary mechanism within the survival process. For those who decide to face the threat, killing becomes one of the possible aggressive acts that result. In essence, a counterposture is initiated as a defense mechanism. Killing now becomes a survival tool, reflecting a cognitive state of mind that is innate and instinctive in everyone.

KILLING IN COMBAT

The concept of killing in combat is extremely controversial by its very nature. While those serving in the military are trained to kill, the "killing response" varies in each person. Killing is a by-product of combat, but training does not eliminate the emotional accountability of a person's actions; and each must deal with their own actions. Grossman (1995) described this process as "The Killing Response Stages":

1. Concern about being able to kill;
2. Killing circumstance;
3. Exhilaration from kill;
4. Remorse and nausea from kill; and
5. The rationalization and acceptance process. (pp. 232–240)

Grossman noted that "some individuals may skip certain stages, or blend them, or pass through them so fleetingly that they do not even acknowledge their presence" (p. 231).

Grossman (1995) also discussed how combat personnel may create a psychological capability to kill as a by-product of "dehumanizing" their enemy. This is a common practice whose origins can be traced back in modern times to World War II. It is not natural for one human being to have a psychological drive to kill another, let alone kill a complete stranger. Creating a psychological profile that tears away at the humanity of another type of person is possible when applied throughout the daily training of military personnel. While this might appear inhumane to some, we suggest that war and killing is inhumane to begin with.

For some, killing in combat appears as a "normal" behavior. However, this is not necessarily the case. Under most circumstances, killing in combat becomes a responsive action. An action that is voluntary in nature because of training, but facilitated as a need for survival through a self-defense mechanism (fight or flight). In many ways, this is reminiscent of Abraham Maslow's hierarchal theory of human motivation and behavior.

Maslow's first level of needs—"self-actualization"—points to morality, problem solving, and acceptance of facts, among other things. The second step—"esteem"—introduces self-esteem and confidence. The third level—"love/belonging"—of the pyramid mentions friendship, and the fourth level—"safety"—security of the body. Killing in combat is a defense mechanism; a means to an end when faced by an enemy focused on killing others. The ability to kill is ingrained through training; however, the willingness, or necessity, to kill is relative as it relates to Maslow's theoretical approach to human motivation and behavior and the need to survive.

RELIGIOUS AND POLITICAL IDEOLOGICAL MOTIVATIONS

For a select few, killing is driven by religious or political ideologies. Persons belonging to domestic terrorist organizations such as the Ku Klux Klan or the Christian Identity Movement are motivated by religious fervor based on their interpretation of biblical writings. In some noted cases, people belonging to organizations such as militias kill based on political ideologies. They are motivated by their disdain for government actions and commit terrorist acts in an effort to change political policies. Timothy McVeigh's bombing of the Murrah Federal Building in Oklahoma City is a clear example of the militia mentality and the desire to kill based on the desire to change political policy.

Some kill for both religious and political ideologies, as in the case of Islamist terrorism. Jihad (holy war) is a way of life driven by a religious belief structure. According to Shoebat (2007),

A jihadist's view breaks the world into two classes, the Muslim class and the non-Muslim class which, in turn, will survive under the rule of Islam.... For the Jihadists of today the goal is to regain the glory days of the 7th century. (pp. 27–28)

Individuals who are recruited, trained, and execute terroristic acts do not necessarily believe that killing infidels (nonbelievers) is unjust; rather they believe infidels present a danger through their teachings and perceived understanding of Islam. As is the case of those who commit terrorist acts under the Christian Identity Movement, this religious belief structure is not adhered to by all within the religion of Islam (Hughbank, Niosi, & Dumas, 2010).

THE HUMAN NATURE PARADIGM

Withdrawal from society may play a pivotal role in psychosocial aspects of killing as an outlet. Many consider those who isolate themselves from the rest of society as outsiders. Wilson (1967) described the outsider as follows:

At first sight, the Outsider is a social problem,... [and] the outsider tends to express himself in Existentialist terms. He is not very concerned with the distinction between body and spirit, or man and nature; these ideas produce theological thinking and philosophy; he rejects both. For him, the only important distinction is between being and nothingness. (pp. 11 and 27)

Based on this concept of isolation and Wilson's analysis of the outsider's social issues, a correlation could be drawn between the alienation from societal norms and the desire to express oneself through the act of killing. Accordingly, "the problem of death, and of meaning in life, is completely dissociated from human cruelty and 'man's inhumanity to man'" (Wilson, p. 149).

The common denominator in this case now becomes the "question of identity" within an outsider as "the outsider is not sure who he is. 'He has found an 'I', but it is not a true 'I'.' His main business is to find his way back to himself" (p. 147). This is especially true in the cases of Klebold, Harris, and Cho. Klebold and Harris felt ostracized from the rest of the kids in Columbine High School, which pushed them to create the "trench coat mafia"; and Cho clearly identified "those rich kids" on multiple occasions in the video he sent to NBC news before he began his killing spree at Virginia Tech University. These questions of identity lead to a radicalization process ending in a killing spree.

For researchers, the process of radicalization has quickly become an area of interest, especially as it pertains to the various forms of terrorism. Bhatt and Silber (2007) defined four phases of the radicalization process:

1. Preradicalization;
2. Self-identification;
3. Indoctrination; and
4. Jihadization. (p. 6)

While we are not necessarily focusing on terrorism and jihadization, Bhatt and Silber's first three phases identify psychological transitions to a transformational state enabling an individual to recreate whom they want (or believe they want) to become.

They note that this gradual transformation could occur over a period of 2 to 3 years, and suggest that this "transnational phenomenon of radicalization in the West is largely a function of the people and the environment in which they live" (p. 7). Additionally, Bhatt and Silber suggest this "is a phenomenon that occurs because the individual is looking for an identity and a cause" and that "the individuals who take this course begin as 'unremarkable' from various walks of life" (p. 8). Their work clearly suggests that human nature is directly involved in the radicalization process of individuals, whether serial killers or terrorists.

The radicalization process is not necessarily predetermined by demographics, economic status, or profession. Therefore, it is imperative that we look more closely at human nature and psychological motives to identify how a person emotionally processes acceptance and denial from their peers and society-at-large (Hughbank et al., 2010). Through this psychosocial lens, we might gain a greater understanding of how persons such as Klebold, Harris, Cho, and Kazmierczak felt a need to commit murder within their academic communities.

FEMALES AND KILLING

From a cultural perspective, we have become immune to the concept of a man portrayed as a killer. However, when a woman has been identified as a killer, society generally takes a different ethical perspective. The fact that a person can kill in order to survive or as a by-product of a psychological disorder is a powerful reality; one that is not generally attributed to a woman. This poses the question—does a killing response exist equally in women as it does in men?

There are multiple explanations as to the phenomenon of females killing, especially in combat or in acts of terrorism as martyrs:

- Females create a greater psychological element of surprise to their potential victims;
- When an attack is carried out by a female, the media coverage becomes more extensive;
- Women create a force multiplier for organizations;
- They tend to draw less attention from the general public, thereby making movement easier; and
- In the Muslim culture, women are generally looked down upon and seen as expendable assets. (Hughbank et al., 2010, pp. 148–149)

Aside from these reasons, a female may be willing to align or join a professional military organization or terrorist group because it offers a unique opportunity to become an equal in a male-dominated culture.

Regardless of the psychological motivations or practical reasons for using females as a vehicle for killing, there are also some noted drawbacks:

- The use of females can fracture a terrorist group based on religious and or ideological grounds;
- Use of females can inadvertently aid the government media machine; and
- The act of using females can send the message that a terrorist organization is weak or unable to attract male volunteers. (Hughbank et al., 2010, p. 149)

In American culture, female criminals may kill—commit murder—as a consequence of an emotional disconnect from society. According to Hughbank et al. (2010), "causes will inevitably vary for this mental breakdown, but the common denominators for such horrific actions in our culture would mostly occur as a byproduct of either fear or revenge, which are directly attributed to the acts of terror they carry out as a byproduct of their emotions" (pp. 153–154).

LEGITIMATE AUTHORITY

Conscience is a driving force in the decision-making process. Stanley Milgram, a former Yale University professor, designed and conducted an unparalleled psychological experiment in 1961–1962, studying human tendencies to obey authority against individual conscience based on obedience and aggression. Findings indicated that more than 65% of Milgram's subjects could be easily manipulated into inflicting a (seemingly) lethal electrical charge on a person they had never met. The test subjects believed they were causing physical pain, and despite their test victim's pleas to stop the experiment, 65% still obeyed their orders to increase the voltage of the shocks (Grossman, 1995).

One conclusion drawn from his experiment was the concept that many people act on the words of others regardless of the content of the action and without limitations of conscience. Grossman (1995) believed "even when the trappings of authority are no more than a white lab coat and a clipboard, this is the kind of response that Milgram was able to elicit" (p. 142). These actions are carried out because people believe the command to act originates from a legitimate authority. What if this "legitimate authority" comes from a person wanting to cause harm to others? This authority could vary from a relative, to a colleague, or an outsider with a strong personality such as Jim Jones, David Koresh, or Charles Manson. A person seeking guidance or direction in their life can become an easily accessible target of opportunity presenting a strong willingness to follow these maladjusted "authorities" without issue or question (Hughbank et al., 2010). The understanding of Maslow's theory and Milgram's study can provide valuable insights into the process of understanding human nature and pathology as they pertain to the killing process.

References

Bhatt, A., & Silber, M. (2007). *Radicalization in the west: The homegrown threat.* New York, NY: New York Police Department. Retrieved from http://www.nypdshield.org/public/SiteFiles/documents/NYPD_Report-Radicalization_in_the_West.pdf

Grossman, D. A. (1995). *On killing: The psychological cost of learning to kill in war and society.* Boston, MA: Little, Brown.

Hughbank, R. J., Niosi, A. F., & Dumas, J. C. (2010). *The dynamics of terror and creation of homegrown terrorism.* Mustang, OK: Tate.

Shoebat, W. (2007). *Why we want to kill you: The Jihadist mindset and how to defeat it.* New York, NY: Top Executive Media.

Taylor, S. E., Klein, L. C., Lewis, B. P., & Gruenewald, T. L. (2000). Biobehavioral responses to stress in females: Tend-and-befriend, not fight-or-flight. *Psychological Review, 107*(3), 411–429. doi:10.1037///0033-295X.107.3.411. Retrieved from http://scholar.harvard.edu/sites/scholar.iq.harvard.edu/files/marianabockarova/files/tend-and-befriend.pdf

Wilson, C. (1967). *The outsider.* New York, NY: Penguin.

54 MILITARY SEXUAL TRAUMA

Elizabeth H. Anderson and Alina Surís

DEFINITION AND UNIQUE CHARACTERISTICS

The Veterans Health Care Act of 1992 (Public Law 102-585) defines military sexual trauma (MST) as "psychological trauma, which in the judgment of a mental health professional...resulted from a physical assault of a sexual nature, battery of a sexual nature, or sexual harassment, which occurred while the veteran was serving on active duty." The term "sexual harassment" refers to "repeated, unsolicited verbal or physical contact of a sexual nature which is threatening in character" (Section 1720D). The term "MST" as used in this chapter will refer to attempted and completed sexual assaults only and will not include sexual harassment, as the majority of the published literature available does not include sexual harassment.

As summarized by Surís and Lind (2008), unique characteristics of sexual trauma associated with military service differentiate it from civilian sexual trauma. Because MST occurs while a person is on active duty, it typically occurs where they live and work. The result is that they often may have to continue to working and living (e.g., on base) with the perpetrator. The perpetrator of military sexual trauma may be another service member, supervisor, or higher ranking official, and the victim may rely on him or her for services, security, evaluations, or promotions. This situation is not as common in the civilian world, and when it does occur, the individual has the option to immediately quit his or her job, something not possible in the military. Strong unit cohesion may contribute to a feeing that the victim cannot speak up about the assault or harassment for fear of being ostracized. Often, victims of MST report that they were told to stay quiet and not cause trouble. Individuals may also not speak out or seek mental health services because they fear it will negatively influence potential for deployment and career advancement.

Research has consistently shown that sexual assault in the military is associated with the development of psychiatric sequelae, physical health issues, and poorer quality of life (Surís & Smith, 2011). For example, victims of MST are two to three times more likely to have a mental health diagnosis, with PTSD having the strongest association with MST (Kimerling, Gima, Smith, Street, & Frayne, 2007). In fact, women veterans with MST were found to be nine times more likely to develop PTSD compared with women veterans with no history of sexual assault (Surís, Lind, Kashner, Borman, & Petty, 2004). Other diagnoses found to have a strong association with MST include depression, alcohol abuse, anxiety disorders, dissociative disorders, eating disorders, and personality disorders (Kimerling et al., 2007).

The few MST studies that have included male veterans have found gender differences in reported psychiatric symptoms. O'Brien, Gaher, Pope, and Smily (2008) found that men with MST describe significantly more trauma symptoms and experience more lasting sexual problems than women with MST. However, the same study found no significant differences between men and women with MST in levels of anxiety, depression, dissociation, and sleep disturbance. Surís and Smith (2011) reported that men are typically reluctant to acknowledge that they have experienced MST for many reasons. Many of them are unaware that sexual assault has more to do with power and dominance, not sexuality or sexual orientation. Consequently, they view MST as something that "happens to women" and acknowledging they are survivors of sexual assault negatively impacts their self-perceptions of what it means to "be a man." In attempts to "prove their masculinity" to themselves and others, male MST survivors may become promiscuous and, if in a relationship, unfaithful to their significant other. Another problem seen in male veterans with MST is confusion about their sexual identity and orientation. Male MST survivors may wonder if they were targeted because they were viewed as homosexual by their attackers. Other male survivors may develop extreme hatred and distrust toward homosexuals because they assume that the perpetrators of their assaults were gay men, when that is often not accurate.

In their review of MST literature, Surís and Lind (2008) found that female veterans with MST experience more physical symptoms, including headaches, chronic fatigue, pelvic pain, menstrual problems, and gastrointestinal symptoms, than female veterans who have not experienced MST. Medical conditions such as liver disease and chronic pulmonary disease have shown moderate association with MST in both women and men (Kimerling et al., 2007). Obesity, hypothyroidism, and weight loss were significantly associated with MST in women, while AIDS was significantly associated with MST in men (Kimerling et al., 2007).

PREVALENCE

The reported prevalence rates of military sexual trauma vary greatly due to factors such as the definition of MST used, the manner of obtaining the data (e.g., database, mailed survey, telephone survey, in-person), the purpose of the study for which the data was gathered (e.g., descriptive, diagnostic, etc.), and the respondent sample (e.g., treatment seeking, compensation seeking, service era, active duty vs. veteran status, etc.) (Surís & Smith, 2011). In their review of the literature, Surís and Lind (2008) found prevalence rates that ranged from 0.4 to 71%, however the majority of studies reviewed reported prevalence rates between 20 and 43%. It should be noted that although the average lifetime prevalence rate of civilian sexual assault is 25%, MST prevalence rates are usually based on a time period of 2 to 6 years, during which service members are on active duty. Rates of sexual assault during these limited years are considerably higher than civilian lifetime rates, suggesting an increased risk for sexual assault among Active Duty military personnel (Surís & Smith, 2011).

The Veterans Health Administration began national MST screening in VA facilities for both men and women in 2002. In the 2011 fiscal year, 2.5% (118,703) of the 4.8 million veterans screened in VA Medical Centers nationwide endorsed MST. The rate of endorsement for women was 23% (65,796), while the rate for males was 1.2% (52,907). Of the 2.6 million veterans screened at community-based outpatient clinics in the 2011 fiscal year, 2.3% (59,218) endorsed MST. The endorsement rates were 22.5% (31,393) for women and 1.1% (27,825) for men. It is also important to note that we do not know the prevalence rates for MST for men and women who do not access their health care through the Veterans Health Administration.

The majority of studies examining the prevalence of MST have utilized female-only veteran samples. The few studies that have investigated the prevalence of MST in men typically find rates of around 1%. Despite the lower prevalence rate among men, the total number of male and female survivors of MST is approximately equal in the VA due to the higher number of men in the military (Surís & Smith, 2011). Despite this fact, men have been largely ignored in MST literature.

Although, technically not labeled as MST, the DoD reported that in 2011, they had a total of 3,192 sexual assaults reported across all service branches (Department of Defense [DoD], 2012). Even if calculated as percentages, these numbers cannot be directly compared to the VA's numbers because the VA includes harassment in its definition of MST, while DoD only includes assaults.

TREATMENT IN VA

The Veterans Health Care Act of 1992 (Public Law 102-585) mandated free counseling for women veterans with MST. The Veterans Health Programs Extension Act of 1994 (Public Law 103-452) extended these services to men and required that the VA screen all veterans for MST. More recently, the Veterans Health Program Improvement Act of 2004 (Public Law 108-422) made the treatment services for MST a permanent benefit and extended MST counseling and treatment to cover active duty for training service members.

There are many possible psychological outcomes that can result from MST, and treatment is therefore based on the resulting symptoms and diagnoses in each individual. The severity and duration of symptoms will vary depending on factors including the victim's previous trauma history, the nature of the victim-perpetrator relationship, the victim's perception of the traumatic event, and the quality of their support system (Surís & Lind, 2008).

Due to the wide-ranging clinical presentations of veterans with MST, it is beyond the scope of this chapter to discuss treatments that may be beneficial for each possible diagnosis or set of symptoms. This chapter will focus primarily on the treatment of PTSD since it is the psychiatric disorder most highly associated with MST. The VA's Uniform Mental Health Services handbook (VA, 2008) required that all veterans with PTSD have access to cognitive processing therapy (CPT) or prolonged exposure therapy (PE). Additionally, the VA/DoD Clinical Practice Guidelines for PTSD recommend these two therapies as first-line treatments for treating PTSD.

CPT is a structured, time-limited cognitive behavioral therapy that focuses on trauma-related beliefs regarding oneself, others, and the world (Resick, Monson, & Chard, 2007). Treatment typically consists of 12 individual or group therapy sessions. These sessions initially focus on education about PTSD symptoms, cognitive theory, and emotional processing with a strong emphasis on the relationship between thoughts and emotions. Sessions then shift to identifying and evaluating the helpfulness and accuracy of cognitions related to trust, safety, intimacy, esteem, and power/control that have changed in the aftermath of the trauma. Clients are then taught to challenge and modify their problematic beliefs, or "stuck points." CPT also helps clients process the trauma by having them write a detailed narrative of the traumatic event, including thoughts and emotions. Clients subsequently read these narratives to themselves and to the therapist in session.

Prolonged exposure therapy (PE) is a manualized treatment approach designed to help clients emotionally process a specific traumatic event and reduce avoidance behavior (Foa, Hembree, & Rothbaum, 2007). Repeated and prolonged exposure to feared and avoided memories and situations teaches clients that they can tolerate their symptoms of anxiety and that, as they become habituated to the memory or situation, their feelings of distress eventually subside. The course of treatment for PE is typically between 9 and 15 sessions. In the initial session, clients are educated about PTSD symptoms and the rationale for using imaginal and in vivo exposure to reduce avoidance symptoms. Clients are also introduced to breathing retraining to assist in relaxation. Subsequent sessions use imaginal exposure, which involves retelling the traumatic event multiple times during a session. During the imaginal exposure, the therapist will ask the client to rate their distress level every 5 minutes and also offer encouragement and support. Afterward, the client and therapist explore and process the exposure experience. Between sessions, clients listen to recordings of their imaginal exposure therapy sessions at least once a day. Clients also take part in in vivo (real life) exposure to feared and avoided situations that remind them of the trauma.

Inherent in the treatment of MST related PTSD is the issue of avoidance, especially when using trauma-focused therapies. As with other traumas, MST victims may seek to avoid the distress of facing the memory of their assault. Avoiding aversive memories or situations will cause a temporary reduction in distress, which can act as a reward for avoidance behavior. Avoidance may become the only way an individual feels they can control their distress and anxiety when, in fact, repeated avoidance exacerbates the distress associated with the feared memory or situation, solidifying the PTSD symptoms. The fear of experiencing the traumatic memory in therapy may be overwhelming for some. Failing to show up to initial appointments or dropping out of therapy is not uncommon. It is important for clinicians to establish a strong therapeutic alliance and monitor the client's level of distress during sessions. In addition, the VA's Uniform Mental Health Services handbook recommends that facilities provide same-sex providers for veterans seeking treatment for MST when clinically appropriate. However, male MST survivors typically do not want male therapists.

There is little published research specifically focusing on treating the psychological sequelae of sexual assaults that occurred during active duty. In our review of the literature, we found only one recently published study that has examined the effectiveness of an evidence-based therapy for treating PTSD resulting from military sexual assault. This randomized clinical trial compared the effectiveness of CPT to present-centered therapy (PCT) in treating PTSD related to MST and found that male and female veterans who received CPT had a significantly greater reduction in self-reported, but not clinician assessed, PTSD symptoms (Surís, Link-Malcolm, Chard, Ahn, & North, 2013). Although both treatment groups demonstrated significant improvement in clinician-assessed and self-reported symptoms of PTSD and self-reported symptoms of depression, pre-post effect sizes, which were mostly moderate to large, trended larger in the CPT group compared to PCT group across almost all measures. This study provides preliminary evidence supporting the use of CPT to reduce self-reported symptoms of PTSD in veterans with MST.

In cases of depression and anxiety, which are also common sequelae of MST, the VA's Uniform Mental Health Services handbook

(VA, 2008) recommends cognitive behavioral therapy (CBT), acceptance and commitment therapy (ACT), and interpersonal therapy (IT). Treatment for substance use disorders (SUD) may include empirically supported addiction-focused interventions, such as cognitive-behavioral coping skills training, behavioral couples therapy, the community reinforcement approach, motivational enhancement therapy, contingency management/motivational incentives, and 12-step facilitation.

Every VA facility has a designated MST coordinator who acts as a point person for MST-related issues. While all VA medical centers employ clinicians who have been trained to offer PTSD treatment, some sites have specialized PTSD programs or treatment teams, such as the Women's Stress Disorder Team, that are specific to MST. There are currently 13 MST inpatient and residential VA treatment programs, the majority of which serve only women. All Vet Centers provide assessment and referrals for sexual trauma counseling, and most provide onsite MST counseling.

MILITARY AND COMMUNITY SERVICES

The Department of Defense (DoD) provides the same evidence-based treatments for treating PTSD as the VA. The two organizations jointly developed the VA/DoD Clinical Practice Guideline for the Management of Posttraumatic Stress (VA/DoD, 2010) to assist facilities from both organizations in implementing evidence-based clinical procedures for the assessment and treatment of individuals with posttraumatic stress. The guidelines gave the highest recommendation grade to trauma-focused therapies, such as CPT and PE, and stress inoculation training based on quality of evidence and the net benefit of the intervention.

In response to the policy recommendations of a Department of Defense (DoD) task force and the resulting DoD directive, the Sexual Assault Prevention and Response Office (SAPRO) was permanently established in 2005 to oversee the sexual assault policy of the DoD, which includes the Army, Navy, Air Force, and Marine Corps. It should be noted that SAPRO policy does not extend to individuals sexually assaulted by a spouse or intimate partner. This type of assault is covered by the DoD's Family Advocacy Policy (FAP). The SAPRO has developed and implemented programs that provide care to survivors of sexual assault and offer education and training to prevent sexual assault.

There are two types of reports established by the SAPRO: restricted and unrestricted. Restricted reporting is confidential and allows service members who have been sexually assaulted to obtain counseling and medical care without the notification of their commanders or law enforcement. With an unrestricted report, their commanders and law enforcement will be notified, triggering an investigation. A restricted report may be changed to an unrestricted report, but an unrestricted report may not be changed to a restricted report. Sexual Assault Response Coordinators (SARCs) and Victim Advocates (VAs) are knowledgeable of the exceptions and limitations of restricted reporting restrictions and help sexual assault victims determine which type of report would best serve their needs. SARCs oversee the coordination of care for sexual assault victims and monitor victim services from the initial contact and until the victim no longer requests support. They also oversee the training and case assignment of VAs (uniformed and civilian), who provide support, education, and resources to victims of sexual assault.

Regarding other resources, the SAPRO also oversees the Safe Helpline, which offers anonymous and confidential crisis support services for military personnel at all hours via the Internet, phone, or text message. The DoD's Military OneSource provides counseling services face-to-face, over the phone, or online for a variety of military issues, including MST.

References

Department of Defense. (2012). *Department of Defense fiscal year 2011 annual report on sexual assault in the military*. Retrieved from http://www.sapr.mil/media/pdf/reports/Department_of_Defense_Fiscal_Year_2011_Annual_Report_on_Sexual_Assault_in_the_Military.pdf

Department of Veterans Affairs, Veteran's Health Administration (VA). (2008). Uniform mental health services in VA medical centers and clinics. In *VHA Handbook 1160.01*. Retrieved from http://www1.va.gov/vhapublications/ViewPublication.asp

Department of Veterans Affairs & Department of Defense (VA/DoD). (2010). *VA/DoD clinical practice guideline for management of post traumatic stress*. Retrieved from http://www.healthquality.va.gov/Post_Traumatic_Stress_Disorder_PTSD.asp

Foa, E. B., Hembree, E. A., & Rothbaum, B. O. (2007). *Prolonged exposure therapy for PTSD: Emotional processing of traumatic experiences*. Oxford, UK: Oxford University Press.

Kimerling, R., Gima, K., Smith, M., Street, A., & Frayne, S. (2007). The Veterans Health Administration and military sexual trauma. *American Journal of Public Health, 97*(12), 2160–2166.

O'Brien, C. O., Gaher, R. M., Pope, C., & Smily, P. (2008). Difficulty identifying feelings predicts the persistence of trauma symptoms in a sample of veterans who experienced military sexual trauma. *Journal of Nervous and Mental Disease, 196*, 252–255.

Resick, P. A., Monson, C. M., & Chard, K. M. (2007). *Cognitive processing therapy: Veteran/military version*. Washington, DC: Department of Veterans Affairs.

Surís, A., & Lind, L. (2008). Military sexual trauma: A review of prevalence and associated health consequences in veterans. *Trauma, Violence, and Abuse, 9*, 250–269.

Surís, A., Lind, L., Kashner, T. M., Borman, P., & Petty, F. (2004). Sexual assault in women veterans: An examination of PTSD risk, health care utilization, and cost of care. *Psychosomatic Medicine, 66*, 749–756.

Surís, A., Link-Malcolm, J., Chard, K., Ahn, C., & North, C. (2013). A randomized clinical trial of cognitive processing therapy for veterans with PTSD related to military sexual trauma. *Journal of Traumatic Stress Studies*. doi: 10.1002/jts.21765. [Epub ahead of print].

Surís, A., & Smith, J. (2011). PTSD related to sexual assault in the military. In B. Moore & W. Penk (Eds.), *Treating PTSD in military personnel: A clinical handbook* (pp. 255–269). New York, NY: Guilford.

55 PRESCRIPTION OPIOID ABUSE IN THE MILITARY

Jennifer L. Murphy and Michael E. Clark

SCOPE OF THE PROBLEM

The White House, military services, and general public recognize that increases in opioid use and abuse in the military are highly concerning. Prescription drug abuse among US military personnel doubled from 2002 to 2005 and almost tripled between 2005 and 2008, largely due to the nonmedical use of prescription medications (Executive Office, 2010). While the rates for use of illicit "street" drugs such as cocaine, methamphetamines, and marijuana have remained low in the last decade at about 2%, the nonmedical use of prescription pain relievers has reached almost 12% (Executive Office, 2010). According to a 2008 Pentagon health survey that included almost 30,000 troops, about 22% said they had abused pain medications in the previous year, and 13% said they had done so in the previous 30 days (Bray et al., 2009).

OPIOID USE AND ABUSE IN THE GENERAL POPULATION

Americans are 4.6% of the global population but consume 80% of the global opioid supply and 99% of the hydrocodone supply (Manchikanti, Fellow, Ailinani, & Pampati, 2010). Abuse of prescription pain medications is the fastest growing drug problem in the United States. Retail sales of commonly used opioid medications such as oxycodone, hydrocodone, hydromorphone, morphine, methadone, fentanyl base, and codeine have increased an average of 149% in the last 10 years, and hydrocodone continues to be the most prescribed medication in the United States (Manchikanti et al., 2010). Data from the most recent 2010 National Survey on Drug Use and Health Studies showed that 1 in 20 people (aged 12 or older) reported nonmedical use of painkillers in the past month, twice the number who used cocaine, hallucinogens, inhalants, and heroin combined (Substance Abuse and Mental Health Services Administration, 2011). Prescription medications also appear to be replacing marijuana as the top "gateway drug" that leads to further substance abuse.

The increase in unintentional drug-related overdose death rates in recent years, which has been largely driven by increased use of opioids, is highly disturbing. According to the Center for Disease Control and Prevention's National Center for Health Statistics, in 2008 the number of poisoning deaths surpassed motor vehicle–related deaths for the first time since at least 1980 (Warner, Chen, Makuc, Anderson, & Miniño, 2011). Of the overdose deaths in the United States in 2008, almost 55% involved a prescription drug, more than all illicit drugs combined. The number of drug poisoning deaths (89% of total) involving opioid analgesics more than tripled in the last decade, increasing more rapidly than deaths involving any other types of drugs. Opioids were involved in more than 40% of all drug poisoning deaths in 2008, up from about 25% in 1999. Of the opioid-related poisoning deaths, the majority involved natural and semisynthetic opioid analgesics such as morphine, hydrocodone, and oxycodone (Warner et al., 2011).

OPIOID USE AND ABUSE IN THE MILITARY POPULATION

While trends among the general public are alarming, the 2008 DoD Survey of Health Related Behaviors among Active Duty Military Personnel revealed that prescription drug abuse among members of the military exceeded rates observed in the civilian population (Bray et al., 2009). Across all services, 11.5% of military personnel reported prescription drug misuse compared to 4.4% in the civilian population (Executive Office, 2010). The prevalence of prescription drug misuse among females in the military was a staggering 13.1%, more than four times the rate for civilian women. Women in the Army were also more than twice as likely as men in the Air Force, Navy, and Coast Guard to have used any illicit drug in the last 30 days, including prescription drugs used nonmedically (Executive Office, 2010). And according to the *Army Times*, between 2009 and 2011, 72% of drug-related undetermined or accidental deaths involved prescription drugs (Tan, 2012).

Prescription painkiller abuse is likely becoming more widespread among military members, partly because of the continued strains of multiple deployments and continuing combat operations around the globe. Advances in trauma care on the battlefield and immediately following have been significant in the past 15 years, resulting in more than 90% surviving after being wounded. Modern warfare can produce serious but often survivable injuries that cause acute pain and may lead to chronic pain. As a result, some service personnel have returned with substance use disorders related to pain relievers necessitated as a result of battlefield wounds. Opioids may continue to be dispensed on a chronic basis in Warrior Transition Units, which can lead to issues with misuse and addiction.

In general, there are increasingly high rates of uncontrolled pain in Operation Iraqi Freedom (OIF) and Operation Enduring Freedom (OEF) veterans entering the Veterans Health Administration (VHA)—43% in 2007 to 53% in 2009 (Health Profile of the Department of Defense [Health Profile], 2011). Service members report experiencing a myriad of pain symptoms as they move between the DoD and VHA or retire into the VHA system for care. Routinely carrying heavy equipment and body armor coupled with the frequency of multiple deployments has placed the OIF/OEF population at a higher risk for the development of musculoskeletal injuries and pain conditions than in previous eras. In the last decade, the military's greater reliance on National Guard or Reserves personnel has increased the average age of those deployed, which makes the existence of predeployment pain (e.g., low back pain) more common. Furthermore, because these individuals are engaged in the military on a part-time basis they may be in less than ideal physical condition at the initiation of service, which places them at greater risk for injury or development of pain conditions. And while pain can be the result of battle-borne injuries, the commonplace physical wear and tear associated with military life often leads to chronic pain conditions that may decrease quality of life and disrupt career trajectories.

RESPONSE TO THE PROBLEM

In response to the increasing issues related to opioid use as well as the growing epidemic of chronic pain, in 2008 the Office of the Army Surgeon General (OASG) initiated an examination of pain as a distinctive issue for the US Army Medical Command. The Pain Management Task Force (PTF) was created to "develop and implement a comprehensive policy on pain management by the military healthcare system" (Office of Army Surgeon General [OASG], 2010, p. E-1). The PTF consisted of members from all military services, TRICARE, and VHA, and represented the first systematic review of DoD, regional medical commands, and military health care facility clinical policies and regulations from the

perspective of pain management. Twenty-eight site visits were conducted at military hospitals and health clinics as well as VHA and civilian hospitals. In addition, the Military Pain Care Policy Act of 2008, which originated in the House of Representatives (HR 4565), as well as the Veterans Pain Care Policy Act (S 2160), which originated in the Senate, were both passed by Congress in October 2008 and signed into law. Assessment from legislators suggested that, "comprehensive pain care is not consistently provided on a uniform basis throughout the systems to all patients in need of such care" (OASG, 2010, p.1).

In May 2010, the OASG released the Pain Management Task Force final report to, per the subtitle of the document, provide "a Standardized DoD and VHA Vision and Approach to Pain Management to Optimize the Care for Warriors and their Families" (OASG, 2010). According to the PTF, the goals of the plan were to reduce suffering and improve quality of life for those with pain. While results of the PTF report found that pain management practices were generally consistent with standards of care throughout the system, the area of biggest concern noted was "an overreliance on opioid medication which contributes to the increase in opioid abuse and dependence" (OASG, 2010, p. 34). The task force suggested that the lack of a well-developed comprehensive strategy for pain as well as a consistent continuum of care were primary reasons for the opioid issues.

In addition, the VHA and DoD have worked collaboratively to provide initiatives to mitigate the risk of opioid misuse/addiction/diversion (see Table 55.1; Buckenmaier et al., 2011). The

TABLE 55.1. Risk Mitigation Initiatives of the Department of Defense (DoD) and Veterans Administration (VA)

- DoD–VA Chronic Opioid Therapy Clinical Practice Guidelines
- Opioid—High Alert Medication Initiative
  —Opioid Renewal Clinic
  —Collaborative Addiction and Pain (CAP) Program
  —Opioid Decision Support System
- Opioid Therapy Web-based Course
- Medical Marijuana Directive

Source: Buckenmaier et al., 2011, with slight modification.

latest edition of the Clinical Practice Guideline for the Management of Opioid Therapy for Chronic Pain, published in May 2010, provides empirically supported evidence to assist providers in decision-making regarding the prescription of opioid analgesics on a chronic basis (defined as more than 1 month) in an attempt to foster responsible, reasonable, and consistent prescribing practices (VA/DoD, 2010).

Another initiative for better prescribing practices is the Opioid Renewal Clinic (ORC) model, which includes an opioid treatment agreement, frequent visits, prescribing opioids on a short-term basis (e.g., weekly), and periodic urine drug testing, among other options (Buckenmaier et al., 2011). This model has already been effective in aiding physicians in primary care feel more comfortable prescribing opioids and has helped increase patient satisfaction. Other initiatives target areas such as increasing opioid-related education and support, developing programs for addiction and pain, and clarifying medical marijuana policies.

Expanded drug testing for all military services began in May 2012 to include commonly abused prescription drugs such as hydrocodone and hydromorphone, found in painkillers such as Vicodin, Lortab, and Dilaudid (Tan, 2012). An Army report for FY2011 indicated that 21% of soldiers involved in illegal drug use were abusing prescription drugs. Officials hope that the expanded testing will help curb misuse of prescription drugs as it did illicit drugs during the Vietnam era (Tan, 2012). While prohibitions against misusing prescription or over-the-counter medications are already in place, additional testing will support them. No changes in procedure were made such as testing sites or who would be selected, but the expansion clearly reflects the growing concern over opioid abuse issues in the military.

CHALLENGES

Several distinct challenges specific to the military exist in managing the current prescription drug use crisis. The military culture emphasizes toughness and a "no pain, no gain" attitude. While this sentiment may be valuable on the battlefield and praised in training, it can lead to hesitance in seeking medical care for pain problems. Postponing care and opting to "tough it out" can lead to musculoskeletal issues and increased risk of chronicity for issues that may have been resolved if addressed earlier.

Given the prevalence of both acute and chronic pain among military members, pain management is significantly underrepresented in the DoD's research investment strategy—it is slated to receive less than 2%, or $54.6 million of research funding over the next 5 years (Health Profile, 2011) despite the clear need for additional treatment resources as well as current data regarding the prevalence of prescription opioid abuse. In addition, military medical providers often feel a significant pressure to provide the highest amount of pain relief to those facing battle's heaviest burdens (OASG, 2010). Unfortunately, this strain may lead to providers feeling as if they must increase opioid analgesic doses for fear of reprimand from families or leaders even in cases when they do not feel comfortable doing so.

The transient nature of the military population frequently leads to a lack of continuity in care since providers shift on a regular basis. Lack of care coordination within the military health care system can lead to numerous opioid-related issues such as inconsistent prescribing and inadequate monitoring of misuse/abuse behaviors. Lack of care coordination between the DoD and VHA poses additional challenges to service members receiving VHA health services. Coordination of care between these two organizations often is problematic at both the inpatient and outpatient levels given difficulties accessing documentation and variations in pain treatment practices. Improved coordination between DoD and VHA pain providers and better communication between the DoD and VHA health care systems might facilitate reductions in opioid analgesic misuse or abuse, promote more elaborate and effective pain treatment plans, and prevent delays in accessing appropriate care including the immediate availability of appropriate pain medications.

A more general but critical issue faced in both the civilian and military population is the lack of "ownership" for pain care by a specific discipline. Since pain permeates every discipline of medical care and is a complaint of many patients, it may be treated by primary care providers, surgeons, anesthesiologists, neurologists, and physical medicine and rehabilitation specialists, among many others. However, pain does not clearly fall under the domain of a specific discipline; this has often led to various providers playing a part in care with no identified discipline managing or coordinating the team. The PTF has recognized this shortcoming in the current system and shifted the definition of pain to one that, if not managed well, can become a chronic disease which instead must be managed on a long-term basis like diabetes or heart disease.

RECOMMENDATIONS

While some steps have been taken to address the various challenges discussed, additional recommendations for improving opioid prescribing practices and treatment are needed. Monitoring provider adherence to shared clinical practice guidelines for pain management across DoD and VHA would assist in ensuring greater consistency in treatment approaches and in the prescribing and monitoring of opioid analgesics. While DoD/VHA CPGs currently exist, the information and standards that are outlined are used variably among providers.

The standardization of pain assessment and documentation procedures across DoD and VHA would be helpful for several reasons. Particularly for those with chronic pain, being able to recognize their pain levels and treatment needs as they transition between the military and VHA would provide more consistency in evaluation and treatment. A system for better sharing of information across these systems would raise provider awareness of previous treatments, expedite medication prescriptions, and help minimize costs by avoiding unnecessary duplication of tests. Furthermore, a set of standards for pain assessment and reassessment would assist in evaluating treatment effectiveness and in detecting changes in behavior associated with opioid misuse, abuse, or addiction.

The identification of additional treatment options to include specific program options for opioid use disorders would decrease the prevalence of prescription opioid abuse/misuse. An increase in DoD resources for pain management in the interdisciplinary arena is greatly needed. The role of behavioral medicine, particularly pain psychologists and addiction specialists, should be emphasized to assist in such programs, especially since other psychiatric comorbidities are often present. For those with comorbid pain and substance use, treating the substance disorder without providing education and tools for managing pain can be futile if individuals do not feel equipped to properly manage chronic pain without opioids. Furthermore, the potential incorporation of pain level and opioid analgesic use into general suicide risk assessments should be further explored given increasing awareness about the links between pain and suicide, and opioids and suicide. Finally, increasing the availability of complementary and alternative medicine (CAM) pain management options such as acupuncture in both DoD and VHA health care systems also is recommended as they are routinely used in the civilian population but less available to military personnel.

The abuse of opioid analgesics in the military is a significant and growing concern. Enhanced standardization of care as well as increased treatment options for problematic use are needed to minimize opioid abuse across DoD and VHA. While the military faces unique challenges, coordinated efforts will likely result in improved treatment outcomes and quality of life for military personnel.

References

Bray, R. M., Pemberton, M. R., Hourani, L. L., Witt, M., Rae Olmsted, K. L., Brown, J. M.,... Bradshaw, M. R. (2009). *2008 Department of Defense survey of health related behaviors among active duty military personnel.* Report prepared for TRICARE Management Activity, Office of the Assistant Secretary of Defense (Health Affairs) and U.S. Coast Guard under Contract No. GS-10F-0097L. Washington, DC: US Department of Defense.

Buckenmaier, C. C., Gallagher, R. M., Cahana, A., Clark, M. E., Avery Davis, S., Brandon, H.,... Spevak, C. (2011). War on pain: New strategies in pain management for military personnel and veterans. *Federal Practitioner, 28*(2). Retrieved from http://www.fedprac.com/PDF/028060001s.pdf

Executive Office of the President. (2010, February). *Newsletter of the office of national drug control policy, 1*(2). Retrieved from https://www.hsdl.org/?view&did=22082

Health Profile of the Department of Defense. (2011). *Federal Practitioner, 28*(1). Retrieved from http://www.fedprac.com/PDF/028050001s.pdf

Manchikanti, L., Fellow, B., Ailinani, H., & Pampati, V. (2010). Therapeutic use, abuse, and nonmedical use of opioids: A ten-year perspective. *Pain Physician, 13*, 401–435.

Office of the Army Surgeon General. (2010, May). *Pain management task force: Final report.* Retrieved from http://www.amedd.army.mil/reports/Pain_Management_Task_Force.pdf

Substance Abuse and Mental Health Services Administration. (2011). *Results from the 2010 National Survey on Drug Use and Health: Summary of National Finding.* Rockville, MD.

Tan, M. (2012). *Painkillers to be included in Army drug tests.* Retrieved from http:www.armytimes.com/news/2012/02/army-drug-tests-expanding-to-include-painkillers-022212w/

VA/DoD clinical practice guideline for management of opioid therapy for chronic pain. (2010). Retrieved from http://www.healthquality.va.gov/COT_312_Full-er.pdf

Warner, M., Chen, L. H., Makuc, D. M., Anderson, R. N., & Miniño, A. M. (2011). *Drug poisoning deaths in the United States, 1980–2008.* NCHS data brief, no 81. Hyattsville, MD: National Center for Health Statistics.

56 PSYCHOSOCIAL REHABILITATION OF PHYSICALLY AND PSYCHOLOGICALLY WOUNDED

Walter Erich Penk and Dolores Little

PSYCHOSOCIAL REHABILITATION

Psychosocial rehabilitation is defined as processes for restoring functioning in the home, with families, in the community, and at work, using techniques that add skills and build character to function in one's environment (see United States Psychiatric Rehabilitation Association, or USPRA, at www.uspra.org, for details). Types of functioning include: physical, nutritional, mental and dental, behavioral, psychological, social, environmental, and spiritual (see Real Warriors at

webmaster@realwarriors.net for descriptions and training for Total Force Fitness), for example:

- nutrition and becoming as fit as possible;
- enhancing qualities of living in one's community;
- increasing skills in the military and as civilians;
- enjoying leisure activities;
- fostering relationships with families and friends;
- maturing in mind and character and strengthening spirituality.

Psychosocial rehabilitation techniques are designed to strengthen skills for everyday living. Psychosocial rehabilitation doesn't compensate for but rather capitalizes on one's skills. For those surviving physical and psychological wounds, psychosocial rehabilitation is problem-focused, independent of diagnoses.

The psychosocial rehabilitation techniques that one learns to practice must be those whose effects have been validated by evidence-based randomized clinical trials. Both the DoD and VA are committing resources to validate classification, treatment, and psychosocial rehabilitation (e.g., VA's *Journal of Rehabilitation Research and Development,* as well as ongoing clinical trials.)

New models are being developed to improve everyday functioning, using new structures for services in communities (e.g., Kazdin & Rabbitt, 2013).

The World Health Organization's International Classification of Functioning, Handicaps, and Impairments is widely used to classify problems that psychosocial rehabilitation techniques can address. Psychosocial rehabilitation includes training in exercise and nutrition, education on problems, social skills training, supported education, family psychoeducational groups, peer support, case management, vocational rehabilitation, and spiritual and character growth (Veterans Health Administration [VHA], 2008, describing rehabilitation techniques, and VA/DoD Clinical Practice Guidelines for steps in deciding applications at www.healthquality.va.gov). Psychosocial rehabilitation emphasizes that those receiving services should evaluate outcomes (LaBoube et al., 2012). A valuable resource the VA provides on the Internet to support this process is a personal record for tracking health outcomes at: myhealth.va.gov. See also, www.dcoe.health.mil and www.afterdeployment.org for educational resources.

Techniques discussed in this chapter are based on approaches empirically validated to produce positive results from randomized clinical trials (see Glynn, Drebing, & Penk, 2009, for outcomes and effect sizes from PTSD studies). It is important to note that psychosocial rehabilitation techniques have not yielded negative side effects; while not every study is positive, none are negative. The only risks are not using psychosocial rehabilitation when needs exist. Psychosocial rehabilitation promotes self-determination and self-reliance by promoting physical and psychosocial well-being for military service members and veterans who are motivated to live by developing self-competencies and relating with others.

EXERCISE AND NUTRITION: *MENS SANA IN CORPORE SANO*

When asked what people should desire in living, the Roman poet Juvenal answered, "*mena sana in corpore sano.*" Each should create "a mind that is sound in a sound body." Such goals are comparable to missions in the military and in living as a civilian, that focus on building the physical, psychological, social, and spiritual fitness of the person. Building fitness is the key in promoting resiliency, facilitating recovery, and reintegrating into military and civilian communities.

Exercise and nutrition are essential in daily living, including for those who have been wounded. Types of physical injuries are listed below, along with websites that provide actions to learn for improving physical functioning and recovery:

- amputations (www.amputee.coalition.org);
- paralysis (www.pva.org);
- injury to vision (www.bva.org);
- substance-use problems (www.drugabuse.gov);
- burns (www.burn.recovery.org);
- disfigurement (www.faceit.org);
- pain conditions (www.ampainsoc.org).

Types of invisible psychological wounds include:

- posttraumatic stress disorder (PTSD, www.ncptsd.org);
- traumatic brain injury (TBI, www.dvbic.org);
- depression (www.nimh.nih.gov/health/publications/depression);

- suicide (www.preventionlifeline.org);
- other anxieties (www.dcoe.health.mil/for Warriors.aspx).

Services available are described in more detail in the chapter prepared by Bates, Bowles, Kilgore, and Sorush (2008) in their discussion of fitness, recovery, and return to service. Daily life for military service members and veterans should include actions to promote strength, endurance, power, flexibility, and mobility. Nutrition involves learning qualities about food, nutritional requirements, and food choices (www.realwarriors.org) Research demonstrates that learning to be fit and lean (i.e., not overweight) indeed does improve quality in living, specifically for those recovering from physical and psychological wounds (Moore & Penk, 2011).

PROBLEM EDUCATION

Coping with one's physical and psychological wounds, likewise, must be learned. Programs within the military and for veterans are needed for recovery and for rebuilding resilience (e.g., Bowles & Bates, 2010. See also special issue of *American Psychologist*, January, 2011, volume 66, "Comprehensive Soldier Fitness."). Research is demonstrating warriors can perfect skills for coping with symptoms and with problems arising from physical and psychological wounds, using technological applications (Apps).

The Defense Centers of Excellence for Psychological Health and Traumatic Brain Injury offers Toolkits and Apps that warriors can use to train themselves to cope with physical and psychological problems. Examples are Apps to improve disturbances in sleep (e.g., break-through pain, nightmares, ruminations, early awakening), mood disturbances, interferences in attention and memory associated with concussions, headaches, PTSD, acute stress disorder, depression and suicidal ideation, chronic pain, substance use disorder, and neuropathic and musculoskeletal chronic pain. But Apps do not do the work; warriors must use the Apps. Furthermore, the military psychologist may need to assist the warrior in finding Apps related to his or her situation and provide assistance with using them properly.

SOCIAL SKILLS TRAINING

Avoidance is a major symptom associated with physical and psychological wounds. Hence, wounded warriors must learn skills of engagement, re-creating the equivalent roles for themselves as civilians similar to those they held in their military units. Social skills training improves social support and cohesiveness in everyday living (e.g., Glynn, Drebing, & Penk, 2009). Military and civilian organizations are already in operation and available to build resiliency (Bowles & Bates, 2010). For example, the Technical Cooperation Program, combining resources from five nations, shares information about services to improve and evaluate mental health support during multiple deployments, as well as services to mitigate stigma and barriers to mental health care arising from stresses associated with multiple deployments (see: www.militaryone source.mil).

SCHOOL AND SUPPORTED EDUCATION

Regaining mastery of skills in careers likewise is essential for recovery and resiliency. Returning to education is a time-honored form of psychosocial rehabilitation that began as "moral therapies" in 19th-century New England and later was adapted by Edith Nourse Rogers, who wrote the GI Bill in 1944. World War II veterans used the GI Bill to become what Tom Brokaw later called the "Greatest Generation." Now, current military and veterans are building the next Greatest Generation, using the new GI Bill and state financing.

FAMILY PSYCHOEDUCATION

Psychosocial rehabilitation techniques are being developed to improve functioning within families (Ainspan & Penk, 2012a). Studies are underway validating the efficacy of family psychoeducation about combat-related disorders, demonstrating that such techniques indeed are beneficial for enhancing family support and cohesiveness (Glynn, Drebing, & Penk, 2009). Multiple deployments have been shown to disrupt unity in family, not only straining relationships among spouses but likewise upsetting children in families of military members and veterans. Manuals now guide families of military members and veterans to improve relationships (see www.ouhsc.edu/REACHProgam). Outreach societies carried out by military members and veterans are likewise available to inform and to support families regarding resources (http://defenselink.mil/prhome/mcfp.html; http://yellowribbon.mil).

PEER SUPPORT

Peer support specialists are expanding resources, particularly as many recent military members are transitioning home from combat (Ainspan & Penk, 2012b). Peer support specialists provide many services, some practical (e.g., housing, transportation, and employment) and others focused on health (e.g., developing wellness strategies). Peer specialists are now integrated into VA treatment teams. Manuals train veterans about roles and responsibilities for peer specialists, how recovery and resiliency techniques work, teaching about mental and physical disorders, training in coping with depression, grief, loss, suicidal ideation, and others. One excellent resource, the *Peer Specialist Training Manual* (Harrington, Dohoney, Gregory, O'Brien-Mazza, & Sweeny, 2011), is available from Daniel O'Brien-Mazza, Director, Office of Peer Specialists, Department of Veterans Affairs at: daniel.o'brien-mazza@va.gov.

CASE MANAGEMENT

Case management is an approach to psychosocial rehabilitation in which a trained mental health professional supports community-focused recovery services for those undergoing treatment for medical and mental treatment. Examples of services by professionals with expertise are: transportation, housing, education about medical and psychological conditions, assistance in obtaining federal and state support for medical conditions, and consultation in regaining employment (e.g., VHA, 2008). Hence, roles and responsibilities for case managers are to maintain contact with the professionals as part of follow-up during a client's treatment for medical and mental health conditions, removing barriers that may interfere with results from treatments, and increasing access to services needed to support treatment outcomes. Clinical case findings and program evaluation reports (e.g., DVA's North East Program Evaluation Center, NEPEC) demonstrate that assertive and intensive case management indeed benefits veterans by maintaining supports for those in recovery from physical and psychological wounds (e.g., Glynn, Drebing, & Penk, 2009).

SPIRITUAL AND CHARACTER GROWTH

Growing in character and spirituality are goals of psychosocial rehabilitation. Techniques to promote growth in character are fostered by "positive psychology," as designed and developed by Martin E. P. Seligman. Seligman's techniques were based on his discovery of "learned helplessness," persistent and pervasive attitudes created by conditions in which individuals learn to believe they are helpless, often after experiencing uncontrollable horrifying negative events such as those in combat. Seligman's techniques center on learning optimism to include learning that negativities from experiences in war do not generalize to civilian status. Nor does remembering war mean such reminders are permanently embodied in the person. Finally, horrors of war are not pervasive and need not generalize into relationships at home and in one's community. Treatment techniques teach new skills to control negative emotions in memory. And treatments are being integrated into psychosocial rehabilitation as military and veterans return home.

Psychosocial rehabilitation goes beyond targeting symptoms and the internal life of warriors. Psychosocial rehabilitation focuses on functioning, at the nexus where the person interacts with family, friends, work, and community. Grounded in theories of learning and in building of character to cope with community, based on empirically validated techniques delivered through training and manualized instructions, psychosocial rehabilitation is designed to foster the functioning of wounded warriors to overcome barriers and access resources that improve quality of living within self, for families and friends, with peers and coworkers, for one's community.

References

Ainspan, N. D., & Penk, W. E. (2012a). *Proceedings from Workshop #131: Advancing skills for brief therapeutic treatments to address the needs of returning combat veterans.* Washington, DC: American Psychological Association.

Ainspan, N. D., & Penk, W. E. (2012b). *When the warrior comes home.* Annapolis, MD: Naval Institute Press.

Bates, M. J., Bowles, S. V., Kilgore, J. A., & Sorush, L. P. (2008) Fitness for duty, recovery, and return to service. In N. D. Ainspan & W. E. Penk (Eds.), *Returning wars' wounded, injured, and ill: A reference manual* (pp. 67–101). Westport, CT: Greenwood.

Bowles, S. V., & Bates, M. J. (2010). Military organizations and programs contributing to resilience building. *Military Medicine, 175*(6), 382–385.

Glynn, S., Drebing, C., & Penk, W. (2009). Psychosocial rehabilitation. In E. Foa, T. Keane, M. Friedman, & J. Cohen (Eds.), *Effective treatments for PTSD* (2nd ed.). New York, NY: Guilford.

Harrington, S., Dohoney, K., Gregory, W., O'Brien-Mazza, D., & Sweeny, P. (2011). *Peer Specialist training manual* (First DVA Instructor Edition). Washington, DC: Department of Veterans Affairs.

Kazdin, A. E., & Rabbitt, S. M. (2013). Novel models for delivering mental health services and reducing the burden of mental illness. *Clinical Psychological Science, 1,* 170–191.

LaBoube, J., Pruitt, K., George, P., Mambi, D., Gregory, W., Allen, B.,… Klocek, J. (2012). Partners in change: Bringing people in recovery into the process of evaluating recovery criteria. *American Journal of Rehabilitation Psychiatry, 15,* 255–273.

Moore, B. A., & Penk, W. E. (2011). *Treating PTSD in military personnel: A clinical handbook.* New York, NY: Guilford.

Veterans Health Administration (VHA). (2008). *Uniform mental health policies handbook 1106.* Washington, DC: Department of Veterans Affairs.

57 WORKING WITH MILITARY CHILDREN

Michelle D. Sherman and Jeanne S. Hoffman

Military family life in the 21st century is unique. Large numbers of families are recalibrating themselves in the aftermath of service members' combat deployments to Iraq and Afghanistan, and each family member can be greatly affected. Fortunately, most individuals are resilient and do well after a period of readjustment; however, concerning numbers are experiencing difficulties,

inviting communities and organizations at all levels, from grassroots organizations to the Veterans Administration (VA) and Department of Defense (DoD), to intervene and provide much-deserved support and appreciation.

There are over 2 million military-connected youth today. According to the DoD, almost half of service members have children, and these youth span the entire developmental age range. Of children of Active Duty personnel, approximately 41% are ages 0–5, about 31% are 6–11, about 24% are 12–18, and 4% are 19–23; children of Reserve/National Guard personnel are somewhat older (Department of Defense, 2008). It is imperative to be both proactive and responsive to the needs and experiences of military children, a group whose needs historically have been somewhat neglected.

Young people today face a range of general life challenges, including peer pressure to engage in risky behavior, bullying, worry about family finances, dealing with puberty and sexual issues, concerns about appearance, and academic/school struggles (American Psychological Association, 2009). In addition, military children face a host of additional issues that are noteworthy. As a large majority of these kids attend private/public schools in the community (instead of schools on military installations), it is very important that school and community personnel be cognizant of these unique stressors.

Military families move an average of six to nine times during the course of a child's elementary through high school career. The challenges of making new friends, joining new activities (e.g., church class, soccer team, cheerleading squad), making academic transitions, and starting over can be stressful, even for well-adjusted youth. Children may avoid becoming too "connected" in a community knowing they will be moving again within a few years. This transient lifestyle can interfere with social development and community integration, and can create unique academic challenges. However, some children enjoy the opportunities to live in various parts of the world and meet interesting people, and today's social media and technology can improve youths' abilities to remain connected to friends they have left behind.

Dealing with parental and sibling deployments to combat zones are significant, unique challenges for military youth. The current military environment is marked by a high operational tempo, short "dwell times" between deployments, a norm of multiple deployments, high rates of exposure to trauma, and considerable rates of emotional/physical/spiritual injuries after war, all of which often coalesce to make family life stressful. Children often don't wonder "if" their parent will be deployed, but "when" and "when again." Due to the young ages of many children and the repeated deployments, many parents have spent more time physically apart from their children than in the same home. In essence, many young children are experiencing significant periods of their upbringing in a "single parent" home, one in which the at-home caregiver is often overwhelmed and worried about the deployed individual as well.

With the long duration of the conflicts in the Middle East and the high levels of media coverage of many gruesome military events, many children are aware of the dangerous environments in which our deployed service members are stationed. Not surprisingly, children worry about their parents, fearing for their safety/well-being, and wondering if they will return home. Children whose parents struggle with reintegration and psychological issues may further experience a range of emotions as they face the loss of the parent they knew before deployment; children may experience ripple effects of parental depression/anxiety/PTSD, which can be confusing, sad, and painful. Parental emotional numbing and withdrawal can be especially distressing to children and detrimental to the parent-child relationship.

Before reviewing the growing research base on the experience of military youth, it is important to note that military children also have a range of unique buffers and supports that their civilian peers do not. Especially for Active Duty youth and families living on or near a military installation, a range of programs and resources are available including child care, summer camps, educational opportunities, and access to health care. Especially in this challenging economy, military children are well served by parents having stable jobs and incomes.

RESEARCH ON THE EXPERIENCE OF MILITARY YOUTH

In light of these numerous challenges, it is not surprising that some military children are experiencing emotional distress. Research examining both parent and child reports of youth functioning reveal increased levels of child anxiety and emotional/behavioral problems (Chandra, 2011; Lester et al., 2010). Children whose parents are deployed for longer periods of time are at greater risk for emotional challenges (Lester et al., 2010); it appears to be the cumulative length of time that the parent is away rather than the number of deployments that is related to child functioning.

Similarly, lengthier deployment-related separations are related to lower school achievement scores, with every month away seeming to hurt student achievement (Richardson, 2011). Having one parent/sibling deployed to a dangerous situation and living with caregivers who are often emotionally taxed themselves likely contribute to child distress, children's decreased ability to focus on schoolwork, and reduced caregiver time available to help children with homework and other academic-related activities.

These overwhelmed at-home caregivers appear to be at increased risk for perpetrating physical child abuse and neglect (e.g., Gibbs et al., 2007), with elevated rates during deployment. Higher levels of distress among at-home caregivers appears to be a risk factor for disturbed child functioning. Therefore, supporting the entire family is imperative when considering child functioning and outcomes. Helping a distressed parent access appropriate services for him/herself may be very helpful for the entire family system.

AVAILABLE RESOURCES

Fortunately, as stresses for military families have increased, so have resources. Many preventive and treatment services are available for children for a wide range of needs. Specific programs and resources may differ across installations.

General Resource Summaries for Providers

The *Health Care Providers Resource Guide* (available on the website of the Center for Deployment Psychology: www.deploymentpsych.org/) provides general information and references about the effects of deployment, resources for providers, and tools for assessing and managing deployment-related problems in families and children.

The *Children of Military Service Members Resource Guide* (Defense Centers of Excellence for Psychological Health and Traumatic Brain Injury: http://www.militarychild.org/public/upload/files/DCoE_Children_of_Military_Service_Members_Resource_Guide.pdf) is a compilation of resources (books, activities, films, and websites) that is organized by age groups and highlights topics such as deployment and emotional well-being.

The *Resource Guide for Providers Who Work with Military and Veteran Families* (Alliance of Military and Veteran Behavioral Health Providers: http://deploymentpsych.org/pdf/Family%20Support%20Resource%20Guide.pdf) provides DoD-level information and resources, service-specific resources, and a listing of resources for family, caregivers, and children.

TRICARE

All military families are beneficiaries of this comprehensive insurance system. TRICARE is administered by geographic regions of the United States with separate coverage OCONUS through Tricare Overseas Program (TOP). Both inpatient and outpatient services are covered. General information about behavioral health benefits can be accessed at www.tricare.mil/mentalhealth. The listings of participating mental health providers are not categorized by specialty (e.g., child or family), so beneficiaries need to contact specific providers for this information. No referrals are required for the first eight outpatient behavioral health visits to participating providers per fiscal year; specific information is provided on the TRICARE website.

Extended Health Care Option (ECHO)

ECHO provides services to Active Duty family members who qualify based on specific mental or physical disabilities. ECHO offers integrated services and supplies beyond basic TRICARE benefits. Special education services, medical and rehabilitative services, respite services for individuals with qualifying diagnoses (e.g., autism spectrum disorders) may be funded (http://www.tricare.mil/echo).

Enhanced access to autism services demonstration

This ECHO program provides additional TRICARE benefits to individuals over age 18 months with autism. Empirically supported educational interventions for autism spectrum disorders (EIA) such as functional behavioral analysis (FBA) and applied behavioral analysis (ABA) are authorized. Parent training is mandated, as are meetings between families and those designing and implementing the intervention program. Parents can be referred to http://www.tricare.mil/echo. A related Web-based resource for military families is www.operationautismonline.org, which provides families with an introduction to autism spectrum disorders as well as resources, services, and support.

Exceptional Family Member Program (EFMP)

Each uniformed service has its own Exceptional Family Member Program. Enrollment is mandatory for military families that have a member, adult or child, with special needs (e.g., chronic medical condition, a mental health condition, the need for specialized educational services). The program provides comprehensive medical, educational, housing, community support, and personnel services. It also coordinates services between the military medical systems and community services. Its mission is to identify and document the special medical and/or educational needs of a service member's family members and to ensure that a service member's family will be stationed where needed services are available.

School Behavioral Health (SBH)

School Behavioral Health is a project of the Child, Adolescent, and Family Behavioral Health Office (CAF-BHO), located at Joint Base Lewis-McCord, which provides behavioral health care at military impacted schools at specific sites. Services are provided by child and adolescent psychiatrists, clinical child psychologists, and clinical social workers and include evaluation and treatment of children using individual, family, and group therapy, medication management, and coordination and integration of services within the military system and the community.

Families OverComing under Stress (FOCUS)

The FOCUS Project (www.focusproject.org) provides resiliency training for children and families facing the challenges of deployment. These services are offered at a limited number of installations for Navy, Marine Corps, Air Force, and Army Families. In this program families are taught skills such as communication and problem-solving skills, stress-reduction techniques, and emotional awareness and regulation. Both single-family and group-based programs are provided for parents and children via face-to-face trainings and online classes. Services vary by location (see their website for up-to-date details).

Family Advocate Programs (FAP)

Each Service and each installation has its own FAP that plays a central role in addressing child abuse and neglect in military families. These programs receive reports of child abuse and neglect and notify appropriate law enforcement agencies and state Child Protective Services. In addition, the FAPs provide assessment, crisis intervention, counseling for both victims and abusers, and prevention services (e.g., New Family Support Program for families with infants). FAP also maintains a Central Registry of perpetrators. FAP serves all Active Duty families in addition to Active Reserve and Active Guard families. General information as well as contact information for each installation can be accessed through http://militaryhomefront.dod.mil/sp/fap.

OTHER HELPFUL RESOURCES

Military One Source: www.militaryonesource. mil. This is a comprehensive source of information for military and their families. Adult counseling is also offered 24/7 through 1-800-342-9647.

Military Home Front: www.militaryhomefront. dod.mil is a website of the Department of Defense with extensive information for service members, their families and providers who serve this population. They assert that "you'll find what you need" on this site.

Military Child Education Coalition (MCEC): www.militarychild.org. This organization provides information and support for parents as well as information and training for professionals working with military children and youth.

National Child Traumatic Stress Network: www.nctsnet.org/resources/topics/military-children-and-families. This site offers information, training, and products to assist children dealing with stress. They have materials specific to traumas experienced by military children and provide a comprehensive list of resources for children, parents, and professionals.

Zero to Three: www.zerotothree.org/about-us/ funded-projects/military-families. Focused on young children, this website has materials for families and for professionals working with military families. Specific training for professionals is offered.

National Military Family Association: www. militaryfamily.org. This organization sponsors Operation Purple (network of camps exclusively for military children), provides information for families, and supports research about military families.

Military Kids Connect: www.militarykidsconnect.org. Created by the Department of Defense's National Center for Telehealth & Technology, this website provides an online community for military children (ages 6–17) with age-appropriate resources to support families throughout the deployment cycle.

Psychologists play an important role in understanding and addressing the needs of military youth and families through research, prevention, and treatment efforts. Although psychologists can draw on their generalist clinical competencies in serving these children, obtaining specialized training and consultation about military culture, challenges, resources, and available services can be useful in serving and meeting the needs of this unique population.

References

American Psychological Association. (2009). *Stress in America survey*. Retrieved from http://www.apa.org/news/press/releases/stress-exec-summary.pdf

Chandra, A., Lara-Cinisomo, S., Jaycox, L. H., Tanielian, T., Han, B., Burns, R. M., & Ruder, T. (2011). *Views from the homefront: The experiences of youth and spouses from military families*. Santa Monica, CA: RAND Corporation. Retrieved from http://www.rand.org/pubs/technical_reports/TR913.html

Department of Defense. (2008). *Demographics 2008: Profile of the military community*. Washington, DC: Office of the Deputy Under Secretary of Defense (Military Community and Family Policy). Retrieved from http://www. militaryhomefront.dod.mil/12038/Project%20 Documents/MilitaryHOMEFRONT/ Reports/2008%20Demographics.pdf

Gibbs, D. A., Martin, S. L., Kupper, L. L., & Johnson, R. E. (2007). Child maltreatment in enlisted soldiers' families during combat-related deployments. *Journal of the American Medical Association, 298*(5), 528–535. doi:10.1001/jama.298.5.528

Lester, P., Peterson, K., Reeves, J., Knauss, L., Glover, D., Mogil, C.,… Beardslee, W. (2010). The long war and parental combat deployment: Effects on military children and at-home spouses. *Journal of the American Academy of Child and Adolescent Psychiatry, 49*, 310–320. doi:10.1097/00004583-201004000-00006

Richardson, A., Chandra, A., Martin, L. T., Setodji, C. M., Hallmark, B. W., Campbell, N. F.,… Grady, P. (2011). *Effects of soldiers' deployment on children's academic performance and behavioral health*. Santa Monica, CA: RAND Corporation. Retrieved from http://www.rand.org/content/dam/rand/pubs/monographs/2011/RAND_MG1095pdf

58 IMPACT OF PSYCHIATRIC DISORDERS AND PSYCHOTROPIC MEDICATIONS ON RETENTION AND DEPLOYMENT

David S. Shearer and Colette M. Candy

ARMED SERVICES BEHAVIORAL HEALTH

Military behavioral health resources throughout the Armed Services include Active Duty and civilian psychologists, psychiatrists, social workers, behavioral health specialists, case managers, and others. A full range of behavioral health services are provided in a wide variety of locations. Services can range from consultation, individual or group psychotherapy, psychotropic medication management, to inpatient psychiatric treatment. In this context a wide array of psychiatric diagnoses are made, individual and group therapies are provided as appropriate, and many types of psychotropic medications may be prescribed. However, use of psychotropic medications and the presence of psychiatric disorders may require additional evaluation in light of the many missions of the Armed Services. As a result, guidelines and policies have been put in place to assist medical and behavioral health providers in determining suitability and fitness for continued service in the Armed Forces or for deployment to areas of operation. Although obviously related, psychiatric diagnoses and psychotropic medications are often considered separately in these guidelines (DoD 2006; DoD, 1996b; DoD, 1996a; USCENTCOM 021922Z Mod 11, 2011a).

A service member (SM) who may not be medically fit for continued service in the Armed Forces is referred into the service-specific Disability Evaluation System (DES). As behavioral health providers to the military community, our role includes treatment as well as continual assessment of medical readiness. The goals of behavioral health treatment of SMs are to ameliorate symptoms, assist the SMs in recovery, and reduce overall behavioral impairment. With early identification and intervention, many psychiatric disorders are treatable (DoD, 2006).

RETENTION AND SEPARATION: BEHAVIORAL HEALTH CONDITIONS

When a SM has a behavioral health condition that impacts military performance he or she can be separated or retired from the military for medical or administrative reasons. Both of these separations usually involve a psychological evaluation. The type of separation is based on the diagnosis, which will be discussed below.

Medical Retirement/Separation

Medical retirement and separation are based on authority from DoD Directive 1332.18, Separation or Retirement for Physical Disability (DoD, 1996a) and DoD Instruction 1332.38, Physical Disability Evaluation (DoD, 1996b, incorporating Change 1, July 10, 2006.). This process is guided by the DES and has two component parts; a Medical Evaluation Board (MEB) and a Physical Evaluation Board (PEB). Referral is made initially to the MEB by a military service provider. If the MEB finds the SM does not meet medical retention standards due to a behavioral health condition the case is sent to the PEB. The PEB determines one of four dispositions: return to duty with/without duty limitations, temporary disabled retired list (TDRL), medical separation from active duty, or medical retirement. The SM's deployability may be used as a variable in the DES when determining fitness for duty. An appeal process is available, and SMs may have legal counsel at an appeal hearing. There are some differences in how each branch of the Armed Services may administer this policy (DoD, 1996a, 1996b). Interested readers should consult with service specific regulations and DES programs for specific details. Listed below are psychiatric conditions that may result in referral into the DES (from: DoD, 1996b, Enclosure 4: Guidelines regarding medical conditions and physical defects that are cause for referral into the Disability Evaluation System (DES)).

4.13 Psychiatric Disorders:

E4.13.2. **Disorders with Psychotic Features (Delusions or prominent Hallucinations).** One or more psychotic episodes, existing symptoms or residuals thereof, or a recent history of a psychotic disorder.

E4.13.3. **Affective Disorders (Mood Disorders).** When the persistence or recurrence of symptoms requires extended or recurrent hospitalization, or the need for continuing psychiatric support.

E4.13.4. **Anxiety, Somatoform Dissociative Disorders (Neurotic Disorders).** When symptoms are persistent, recurrent, unresponsive to treatment, require continuing psychiatric support, and/or are severe enough to interfere with satisfactory duty performance.

E4.13.5. **Organic Mental Disorders.** Dementia or organic personality disorders that significantly impair duty performance.

E4.13.6. Eating Disorders. When unresponsive to a reasonable trial of therapy or interferes with the satisfactory performance of duty.

(Additional Note: **E4.13.1.5.** Any MEB listing a psychiatric diagnosis must contain a thorough psychiatric evaluation and include the signature of at least one psychiatrist (identified as such) on the MEB signatory face sheet.)

Administrative Separation

There are administrative separations for behavioral health conditions that the military deems as unsuitable for service when there is impact on the SM's ability to perform the mission. These are also described in DoDI 1332.38 as follows:

E4.13.1.4. Personality, Sexual, or Factitious Disorders, Disorders of Impulse Control not elsewhere classified, Adjustment Disorders, Substance-related Disorders, Mental Retardation (primary), or Learning Disabilities are conditions that may render an individual administratively unable to perform duties rather than medically unable, and may become the basis for administrative separation. These conditions do not constitute a physical disability despite the fact they may render a member unable to perform his or her duties. (DoD, 1996b)

The aforementioned conditions can lead to military performance issues and are typically addressed through a Command Directed Mental Health Evaluation per DoD Instruction 6490.04 (DoD, 2013). If upon completion of the Command Directed Mental Health evaluation any of the above conditions are met the provider can recommend to the Commander an administrative separation. The Commander can then processes a separation as the SM is consider medically fit for duty but not suitable. There are service-specific regulations for behavioral-health-related administrative separations. The reader should be aware that 2013 revisions to this guidance may result in changes to the process or conditions for administrative separation.

DEPLOYMENT-LIMITING PSYCHIATRIC CONDITIONS AND MEDICATIONS

Among the primary missions of military medicine is to provide a fit and ready force for deployment to areas of operation, which include land, sea, and air missions. Deployment typically refers to sending a service member to an operational area that may or may not include combat operations and other activities. Medical readiness is assessed at multiple points in the military lifecycle. Assessment of psychological readiness is conducted at each stage of this lifecycle, which includes sustainment, predeployment, deployment, and postdeployment periods. These are conducted via an annual periodic health assessment (PHA), predeployment health assessment and postdeployment health assessment (PDHA), postdeployment health reassessment (PDHRA), routine health care visits, and during deployment as appropriate (DoD, 2006).

Section 738 of the National Defense Authorization Act for Fiscal Year 2007, Public Law 109-364 (National Defense Authorization, 2006), sets forth requirements to establish policies to determine minimum behavioral health standards for members of the armed forces to deploy to a combat operation or contingency operation. The Assistant Secretary of Defense provided policy guidance for deployment-limiting psychiatric conditions and medications in a November 2006 memorandum to the Secretaries of the Army, Navy, and Air Force (DoD, 2006). In the Attachment to this memo it states:

Serving in the Armed Forces requires the physical and mental fitness necessary to plan and execute missions involving combat as well as Stability, Security, Transition, and Reconstruction Operations .... Any condition or treatment for that condition that negatively impacts on the mental status or behavioral capability of an individual must be evaluated to determine potential impact both to the individual Service member and to the mission. (paragraph 2.1)

US Central Command (CENTCOM) provides additional guidance regarding deployment-limiting psychiatric conditions and medications for CENTCOM Service Members (USCENTCOM 021922Z Mod 11, 2011a). USCENTCOM is one of 10 combatant commands in the US Military. Six of these commands, including CENTCOM, are responsible for a specific geographic area in which

commanders may plan and conduct operations as allowed under the Unified Command Plan. CENTCOM's area of operations includes Afghanistan, Bahrain, Egypt, Iran, Iraq, Jordan, Kazakhstan, Kuwait, Kyrgyzstan, Lebanon, Oman, Pakistan, Qatar, Saudi Arabia, Syria, Tajikistan, Turkmenistan, United Arab Emirates, Uzbekistan, and Yemen.

CENTCOM issued MOD ELEVEN in December 2011 as a modification to USCENTCOM Individual Protection and Individual Unit Deployment Policy (USCENTCOM 021922Z Mod 11, 2011a). Guidance from both the Memorandum on Policy Guidance for Deployment-Limiting Psychiatric Conditions and Medications (ASECDEF Memo, 2007) and CENTCOM's MOD ELEVEN PPG-TAB A (USCENTCOM 021922Z Mod 11, PPG-TAB A, 2011b) are listed below:

Deployment Limitations Associated with Psychiatric Disorders

(From: ASECDEF Memo, Policy Guidance for Deployment-Limiting Psychiatric Conditions and Medications (Attachment), November 7, 2006 [DoD, 2006]):

4.1.4.1. All conditions that do not meet retention requirements or that render an individual unfit or unsuitable for military duty should be appropriately referred through Service-specific medical evaluation boards (MEB) or personnel systems.

4.1.4.2. Psychotic and Bipolar Disorders are considered disqualifying for deployment.

4.1.4.3. Members with a psychiatric disorder in remission or whose residual symptoms do not impair duty performance may be considered for deployment duties.

4.1.4.4. Disorders not meeting the threshold for a MEB should demonstrate a pattern of stability without significant symptoms *for at least 3 months prior to deployment* [emphasis added].

4.1.4.5. The availability, accessibility, and practicality of a course of treatment or continuation of treatment in theater should be consistent with practice standards.

4.1.4.6. Members should demonstrate behavioral stability and minimal potential for deterioration or recurrence of symptoms in a deployed environment, to the extent this can be predicted… This should be evaluated considering potential environmental demands and individual vulnerabilities.

4.1.4.7. The environmental conditions and mission demands of deployment should be considered: the impact of sleep deprivation, rotating schedules, fatigue due to longer working hours, and increased physical challenges with regard to a given mental health condition.

4.1.4.8. The occupational specialty in which the individual will function in a deployed environment should be considered. However, when deployed, individuals may be called upon to function outside their military training as well as outside their initially assigned deployed occupational specialties. Therefore the primary consideration must be the overall environmental conditions and overall mission demands of the deployed environment rather than a singular focus on anticipated occupation-specific demands.

Deployment Limitations Associated with Psychotropic Medication

4.2.1. … Medications prescribed to treat psychiatric disorders may vary in terms of their effects on cognition, judgment, decision-making, reaction time, psychomotor functioning and coordination and other psychological and physical parameters that are relevant to functioning in a an operational environment…

4.2.2. Caution is warranted in beginning, changing, stopping, and/or continuing psychotropic medication for deploying and deployed personnel… use of psychotropic medication should be evaluated for potential limitations to deployment or continued service in a deployed environment.

Medications disqualifying for deployment include:

4.2.3.1. Antipsychotics used to control psychotic, bipolar, and chronic insomnia symptoms; lithium and anticonvulsants to control bipolar symptoms

4.2.3.2. Medications that require special storage considerations, such as refrigeration.

4.2.3.3. Medications that require laboratory monitoring or special assessments, including lithium, anticonvulsants, and antipsychotics.

4.2.4. Psychotropics clinically and operationally problematic during deployments include short half-life benzodiazapines and stimulants. Decisions to deploy personnel on such medications should be balanced with necessity for such medication in order to effectively function in a deployed setting, susceptibility to withdrawal symptoms, ability to secure and procure controlled medications, and potential for medication abuse.

From: USCENTCOM 021922Z Mod 11, PPG-TAB A, 2011b

Use of any of the following medications (specific medications or class of medication) is disqualifying for deployment:

7.I.5. Antipsychotics, except quetiapine (Seroquel) 25 mg at bedtime for sleep.

7.I.6. Antimanic (bipolar) agents. Individual assessment required.

7.I.7. Anticonvulsants used for seizure control or psychiatric diagnoses.

7.I.7.a. Anticonvulsants which are used for non-psychiatric diagnoses, such as migraine, chronic pain, neuropathic pain, and post-herpetic neuralgia, are not deployment limiting as long as they meet the criteria in MOD ELEVEN and PPG-TAB A. No waiver required. (Exception: use of valproic acid or carbamazepine for non-psychiatric diagnoses are deployment limiting).

Waivers

Under certain circumstances, as noted above, a waiver may be requested to allow a service member to deploy who has been identified as having a non-deployable condition or medication regimen. The CENTCOM Surgeon is the approval authority under these circumstances. All behavioral health waivers within USCENTCOM (as defined above) must be obtained from the CENTCOM Command Surgeon or CENTCOM Component (e.g., Army, Navy, Air Force, Special Forces) Surgeon (USCENTCOM 021922Z Mod 11, 2011a).

From: USCENTCOM 021922Z Mod 11, PPG-TAB A, 2011b

Individuals with the following conditions should not deploy (unless a waiver is approved):

7.H.2. Clinical psychiatric disorders with residual symptoms, or medication side effects, which impair duty performance.

7.H.3. Mental health conditions that pose a substantial risk for deterioration and/or recurrence of impairing symptoms in the deployed environment.

7.H.4. History of the following: psychiatric hospitalization; suicide attempt; substance (medication, illicit drug, alcohol, inhalant, etc.) abuse or treatment for such abuse. Such history does not necessitate a waiver request, but requires a pre-deployment evaluation by a BH practitioner authorized to write profiles… Waiver requests should include the results and recommendations from this evaluation, as well as the documentation of completion of any formal substance abuse classes or instruction.

7.H.4.a. Substance abuse disorders (not in remission), (for service members) actively enrolled in Service Specific substance abuse programs require an Individualized Assessment.

7.H.5. Psychiatric disorders with fewer than three months of demonstrated stability from the last change in treatment regimen (medication, either new or discontinued, or dose change). Note: Disorders that HAVE demonstrated clinical stability for three months or greater, without change in therapy, do not require a waiver to deploy (specific exceptions noted elsewhere such as bipolar disorder or use of antipsychotics).

7.H.5.a. Psychiatric disorders newly diagnosed during deployment do not immediately require a waiver or redeployment. Disorders that are deemed treatable, stable, and having no impairment of performance or safety by a credentialed mental health provider do not require a waiver to remain in theater.

SPECIAL CONSIDERATIONS

As might be expected, some specific occupational specialties within the Armed Forces may necessitate more restrictive limitations or a more detailed evaluation. "Special consideration must be given to limitations affecting those under the Personnel Reliability Program and specific operational standards such as aviation, submarines, special operations or other high risk occupational categories" (DoD, 2006, paragraph 4.3.1). For example, aviators and aviation personnel in each Military Service must be specifically cleared by a flight surgeon (Department of Air Force, 2009; Department of the Army, 2008; Department of the Navy, 2010).

References

Department of Air Force. (2009). *Air Force Instruction 48-123: Aerospace medicine, medical examinations, and standards.* Washington, DC: Author.

Department of Defense (DoD). (1996a). *Department of Defense Directive 1332.18: Separation or retirement for physical disability.* Washington, DC: Author.

Department of Defense. (1996b, incorporating Change 1, 2006). *Department of Defense Instruction 1332.38: Physical disability evaluation.* Washington, DC: Author.

Department of Defense. (2013). *Department of Defense Instruction 6490.04: Requirements for mental health evaluations of members of the armed forces.* Washington, DC: Author.

Department of Defense. (2006). *Assistant Secretary of Defense, memo: Policy guidance for deployment-limiting psychiatric conditions and medications and attachment.* Washington, DC: Author.

Department of the Army. (2008). *Flight surgeon's aeromedical checklists, revision.* Retrieved from http://usasam.amedd.army.mil/_AAMA/files/ArmyAPLs.pdf

Department of the Navy. (2010). *U.S. Navy aeromedical reference and waiver guide.* Retrieved from http://www.med.navy.mil/sites/nmotc/nami/

arwg/Documents/WaiverGuide/Waiver%20
Guide%20-%20Complete%20120215.pdf

National Defense Authorization Act for Fiscal Year
2007, Enhanced mental health screening and
services for members of the Armed Forces,
Pub. L. no. 109-364, Sec 738, 120 Stat 2083
(2006).

USCENTCOM 021922Z Dec 11 Mod Eleven to
USCENTCOM Individual Protection and
Individual-Unit Deployment Policy. (2011a).
Retrieved   from   www.cpms.osd.mil/.../
MOD11-USCENTCOM-Indiv-Protection-I
ndiv-Unit-Deployment-Policy-Incl-Tab-A-
and-B.pdf

USCENTCOM 021922Z Dec 11 Mod Eleven to
USCENTCOM Individual Protection and
Individual-Unit Deployment Policy. (2011b).
PPG-TAB A: Amplification of the Minimal
Standards of Fitness for Deployment to the
CENTCOM AOR: To Accompany MOD Eleven
to USCENTCOM Individual Protection and
Individual-Unit Deployment Policy. Retrieved
from www.cpms.osd.mil/.../MOD11-USCEN
TCOM-Indiv-Protection-Indiv-Unit-Deploym
ent-Policy-Incl-Tab-A-and-B.pdf

59 TECHNOLOGY APPLICATIONS IN DELIVERING MENTAL HEALTH SERVICES

Greg M. Reger

Service members returning from military deployments are at increased risk of mental health disorders to include depression, posttraumatic stress disorder (PTSD), and a range of behavioral or substance abuse problems. With increasing numbers of military personnel returning from deployment with behavioral health needs there have been calls to increase access to psychological resources and specialty mental health treatment. Increased access is necessitated by the remote geographic location of many service members, particularly National Guard and reservists, who may live far from military medical centers and VA mental health treatment facilities. Many military personnel also have concerns with treatment stigma and peer or leader perceptions of help seeking, resulting in calls for access to anonymous and less stigmatizing forms of help.

Improvements in treatment outcomes are also of concern. Although effective psychotherapies and medications exist to treat many of the postdeployment concerns of some service members, the efficacy of most treatments with military populations is unknown, and some research has found decreased efficacy with military veterans. Technology developments provide a range of opportunities to improve access to resources and to potentially improve the quality of care delivered.

TECHNOLOGIES TO SUPPORT MILITARY PERSONNEL

Technologies have continued to evolve simultaneously with the military operations in Iraq and Afghanistan. These technologies provide novel opportunities to support the psychological health of service members. This section will briefly review several key technology capabilities currently available to military psychologists.

Internet and Web Resources

With nearly ubiquitous service member access to the Internet, websites provide a readily accessible source of information, self-assessment, and support. Web resources can be accessed anonymously, which may mitigate the stigma of presenting at a brick and mortar military mental health clinic for information and initial screening. A wide range of websites exist, a number of which offer high quality, evidence-based information and resources. Although a comprehensive review of psychological health websites is beyond the scope of this chapter, a review of representative quality examples will elucidate the type of capabilities these technologies can make available. Afterdeployment.org is a Congressionally mandated Department of Defense (DoD) website that provides a self-care solution for service members with preclinical problems. The site includes self-assessments and multisession self-guided workshops for 18 content areas such as anger, depression, alcohol and drugs, and life stress. These workshops provide evidence-supported skills and information that can be learned independent of face-to-face care. Although websites such as these do not serve as a replacement for mental health treatment when indicated, they can provide significant support to military personnel seeking to understand their difficulties or provide key self-management strategies to those whose problems do not rise to a clinical level of intensity. Some providers are also beginning to leverage website capabilities like these to support relevant patient homework between psychotherapy sessions. Others are using web content live during group therapy sessions.

Social networking is a special case of Internet capability that has a primary opportunity for users to offer and obtain social support. Many mental-health-related organizations also use these capabilities to disseminate strategic information to stakeholders. A range of complex issues must be considered by any clinician considering the use of social networking for professional purposes. A recent paper, which focused on the topic of suicide and social networking, provides a thorough overview of some relevant issues (Luxton, June, & Fairall, 2012).

Smartphones and Mobile Computing Platforms

Smartphones are increasingly in the pockets of our military personnel. These mobile computing platforms offer a range of capabilities including impressive computer processing speeds, local device storage of data, access to cloud-based data storage and resources, GPS, 2-way camera/video viewing, accelerometers, phone, and compass functionality. The size of these devices makes them highly portable and accessible to users throughout their day. Specialized computer software applications, or apps, are routinely used by smartphone owners for a wide range of purposes.

Many are beginning to think about how to leverage these capabilities to support the psychological health of Warriors. Smartphones provide ongoing, instant access to web content and apps in a surreptitious and potentially nonstigmatizing format. Collaborations between the DoD and the Department of Veterans Affairs (VA) have resulted in apps to support: (1) self-assessment and population surveillance of warriors and veterans, (2) self-care and symptom management, and (3) the delivery of evidence-based treatments. Regarding self-assessment and surveillance, the T2 Mood Tracker provides one example. This app is designed to support the ecological momentary assessment of service members. Users can track their moods daily (or multiple times a day) on a range of visual analog scales. Service members or veterans can log events and circumstances related to mood changes to help understand their difficulties and support behavior change efforts.

Tracking mood, in and of itself, can be helpful to many people. Others are using mood tracking apps as an adjunct to therapy. Logs from mood tracking apps can be reviewed in therapy sessions to discuss the successes and barriers to success for implemented interventions. An app currently in development is designed to support the ongoing assessment of symptoms or to support population surveillance. This app would allow a patient to complete any validated self-report measure inserted into the platform, provide their responses, and securely transmit their data to a provider or a secure data base. Apps such as these could provide future support to the assessment and ongoing symptom management associated with the delivery of psychotherapy. Alternatively, apps of this type could efficiently support large-scale assessments or surveillance efforts, such as that conducted by entire military units following operational deployments. At risk populations could provide ongoing secure self-assessments to providers to assist in the identification of increased risk between psychotherapy sessions.

A number of apps have also been developed to support the psychoeducation and self-care of service members and veterans. VA and DoD collaborated on PTSD Coach, a smartphone app that provides information about PTSD, self-assessment, and symptom management strategies for military personnel and veterans. If veterans self-rate their distress at a high level, the app leverages the phone capabilities of the device and the user is provided one-touch access to a crisis-line, should they choose to do so.

Apps have also been developed to support the tasks of evidence-based psychotherapy. These apps are not designed to be used as self-help, but rather to support the work of a patient and provider engaged in a manualized treatment. For example, prolonged exposure (PE) is an evidence-based treatment for PTSD. PE Coach was designed by the National Center for Telehealth and Technology (T2), the VA National Center for PTSD, and the Center for Deployment Psychology to improve the implementation, treatment fidelity, and adherence of patients and providers engaged in PE. The app provides a range of capabilities necessary to the treatment protocol to make participation in treatment more convenient. The app supports audio recording of PE therapy sessions directly onto the patient's phone, logging of imaginal and in vivo exposure homework in the app (instead of paper worksheets), tracking and graphical display of trauma-related distress and PTSD symptoms over time, and device calendar reminders of PE sessions and homework. At each session, the therapist can review the patient's homework adherence based on actual app usage supporting the identification of barriers and problems with homework adherence. Obviously, these descriptions of a few apps do not begin to summarize the broad range of apps available for clinical use. Although apps like PE Coach have the ability to transform our delivery of evidence-based treatments in helpful ways, clinicians must carefully judge the quality of the content of apps and research is needed to evaluate the effectiveness of apps like these to determine their value and any contributions to treatment outcomes.

Virtual Reality and Virtual Worlds

Virtual reality (VR) is a more innovative technology available to some military mental health providers. VR leverages computers and a range of peripherals to give the user the psychological experience of participating in a computer-generated environment although they are physically located elsewhere. Head-mounted display systems or immersive visual display systems, vibro-tactile platforms, 3D audio and visuals, haptic devices, and delivery of relevant manufactured olfactory stimuli are common components of VR.

The ability to psychologically transport a user to an alternate location may be relevant to certain military behavioral health goals. Distraction is an effective form of nonpharmacological pain management and there is evidence that VR may be useful for some patients undergoing painful medical procedures. VR-based assessment is another area of interest. Ecologically valid assessment of attention processes and other cognitive functions may be very helpful in the future to providers wanting to answer real-world questions about fitness for duty.

However, the most broadly researched area of VR relevant to military clinicians is probably the potential to use VR to help activate the fear structures of patients engaged in an exposure therapy for PTSD. During VR exposure, multisensory VR stimuli are modified in real time during imaginal exposure to help patients activate their memory and modulate therapeutic levels of emotional engagement. The VR system is not a replacement for formal training in exposure therapy and it does not replace the role of the clinician. However, it may prove to be a useful tool for a skilled clinician to use with military personnel who have developed strong emotional detachment and have difficulty achieving adequate levels of engagement during imaginal exposure. Research has supported the effectiveness of VR exposure, but quality randomized controlled trials are needed to determine the efficacy of this form of exposure therapy.

Shared computer-generated environments, referred to from here forward as virtual worlds (VW), are also being explored for supporting the psychological health of military personnel. VWs typically involve use of a digital representation of oneself, called an avatar, to navigate through and interact with other users and the 3D computer generated environment. In many cases, access is available to anyone with a broadband Internet connection. Users can typically communicate with one another through text-based chat or can use a digital microphone to speak directly through Voice Over Internet Protocol (VOIP). Software capabilities are often incorporated into these spaces to enhance user experience. For example, some VWs allow incorporation of Microsoft Office products to support collaborative work and meetings. In an era of increased calls for efficiency, one can imagine the potential utility of a VW that is approved for use on the military network to increase collaborative DoD meetings while reducing costs associated with travel. Some are considering whether VWs could effectively replace inefficient classroom gatherings of mental health providers for training on evidence-based treatments.

Of all the VW uses considered, experiential learning is one clear use case. The Virtual PTSD Experience is one example. This space is located in the VW called Second Life and takes the user through an asynchronous, stand-alone, interactive educational experience. It teaches the user about the causes, common symptoms, and help available for deployment-related PTSD. Users proceed through a series of "exhibits" that attempt to leverage gaming motivation to help users learn by doing. Like an interactive museum, the Virtual PTSD Experience is an example of using a VW space to deliver psychoeducation in a manner unlike typical didactic methods. This space is available at no cost, and users can name their avatar anonymously, potentially mitigating the stigma related to seeking information about mental health issues.

CONSIDERATIONS IN THE CLINICAL DECISION TO USE TECHNOLOGY

The decision to use a technology to support clinical practice requires deliberate thought to ensure the solution being considered is a good fit. A framework that was previously described for considering the design of virtual environments to support patients with central nervous system dysfunction (Rizzo, Buckwalter, & van der Zaag, 2002) can be adapted for considering questions relevant to a range of technologies. First, can the same benefits be achieved without the technology approach? Gadgets for gadgets' sake do not support military personnel. An honest appraisal of how the technology capabilities are helpful is needed to ensure good clinical decision making. Second, how well do the capabilities of the technology fit the clinical goals? The mere insertion of technology into the treatment plan of a clinical issue does not make treatment better. However, technologies that provide capabilities that help the clinician address a logistic or clinical problem can provide dramatic improvements. Third, consideration must be given to how well a technology solution fits the characteristics of the patient population. Young, technologically experienced "digital natives" make up a sizable proportion of today's Active Duty military. The integration of appropriate technologies into clinical practice can be a successful fit for many. However, service members are not a homogeneous group. Users must have access to the relevant technologies to take advantage (e.g., the Internet or smartphone devices). In addition, certain clinical populations may not be appropriate for certain technology solutions. For example, patients with a lower threshold for seizure activity may not be appropriate to use certain low frequency visual displays. Patients with vestibular problems may need to avoid technologies with the potential for balance or dizziness side effects (e.g., immersive virtual environments). Clinicians must give careful consideration to the specific clinical population and their fit with the targeted technology.

ETHICAL ISSUES

Several sections of the *Ethical Principles of Psychologists and Code of Conduct* (American Psychological Association [APA], 2010) are relevant to the discussion of the clinical use of technology. First, psychologists are expected to practice within the limits of their competence. "Psychologists planning to provide services, teach, or conduct research involving populations, areas, techniques, or technologies new to them undertake relevant education, training, supervised experience, consultation, or study" (APA, 2010, p. 5). Similarly, in emerging areas where agreed upon standards and training do not yet exist, psychologists are expected to take reasonable steps to ensure their competence. Continuing education is increasingly available to support skill development in the use of certain technologies, and some professional societies are beginning to give significant thought to the appropriate use of a range of technologies in clinical practice (e.g., American Telemedicine Association, accessed August 20, 2012). Information should be obtained from guidelines such as these and professionals with relevant experience should be consulted.

Mental health providers seeking to apply technologies in practice should ensure they are current on the available scientific literature supporting the use of the selected technology approaches, as well as the limits of what can currently be concluded from that literature. Doing so supports compliance with the APA ethical requirement to obtain informed consent. If a selected technology treatment is judged to be one for which recognized techniques and procedures have not been established, providers must inform Service Members "of the developing nature of the treatment, the potential risks involved, alternative treatments that may be available, and the voluntary nature of their participation" (APA, 2010, p. 13).

Some technologies increase the risk of challenges to professional boundaries. Communication technologies that include e-mailing or texting provide instant delivery of messages and the expectations and management of such communications during nonbusiness hours can be complicated. It is possible that the APA Ethics Code's Standard 3.05 on Multiple Relationships could be relevant to the use of some technologies in clinical practice. Finally, some technology solutions support the face-to-face delivery of care. If a technology solution obtains or stores confidential information, psychologists have an ethical obligation to take reasonable steps to protect that information (Standard 4.01).

A range of technologies are emerging with interesting and potentially useful capabilities to support some of the goals of the military psychologist. With these capabilities comes the obligation to think carefully about when and where such technology applications are appropriate. A thoughtful and ethical clinical implementation of technologies has the potential to dramatically impact and improve our future care of Warriors.

References

American Psychological Association. (2010). *Ethical principles of psychologists and code of conduct.* Retrieved from http://www.apa.org/ethics

American Telemedicine Association. (2012, August). *Telemedicine standards and guidelines.* Retrieved from http://www.americantelemed. org/i4a/pages/index.cfm?pageid=3311

Luxton, D. D., June, J. D., & Fairall, J. M. (2012). Social media and suicide: A public health perspective. *American Journal of Public Health, 102*(Suppl. 2), s195–s200.

Rizzo, A. A., Buckwalter, J. G., & van der Zaag, C. (2002). Virtual environment applications in clinical neuropsychology. In K. M. Stanney (Ed.), *Handbook of virtual environments: Design, implementation, and applications* (pp. 1027–1064). Mahwah, NJ: Erlbaum.

60 WHAT WE HAVE LEARNED FROM FORMER PRISONERS OF WAR

Brian Engdahl

American former prisoners of war (POWs) are exceptional individuals. As many as 110,000 POWs were alive in the mid-1950s; at last estimate, less than 30,000 remained (US Department of Veterans Affairs, 2006). Roughly 25,000 were from World War II (WWII), 2,000 from Korea, 550 from Vietnam, and approximately 40 from the Cold War, Desert Storm, and subsequent conflicts combined. Very few remain on active duty. Nearly all were subjected to a spectrum of harsh abuse and suffered a myriad of insults, including malnutrition, exposure to environmental extremes, infections, and physical and emotional injuries (for further detail, see Skelton, 2002).

Postrepatriation treatment has always placed priority on restoring lost weight and treating medical illnesses and physical injuries. Following WWII and the Korean War, it was not uncommon during repatriation examinations for POWs to be told that their ordeal would shorten their life span, although the studies summarized by Beebe (1975) and Page (1992) showed increased morbidity and mortality only in the first postwar years. For all too many, their psychological injuries were only recognized in hindsight. Beebe proposed a model to explain negative captivity effects. They stem from two types of trauma: one is physical and primarily short-term, caused by malnutrition, infection, and physical injury; the other is psychological and essentially permanent, leading to a loss of "ego strength" and lowered thresholds for both physical and psychological distress.

Many POWs suffer from what we now know as posttraumatic stress disorder (PTSD), almost entirely traceable to combat and prison camp trauma. The psychological challenge of being held captive cannot be overemphasized; see Farber, Harlow, and West (1957) for their insightful discussion of the "three D's" faced by POWs—debility, dependency, and dread.

PTSD is defined by an enduring set of maladaptive symptoms that arise after exposure to one or more potentially life-threatening events (see elsewhere in this volume). These symptoms include the unwanted reexperiencing of painful trauma memories—nightmares, daytime intrusive memories, and psychological distress and/or physiologic arousal when reminded of the trauma. Other symptoms include avoidance of trauma reminders, withdrawing from one's environment, plus a numbing of responsiveness. Persistent arousal—sleep disturbances, irritability, exaggerated startle responses, and/or hypervigilance—also contributes to the functional impairments accompanying PTSD.

In a community sample of POWs (N = 262) from WWII and the Korean War, over half met lifetime criteria for PTSD, and 30% met criteria for current PTSD, 40 to 50 years after repatriation (Engdahl, Dikel, Eberly, & Blank, 1997). The most severely traumatized group—POWs held by Japan—had lifetime PTSD rates of 84% and current rates of 58%. In a further study of this group (Dikel, Engdahl, & Eberly, 2005), over 50% of the variance in current PTSD severity was accounted for by a combination of prewar, wartime, and postwar factors. POW camp trauma was most predictive of PTSD severity, followed by a narrowly defined postwar social support variable: interpersonal connection. Prewar conduct disorder behavior positively predicted PTSD and negatively predicted interpersonal connection. Combat exposure and being younger at capture also predicted PTSD. Prewar family closeness did not predict PTSD, but predicted postwar interpersonal connection. This study provides strong evidence that trauma is by far the most significant predictor of PTSD severity and chronicity among POWS. Few of these men had ever sought mental health treatment. This is changing as DoD and the Department of Veterans Affairs continue their efforts to destigmatize mental health problems and encourage service members and veterans to seek treatment.

The long-lasting effects of combat and imprisonment are not universally negative. A narrow focus on negative effects blinds us to the complexity of responses to trauma, and the resilience that survivors exhibit. Posttraumatic growth is an important but often overlooked aspect of functioning among trauma survivors. World War II and Korean War POWS learned through their combat and prison camp trauma that they were stronger than they thought they were. They also developed a greater appreciation of life and personal relationships; many had positive growth in their spiritual lives (Erbes et al., 2006).

Research findings coupled with intense and persistent congressional lobbying have established a list of "presumptive" service-connected disabilities for POWs. If found, they are automatically presumed to be related to the POW's military experience, without requiring historical written proof, qualifying the POW for care and disability benefits (Skelton, 2002). Many medical records were never generated, or were lost, leaving the POW with no way to prove a connection between wartime service and present-day illnesses. The "presumptives" initially included arthritis due to injury, any disease due to malnutrition, chronic dysentery, frostbite, helminthiasis (parasitic worms), psychosis, panic disorder, PTSD and other anxiety disorders, depression, peripheral neuropathy (nerve damage), irritable bowel syndrome, and peptic ulcer disease. Subsequent research and lobbying led to the addition of ischemic heart disease, cirrhosis, stroke, and osteoporosis.

OPERATION HOMECOMING AS A MODEL FOR POW REPATRIATION

In contrast to the unorganized, sometimes indifferent reception afforded repatriated POWs from previous wars, Operation Homecoming was conducted in the Philippines for repatriated Vietnam POWs in February–March 1973 (Ursano & Rundell, 1989). Safeguards included carefully balanced diets and insulation from the press. A man of equivalent rank, service branch, and background greeted each POW and served as an escort, a buffer, and a source of support. He provided up-to-date information about the world and the POW's family. Navy Captain Jeremiah Denton's first words upon landing were: "We are honored to have had the opportunity to serve our country under difficult circumstances. We are profoundly grateful to our Commander in Chief and to our nation for this day. God bless America" (Sterba, 1973).

CURRENT RESOURCES

At the service branch level, resources such as the USAF's Family Readiness Edge applies to all deployed service members including POWs, and their families: (http://www.afcrossroads.com/famseparation/pdf/ReadinessFAmily.pdf)

A triservice program, the Robert E. Mitchell Center for Prisoner of War Studies (http://www.med.navy.mil/sites/nmotc/rpow/Pages/default.aspx), Pensacola, Florida, provides follow-up studies of repatriated POWs of all eras, documenting captivity-related physical and mental problems within the context of extensive annual evaluations. The Center also trains medical personnel of all Services assigned to operational billets who might be involved in repatriation.

At the highest level, the Defense Prisoner of War/Missing Personnel Office (DPMO, http://www.dtic.mil/dpmo/) searches for MIAs from past conflicts, and also oversees efforts to account for and recover personnel who have become separated from their units during more recent actions. This includes the rescue, recovery, and reintegration of captured or missing personnel through diplomatic and military means.

APPLYING WHAT WE HAVE LEARNED

DoD continues to incorporate lessons learned into future capabilities, ensuring that personnel are properly trained and accounted for. This includes the increasing numbers of DoD contractors and civilians who accompany the military force. In the absence of future scenarios in which large numbers of Americans are taken captive, efforts have accordingly shifted from personnel recovery and the repatriation process to preparing service members, contractors, and civilians to avoid capture. Training also focuses on proper responses if they are captured. Lessons learned through research and operations such as at the Mitchell Center are used in SERE (survival, evasion, resistance, and escape) training. Improvements in technology, planning, training, and command and control have all combined to form a rapid, organized response to isolating events.

In the words of the recent head of the Mitchell Center, Robert E. Hain, CAPT, MC, USN (Ret),

A question that is on our collective minds deals with the fact that so many of our people went through a truly terrible experience but emerged at the other end a better, stronger person. Answers to this question of why ultimately helps us contribute to the body of knowledge that prepares present day fighters to be deployed to a war zone. (Booher, 2012, p. 20)

References

Beebe, G. W. (1975). Follow-up studies of World War II and Korean War prisoners: II. Morbidity, disability, and maladjustments. *American Journal of Epidemiology, 101,* 400–422.

Booher, A. (2012, March/April). Retiring with a remarkable legacy. *EX-POW Bulletin,* p. 20.

Dikel, T., Engdahl, B., & Eberly, R. (2005). PTSD in former POWs: Prewar, wartime, and postwar factors. *Journal of Traumatic Stress, 18,* 69–77.

Engdahl, B. E., Dikel, T., Eberly, R. E., & Blank, A. (1997). Posttraumatic stress disorder in a community sample of former prisoners of war: A normative response to severe trauma. *American Journal of Psychiatry, 154,* 1576–1581.

Erbes, C., Johnsen, E., Harris, J., Dikel, T., Eberly, R., & Engdahl, B. (2006). Posttraumatic growth among American POWs. *Traumatology, 11*(4), 285–295.

Farber, I., Harlow, H., & West, L. (1957). Brainwashing, conditioning, and DDD (debility, dependency, and dread). *Sociometry, 20,* 271–285.

Page, W. F. (1992). *The health of former prisoners of war: Results from the medical examination survey of former POWs of World War II and the Korean conflict.* Washington, DC: National Academy Press. Retrieved from http://books.google.com/books?hl=en&lr=&id=bz8rAAAAYAAJ&oi=fnd&pg=PA1&dq=Page+National+Academy+press+prisoner+of+war&ots=wwm1vCWpPX&sig=dRP6rAVRENH8I0BB2n_1BiG72A#v=onepage&q=Page%20National%20Academy%20press%20prisoner%20of%20war&f=false

Skelton, W. P., III. (2002). *The American ex-prisoner of war.* Retrieved from http://www.public-health.va.gov/docs/vhi/pow.pdf

Sterba, J. (1973, February 12). First prisoner release completed. *New York Times,* p. 1.

Ursano, R. J., & Rundell, J. R. (1989). The prisoner of war. In F. D. Jones, L. R. Sparacino, V. L. Wilcox, J. M. Rothberg, & J. W. Stokes (Eds.), *War psychiatry.* Washington, DC: Office of the Surgeon General. Retrieved from http://freeinfosociety.com/media/pdf/4569.pdf#page=436

US Department of Veterans Affairs. (2006). American Prisoners of War (POW) and Missing in Action (MIA). Office of the Assistant Secretary for Policy and Planning.

61 CLINICAL RESEARCH IN THE MILITARY

Stacey Young-McCaughan

Research conducted in the military has been instrumental in advancing medical practices both on and off the battlefield. Examples include medical evacuation, vaccine development, and the use of blood product replacement. In mental health, military psychologists have conducted important research in the area of health promotion and more recently in the care of military-related posttraumatic stress. Researchers working in a military setting must follow the same regulations and ethical standards adhered to by researchers working in a civilian setting, as well as additional regulations specific to the military. Navigating the regulations as well as the interpretation of the regulations within a specific organization can be challenging. Having a basic understanding of the key issues can facilitate the review and approval of a research project so that high quality research can be accomplished.

DEFINITION OF RESEARCH

The section of the Department of Defense (DoD) Code of Federal Regulations (CFR) that governs human subjects research in the military (32 CFR 219) defines research as, "a systematic investigation, including research development, testing and evaluation, designed to develop or contribute to generalizable knowledge" [32 CFR 219.102(d)]. There are debates about what is research as opposed to quality assurance (QI), performance improvement (PI), case reports, utilization review, and education. There are similarities between these types of activities; the bullets below compare each of these terms with research.

- QI or PI (terms often used interchangeably) are activities that aim to improve local processes and/or outcomes within one facility or institution as compared to research, which aims to generate knowledge with wider applicability.
- Case Reports review a provider's care of one patient, usually for educational purposes. A case report is commonly an interesting observation others can benefit from knowing about, while research is a deliberate investigation.
- Cases Series are a review of a provider's clinical care of more than one patient with a similar condition. The number of patients included in the series is not a defining factor; rather that one person's clinical experiences are being presented, again usually for educational purposes.
- Utilization Review is an evaluation of the use of resources in a specific health care activity, as opposed to research that aims to generate knowledge with wider applicability.
- Education is the transferring of information from one group to another to spread generalizable knowledge, as opposed to research that aims to generate knowledge.

One of the main differences between research and the activities listed above is the standardization of research methods. While other activities allow for modifications to the procedures while data collection is underway, research methods require that the data collection procedures be held constant and are only changed with good justification and approval from the IRB. Investigators may want an IRB review and approval for a project as research because they want to publish the findings; however, IRB review and approval is not a requirement for publication. Investigators wishing to publish the findings from their nonresearch activity need to submit their findings to an appropriate journal and should describe their processes and findings appropriate to the activity. For example, a QI or PI project should use section headings such as "issue to be addressed" rather than "research question," "imperative for the project" rather than "review of literature," "procedures for collecting and evaluating data" rather than "methods," and so forth. This is an example of why military psychologists are encouraged to speak with their supervisors and local Office of the Institutional Review Board (OIRB) to determine how the regulatory rules and policies apply to a specific activity.

RULES AND REGULATIONS

In the conduct of human subjects research, both civilian and military organizations follow 45 CFR 46, commonly known as "the Common Rule." The section of the Code of Federal Regulations specific to the military is 32 CFR 219, but this document is identical to 45 CFR 46. Additionally, investigators in both civilian and military organizations conducting studies of drugs or devices must follow applicable Food and Drug Administration (FDA) regulations, primarily 21 CFR 50 and 21 CFR 56. All research conducted in an organization is done under an "assurance of compliance." The assurance publically states that the organization will adhere to the Common Rule as well as the ethical principles set forth in the National Commission for the Protection of Human Subjects of Biomedical and Behavioral Research (also known as the Belmont Report). The assurance identifies the "Institutional Official" as the person in the organization responsible for ensuring the ethical conduct of research in the organization. In military organizations this is usually the Commander.

The organizational body that reviews, approves, and monitors biomedical and behavioral research involving humans for the Institutional Official is the IRB. The federal regulations empower IRBs to approve, require modifications in planned research prior to approval, or disapprove research proposals. IRBs are responsible for critical oversight functions for research conducted on human participants to include compliance with scientific, ethical, and regulatory standards and regulations.

The regulations unique to the military are outlined in the Department of Defense Instruction (DoDI) 3216-02, titled "Protection of Human Subjects and Adherence to Ethical Standards in DoD-Supported Research" and published October 20, 2011. The additional requirements beyond the Common Rule outlined in this document are bulleted and briefly described below.

- **Additional Protections for DoD Personnel as Research Participants**. While participation in research is voluntary for military members, service members must follow their command policies regarding the requirement to obtain command permission to participate in research while on duty. Commanders cannot order service members to participate in research. Furthermore, military and civilian supervisors, unit officers, and noncommissioned officers (NCOs) are prohibited from influencing the decisions of their subordinates to participate in a research study. For research determined to be greater than minimal risk and when recruitment occurs in a group setting, the IRB must appoint an ombudsman. The ombudsman cannot be associated with the research. He or she must be present during the recruitment in order to monitor that the voluntary involvement or recruitment of the service members is clearly and adequately stressed and that the information provided about the research is clear, adequate, and accurate.

- **Requirement for a Research Monitor**. For research determined by the IRB to be greater than minimal risk, the IRB must appoint a research monitor. The Research Monitor is generally asked at a minimum to review adverse events and unanticipated problems involving risk to subjects or others (UPIRTSO) reports prior to submission to the IRB for any concern for protection of human subjects. Additional duties of the research monitor are determined based on the specific risks or concerns about the research.

• **Requirement for Intent to Benefit if Consent Is to Be Obtained from a Legal Representative Rather Than the Research Participant Him- or Herself**. If it is anticipated that the research participant cannot consent for him- or herself and that informed consent will be obtained from the participant's legal representative, also known as the legally authorized representative or LAR, the research must intend to benefit the individual subject in accordance with Title 10, U.S.C., Section 980 (10 USC 980). This presents a challenge to military investigators conducting natural history or placebo controlled clinical trials with critically injured individuals and/or minors where a LAR must provide consent. The simple observation of the disease course or use of a placebo does not provide an intent to benefit and is not approvable by a military IRB. If a LAR will provide consent, investigators need to include as part of their protocol how participation in the study will benefit all participants.

• **Requirement for DoD Component (or Service) Review and Oversight (previously known as "second level review")**. An administrative review by the military service component must be conducted before the research involving human subjects can begin to ensure compliance with all applicable regulations and policies, including any applicable laws and requirements and cultural sensitivities of a foreign country if conducted in a foreign country. While this Component review is not intended to be an additional IRB review, the Component office can and does issue stipulations that must be addressed by the investigator and/or the reviewing IRB(s) prior to the start of research.

• **Compensation of Service Members for Participation in Research.** Service members participating in research while on duty may only be compensated up to $50 for each blood draw. However, service members while off duty may be compensated a reasonable amount according to local reimbursement practices and the nature of the research as approved by the IRB. Again, as outlined in the first bullet in this section, while participation in research is voluntary for military members, service members must follow their command policies regarding the requirement to obtain command permission to participate in research while on duty.

• **Protecting Human Subjects from Medical Expenses if Injured.** The DoD requires the provision of medical care, or compensation for research-related injuries. This is usually not a consideration conducting research within a military organization with military beneficiaries who enjoy medical care as one of their benefits regardless of the cause for need of care.

In addition to the federal and DoD requirements, each of the military services has regulations, directives and/or instructions that guide the conduct of research. The primary documents for each of the major services are listed below.

• Army—AR 40-38 (Clinical Investigation Program) and AR 70-25 (Use of Volunteers as Subjects of Research)
• Navy—SECNAVINST 3900.39D—Protection of Human Subjects (Nov 6, 2006)
• Air Force—AFI 40-402 (Protection of Human Subjects in Biomedical and Behavioral Research)

And finally, the roadmap for how human research is conducted following the federal and DoD requirements in a specific organization is outlined in the organization's "Human Research Protection Program" or HRPP. The HRPP sets forth the structure, policies, and procedures to assure that the rights and welfare of human participants in research are protected and that all activities conform to federal, DoD, and service regulations, policies, and guidelines. The person responsible for ensuring that the HRPP is adhered to and for facilitating the work of the IRB for the Commander is the Human Protections Administrator (HPA).

Often, research projects involve multiple investigators from multiple organizations as well as multiple performance sites for the conduct of the research, necessitating IRB review and approval by multiple boards. In an attempt to decrease the volume of regulatory documents that need to be submitted, an Institutional Agreement for IRB Review (IAIR) can be employed whereby one IRB relies on the review of another IRB. DoDI 3216-02 allows military IRBs to defer to a civilian IRB; however the civilian IRB is responsible for ensuring that all the DoD requirements are met and many civilian IRBs do not have the expertise or inclination to take on these additional responsibilities.

In July 2011 the Department of Health and Human Services issued an Advance Notice of Proposed Rulemaking (ANPRM) proposing to change the Common Rule, 45 CFR 46, that if approved will significantly change the review, approval, and conduct of human subjects research. The DoD has submitted comments on the proposed changes to the regulation. It is not clear at this time whether the DoD will adopt all or even part of the new regulation if it is published, or continue to follow 32 CFR 219 as it is currently written.

CONSIDERATIONS FOR SUBMITTING A PROTOCOL FOR IRB REVIEW

There are three determinations that can be made regarding the conduct of research: (1) research not involving human subjects, (2) research involving human subjects, and (3) research involving human subjects but exempt from review as allowed in 21 CFR 219. The investigator cannot make the determination that the research is not involving human subjects or that the research is exempt from review. Local policy will dictate how these determinations are made. If it is determined that the research does involve human subjects, the protocol

can undergo two different types of review. A full IRB review requires the presentation and discussion of the protocol at a convened IRB meeting. An expedited review can be accomplished by an individual designated by the IRB. Research that falls into one of six categories outlined in the Federal Register and that is no greater than minimal risk can be reviewed expeditiously. Examples of research eligible for expedited review include research collected prospectively using data collected through noninvasive procedures routinely employed in clinical practice and research involving data, documents, records, or specimens that were collected as part of routine clinical care.

In accordance with The Common Rule as with research conducted in civilian institutions, research conducted in military institutions requires informed consent. There are situations where consent can be waived or an alteration of documentation of consent can be accomplished. The IRB can help investigators determine the best approach to consent for a particular project.

In addition to the protocol and consent documents, IRBs are required to ensure that the principal investigator and research staff engaged in research are appropriately trained to specifically conduct the research proposed as well as conduct human subjects research adhering to ethical standards of research. Usually this is accomplished through submission of biographies or a curriculum vitae for each research staff member as well as documentation of training in the conduct of human subjects research. Many institutions use the Collaborative IRB Training Initiative (CITI), www.citiprogram.org, to document training in human subjects research. And finally, the IRB asks that investigators disclose any conflicts they may have in the conduct of the proposed research, or state that there is no conflict.

CONFIDENTIALITY

In addition to the federal regulations on the conduct of human subjects research, investigators must follow the Federal Health Insurance Portability and Accountability Act (HIPAA) regulations 45 CFR 160 and 164 as well as DoD

6025.18-R (DoD Health Information Privacy Regulation). In general, investigators must obtain the research participant's permission to access, use, and/or disclose personally identifiable information, commonly referred to as protected health information (PHI). There are 19 identifiers that can be used to identify an individual; the most common are name, address, phone number, social security number, e-mail addresses, and birth dates. Even though an investigator has a research participant's permission to use his or her PHI, the investigator is still ethically obligated to protect this information. Common ways in which investigators do this are by using code numbers on all data collection forms, securing paper data files in a locked file cabinet, and password protecting electronic research files. The institution's HIPAA Privacy Board can approve a waiver of HIPAA authorization if disclosure/use of PHI involves no more than minimal risk to the privacy of individuals, the research could not practicably be conducted without the waiver, and the research could not practicably be conducted without access to and use of the PHI.

BENEFITS OF USING A MILITARY POPULATION

Military beneficiaries are generally a healthy population with few comorbidities as compared with civilian populations. The military electronic medical record is an incredible resource to follow individuals over time as care is delivered at various health care settings and locations. Military beneficiaries generally have at least a high school education as well as additional specialized skill training and are capable of making an informed decision about whether or not to participate in a research study. Investigators conducting research using military populations have found service members generally willing to participate and compliant with the research procedures and follow-up. The altruistic nature of military service that values individual sacrifice for the welfare of others serves to benefit investigators who conduct research using a military population.

References

Department of Defense Instruction (DoDI) 6025.18, Privacy of Individually Identifiable Health Information in DoD Health Care Programs (2009). Retrieved from http://www.dtic.mil/whs/directives/corres/pdf/602518p.pdf

Department of Defense Instruction (DoDI) 3216-02, Protection of Human Subjects and Adherence to Ethical Standards in DoD-Supported Research (2011). Retrieved from http://www.dtic.mil/whs/directives/corres/pdf/321602p.pdf

Department of Defense (DoD) Title 32—National Defense—Part 219—Protection of Human Subjects (32 CFR 219). Retrieved from http://ecfr.gpoaccess.gov/cgi/t/text/text-idx?c=ecfr&tpl=/ecfrbrowse/Title32/32cfr219_main_02.tpl

Department of Health and Human Services (DHHS) Title 45—Public Welfare—Part 46—Protection of Human Subjects (45 CFR 46). Retrieved from http://ecfr.gpoaccess.gov/cgi/t/text/text-idx?c=ecfr&tpl=/ecfrbrowse/Title45/45cfr46_main_02.tpl

Department of Health and Human Services (DHHS) Title 45—Public Welfare—Part 160—General Administrative Requirements (45 CFR 160) & Part 164—Security & Privacy (45 CFR 164). Retrieved from http://ecfr.gpoaccess.gov/cgi/t/text/text-idx?c=ecfr&tpl=/ecfrbrowse/Title45/45cfr160_main_02.tpl and http://ecfr.gpoaccess.gov/cgi/t/text/text-idx?c=ecfr&tpl=/ecfrbrowse/Title45/45cfr164_main_02.tpl

Food and Drug Administration (FDA). Title 21—Food and Drugs Chapter I—FDA DHHS Subchapter A—General Part 50 Protection of Human Subjects. Retrieved from http://www.accessdata.fda.gov/scripts/cdrh/cfdocs/cfcfr/CFRSearch.cfm?CFRPart=50

Food and Drug Administration (FDA). Title 21—Food and Drugs Chapter I—FDA DHHS Subchapter A—General Part 56 Institutional Review Boards. Retrieved from http://www.accessdata.fda.gov/scripts/cdrh/cfdocs/cfcfr/cfrsearch.cfm?cfrpart=56

62 MEASURING RESILIENCE AND GROWTH

Lynda A. King and Daniel W. King

Much of the literature on highly stressful events encountered by military personnel in threatening situations examines negative outcomes, the problems that can ensue following deployment and exposure to a war zone (e.g., mental distress, risky behaviors, social isolation). But there is increasing attention given to adaptation, functionality, and even positive gains following extreme life events and circumstances. Indeed, military psychologists and others recently have focused on the concepts of *resilience* to stressors and the possibility of *growth* following trauma. In this chapter, we discuss how to operationalize and measure resilience and growth, with attention to two perspectives: one that assesses individual differences in personal trait or state characteristics, and the other that documents individual differences in patterns of change as an index of adaptive functioning.

MEASURING RESILIENCE: TRAIT/STATE PERSPECTIVE

Resilience has traditionally been defined in terms of a process of activating personal assets and resources in times of stress and adversity to facilitate normal functioning. Presumably, the link between the personal assets and resources and some desirable outcome gives insight into the nature of the resilience process. Resilience-promoting attributes include hardiness, sense of mastery, cognitive flexibility, coping ability, spirituality, skill in recognizing and garnering social support, or any other protective factors that maintain well-being (see Connor & Davidson, 2003, and Johnson et al., 2011, for detailed listings). To the extent that these individual difference factors are stable and enduring personal characteristics, they can be considered *traits*; to the extent that they are situationally specific, malleable, or amenable to alteration, they would be designated as *states*. As with the measurement of any psychological variable, the goal is to assign each individual a quantitative value or score that represents the person's standing on the attribute. Accordingly, an individual high on the attribute is hardier, draws more appropriately from a coping repertoire, marshals enhanced support from others, and so forth, than one lower on the attribute. In a resilience research context, such factors are typically treated as independent variables or predictors of some dependent variable representing successful adaptation.

Below are summary descriptions of three exemplary measures of this type: again, those measures aimed at assessing trait or state protective factors that presumably delineate the process of a resilient response to stress and adversity:

- *Connor-Davidson Resilience Scale* (CD-RISC; Connor & Davidson, 2003). Connor and Davidson defined resilience as a general

301

ability to cope with stress and identified qualities of resilient individuals (e.g., hardiness, self-esteem, problem-solving skills, faith) to guide item development. The resulting measure contains 25 items, each of which is a self-descriptive statement (e.g., "I can deal with whatever comes my way," "Even when things look hopeless, I don't give up"). Each item is scored on a scale from 0 (*not true at all*) to 4 (*true nearly all of the time*). The scale development process used multiple samples to document internal consistency and test-retest reliability, convergent and discriminant validity, and sensitivity to change consequent to a targeted intervention. A factor analysis of item content yielded five factors: (1) personal competence, high standards, and tenacity; (2) trust in one's instincts, tolerance of negative affect, and strengthening effects of stress; (3) positive acceptance of change and secure relationships; (4) control; and (5) spiritual influences. This instrument presently is the most widely used measure of resilience.

- *Dispositional Resilience Scale* (DRS-15; Bartone, 1995). Bartone's self-report measure of resilience centers on the construct of hardiness, which he defines as a generalized style of healthy functioning with cognitive, emotional, and behavioral elements. This measure contains 15 items, 5 assessing each of 3 facets: (1) sense of control over one's life ("My choices make a real difference in how things turn out in the end"), (2) commitment in terms of the meaning ascribed to new experiences ("Most of my life gets spent doing things that are meaningful"), and (3) openness to viewing change as challenge ("Changes in routine are interesting to me"). For each item, the response options are 0 (*not at all true*) to 3 (*completely true*). The DRS-15 has a long history of development and use in the study of organizational leadership, including the context of military settings. Psychometric qualities of the instrument have been established via both classical test theory and item response theory perspectives, and scores have been associated with performance and health across a variety of military samples both in the United States and internationally. Normative information for adults and college students is available.

- *Response to Stressful Experiences Scale* (RSES; Johnson et al., 2011). The RSES is a relatively new self-report measure of individual differences in cognitive, emotional, and behavioral responses to stressful life events. Item selection and validation were accomplished using a series of samples drawn from active-duty and reserve component military units, the large majority of which had been deployed either to the Iraq or Afghanistan theater of operations. With systematic attention to content validity, a 22-item scale was developed. Respondents are asked to judge how they think, feel, or act during or immediately after a stressful event. Sample items include, [*During or after life's most stressful events, I tend to…*] "expect that I can handle it," "look for creative solutions to the problem." Response options range from 0 (*not at all like me*) to 4 (*exactly like me*). Johnson et al. provided evidence for internal consistency and test-retest reliability, as well as convergent, discriminant, concurrent, and incremental validity. Factor analysis of the RSES suggested five resilience-promoting factors: (1) meaning-making and restoration, (2) active coping, (3) cognitive flexibility, (4) spirituality, and (5) self-efficacy. Further psychometric study and applications to additional samples are recommended.

MEASURING RESILIENCE: CHANGE PERSPECTIVE

The second approach to documenting resilience focuses directly on the ability to detect longitudinal patterns of growth or decline in an individual's standing on a relevant attribute. Here, resilience is operationalized as a process of dynamic change wherein one's status on an individual difference characteristic (e.g., functional health) is altered over time—or not—in response to one or more forces or potentially explanatory personal (e.g., coping ability) or contextual (e.g., severity of combat exposure) variables. This definition of resilience as *change process* derives from developmental science's attempts to understand how children and adolescents facing highly aversive life circumstances, such as poverty or persistent abuse, arrive at adulthood as stable and relatively healthy individuals. Resilience as change in adults is typically conceptualized as a trajectory of scores on an attribute over time, defining degree of adaptive functioning following exposure to an identified potentially traumatic event and representing a progressive, adaptive reaction to personal threat.

Bonanno (2004) proposed four patterns of change or change trajectories:

1. In the *chronic dysfunction* pattern, initial response on adaptive functioning is quite low and remains so over the full interval during which observations are made; the individual responds negatively to the exposure and fails to improve.
2. A *delayed reaction* is characterized by normal functioning at initial status, followed by a decline after some time; the individual initially appears to be unaffected by the experience, but subsequently shows evidence of deterioration.
3. *Recovery* is just the opposite; the pattern displays an initial negative reaction to the event, followed by recuperation and restoration of adaptive functioning.
4. Finally, *resilience* is denoted by a pattern showing no dysfunction from initial assessment throughout the full observation period; resilient individuals appear uninfluenced by potentially traumatic events.

Masten and Obradovic (2008) set forth a more elaborate change trajectory typology of responses to disaster, taking into consideration the individual's possible preexposure status. For example, they proposed a pattern of persistent unresponsive dysfunction both prior to and following exposure as well as a pattern of maladaptive functioning prior to exposure that shows an initial decline but then returns to the prior (yet still maladaptive) state. Additionally, they introduced the possibility of positive gains in functioning with two patterns of growth over and above preexposure status. Moreover, they viewed both a recovery pattern and a growth pattern as forms of resilience.

The important point is that resilience is no longer equated to personal trait or state, but rather operationalized in terms of how an individual does or does not change on some index of functioning following stress and adversity. The nature of the variable representing adaptive functioning—the one being tracked over time—appears to be rather inconsequential. It could be a positively valenced attribute (e.g., functional health, life satisfaction) or the absence of psychosocial distress (e.g., lack of posttraumatic stress symptoms, absence of marital strife). Importantly, the object is to observe and quantify resilience as individual differences in the intraindividual (within-person) change process. The intraindividual change trajectory is typically viewed as a research outcome or dependent variable.

King et al. (2012) offered a selection of increasingly sophisticated analytic tools that can be used to "measure" or document resilience as within-person change:

- The most logical and straightforward method of documenting change in an individual over time is to compute a *simple difference score*, subtracting the person's score on a variable at an earlier assessment from that person's score on the same variable at a subsequent assessment. Then, that difference score or index of resilience can be related to other variables, risk or protective factors. Historically, some methodologists cautioned against using simple difference scores, claiming that difference scores are unreliable, but other methodologists have argued that simple difference scores should not be dismissed as reasonable representations of within-person change.
- A second method to index intraindividual change is a *residualized change score*, or *partialed change score*: the difference between a person's observed score on a measure at a particular time and the score predicted for that person from a prior score on the same measure via regression analysis. In turn, this residualized change score (say, for functional health) is usually regressed on scores on another variable of interest (e.g., a protective or resilience-promoting factor, such as hardiness). A significant relationship between the

protective factor (hardiness) and the residualized change score (functional health) suggests the efficacy of that factor as a predictor of change. Computationally, this partialing strategy is accomplished in a multiple regression framework, with attention to the significance of the partial regression coefficient. This method assumes stationarity, that there are no unaccounted-for influences on the residualized change score, an assumption that is quite difficult to defend in many research situations.

• Extending the notion of residualized change, another class of procedures, *time series analysis*, uses data from a single individual assessed over many occasions. One could, for example, predict the individual's residualized change occasion to occasion (a potential resilience pattern) from previous or lagged scores on a candidate protective factor. Resulting parameter estimates, partial regression coefficients, calculated over occasions can be considered characteristics of that one person, distinguishable from estimates for another person, just as a score on a measure is intended to differentiate one person from another. These within-person parameter estimates subsequently can be regressed on between-person characteristics (e.g., gender, age) to explore more complex interactions between the within-person factor and the between-person factor, in predicting change, as in *multilevel regression techniques*.

• One may use *growth curve modeling* to document dynamic change, also relying on a multilevel data structure. The goal is to describe a trajectory of scores on an outcome variable over a set time interval, with the score on the outcome being a function of time since some starting point. The procedure uses within-person repeated assessments and the times of assessment to generate a series of parameter estimates that describe a best-fitting regression line for each individual. In the simplest case of straight-linear regression, at least two parameters are estimated for each person: intercept (typically the score at initial assessment) and slope (amount of change in the outcome per unit change in time). Associations between other individual difference characteristics and these intercept and slope variables may be examined, and more intricate models of curvilinear change can be considered.

• *Latent difference score analysis* is used to disaggregate a trajectory of change over time into a sequence of segments, each of which defines a change in the value of an outcome over that particular time interval. Thus, this method accommodates shifts in the rates of change from interval to interval and enables a researcher to pinpoint differential associations between predictors and change along the succession of intervals. Latent difference score models have certain advantages. The difference scores are represented as differences between two perfectly reliable latent variables, thereby maximizing the reliability of the difference scores themselves. The model also accommodates the control of the effects of prior status on the change variable. Finally, the model controls for the influence of extraneous unmeasured variables, eliminating the need to assume stationarity.

More detailed descriptions, relevant citations to each technique, and examples of use of these methods to document change are provided by King et al. (2012).

MEASURING GROWTH

Over the last two decades, there has been increasing interest in the possibility that exposure to highly stressful events might yield positive personal gains, variously referred to as *benefit-finding, stress-related growth*, or *posttraumatic growth*. Conceptually, this phenomenon is typically explained as the attempts of an individual who has undergone life crisis to reconcile that experience in light of a prior-held worldview or global meaning. The individual seeks to think through the implications of the event, process the emotions linked to the event, and recast the event into a broader context with possible constructive and affirmative consequences. Through a process akin to meaning-making, the individual expresses gains or positive changes in the following types of domains: quality of relationships with others, beliefs in one's ability to cope with future adversity, appreciation for life, commitment to personal values, and spirituality.

While the construct references actual or veridical change, the measures of growth have typically relied on self-reported retrospective judgments of prior status in relation to current status. Two examples of these instruments assessing perceived growth are:

- *Posttraumatic Growth Inventory* (PGI; Tedeschi & Calhoun, 1996). The PGI was developed specifically to assess positive benefits in persons who have experienced a trauma. Item selection derived from the content domains of perceived changes in self, changed relationships with others, and changed philosophy of life. This 21-item scale uses a 0 (*I did not experience this change…*) to 5 (*I experienced this change to a very great degree…*). Sample items are "I established a new path for my life" and "I have a stronger religious faith." Tedeschi and Calhoun reported acceptable levels of internal consistency and test-retest reliability as well as evidence for concurrent and discriminant validity. A principle components analysis yielded five components: (1) relating to others, (2) new possibilities, (3) personal strength, (4) spiritual change, and (5) appreciation of life.
- *Stress-Related Growth Scale* (SRGS; Park, Cohen, & Murch, 1996). Similarly, Park, Cohen, and Murch sought to develop a measure of positive outcomes from a stressful event. Item generation proceeded from three general content domains, defined in terms of changes in personal resources, social resources, and coping skills. The SRGS contains 50 self-descriptive item statements, and the response options are 0 (*not at all*), 1 (*somewhat*), and 2 (*a great deal*). Sample items are "I rethought how I want to live my life" and "I learned better ways to express my feelings." Park et al. provided information on internal consistency and test-retest reliability and offered extensive support for concurrent validity. They also showed significant relationships between scores on the

SRGS and actual pre- to poststressor positive change on other indicators of well-being, including optimism, positive affectivity, and satisfaction with social support.

In addition to these self-report measures of perceived change, it is certainly possible to operationalize growth as a change trajectory or a trend toward higher scores on some selected index of personal well-being, sense of self, or worldview (see prior characterization of growth by Masten & Obradovic, 2008). In such a case, one could adopt any of the aforementioned analytic tools (simple difference scores, residualized change scores, etc.), with the emphasis on improvement over and above one's preevent state.

References

Bartone, P. T. (1995, July). *A short hardiness scale.* Paper presented at the annual convention of the American Psychological Society, New York, NY.

Bonanno, G. A. (2004). Loss, trauma, and human resilience: Have we underestimated the human capacity to thrive after extremely aversive events? *American Psychologist, 59,* 20–28.

Connor, K. M., & Davidson, J. R. T. (2003). Development of a new resilience scale: The Connor-Davidson Resilience Scale (CD-RISC). *Depression and Anxiety, 18,* 76–82.

Johnson, D. C., Polusny, M. A., Erbes, C. R., King, D., King, L., Litz, B. T.,…Southwick, S. M. (2011). Development and initial validation of the Response to Stressful Experiences Scale. *Military Medicine, 176,* 161–169.

King, L. A., Pless, A. P., Schuster, J. L., Potter, C. M., Park, C. L., Spiro, A., III, & King, D. W. (2012). Risk and protective factors for traumatic stress disorders. In G. Beck & D. Sloan (Eds.), *Oxford handbook of traumatic stress disorders* (pp. 333–346). New York, NY: Oxford University Press.

Masten, A. S., & Obradovic, J. (2008). Disaster preparation and recovery: Lessons from research on resilience in human development. *Ecology and Society, 13*(1), 9. Retrieved from http://www.ecologyandsociety.org/vol13/iss1/art9/

Park, C. L., Cohen, L. H., & Murch, R. L. (1996). Assessment and prediction of stress-related growth. *Journal of Personality, 64,* 71–105.

Tedeschi, R. G., & Calhoun, L. G. (1996). The Posttraumatic Growth Inventory: Measuring the positive legacy of trauma. *Journal of Traumatic Stress, 9,* 455–471.

63 TRANSITIONING THROUGH THE DEPLOYMENT CYCLE

Sherrie L. Wilcox and Michael G. Rank

Military psychologists have several unique advantages over civilian providers when it comes to helping service members and their families—they are immersed within the military culture, they experience the deployment cycle through their own deployments, and they have access to military personnel and their families to prevent and treat illness across the deployment cycle. Civilian providers are most likely to see military personnel and their families after deployment, where physical and psychological challenges are established and only tertiary prevention can be implemented. For military psychologists, this means being aware of the service and family members' stage of deployment and the associated experiences within their respective stage, and strive to help early stage military families prepare for later stages. This chapter navigates through the deployment cycle and discusses clinical implications for each stage of deployment.

THE DEPLOYMENT CYCLE

Deployments vary in length and location and refer to the time when military personnel are away to train or perform a mission. Military personnel often experience multiple deployments throughout their military career, which affect both military personnel and their families, as military families must also adjust to transitions associated with deployment. The six-stage deployment cycle described in this chapter is derived from existing approaches (Peebles-Kleiger & Kleiger, 1994; Pincus, Leiner, Black, & Ward Singh, 2011). Each cycle refers to a single "round-trip" deployment to an operational or training area and back home—it is an ongoing process. This chapter focuses on the longer-term deployments to operational areas.

Stage 1: Training and Preparation

The Department of Defense provides the forces to deter war and to protect the security of the country, and in order to maintain a military force with a high level of readiness, military personnel are always training across the spectrum of military operations. Although military personnel often have a job with a civilian equivalent, their primary occupation is as a war fighter and their job is mission focused. Stage 1 of the deployment cycle refers to the time before military personnel receive a warning order indicating an upcoming deployment. This is when they engage in usual job duties including the training and preparation for combat.

In this stage, military personnel and their families have established routines. Military personnel are working their normal schedule, which can last from 8 to 12 hours or longer. Depending on their job, military personnel may already be spending long hours in the field training, while others have a schedule that generally resembles that of their civilian counterparts. Family members also have their own existing routines. The spouse/partner will be working, volunteering, and/or engaging in household activities, including caring for children.

At this point, military personnel and families should develop a high level of readiness. The following is a list of a few basic activities that should be completed as early in the deployment cycle as possible: (1) the military family should already be engaging in counseling, briefings, and trainings to prepare for potential deployments and strengthen the family unit; (2) family members should be enrolled in the military benefits system; (3) the record of emergency data (DD93) should be complete; (4) military personnel should have life insurance, an advanced medical directive, last will and testament, trusts, and a power of attorney; and (5) military personnel should be attending any available mental health trainings related to resilience.

From the clinical perspective, it is imperative to assess service and family member readiness for a deployment in this stage. Military personnel are physically, psychologically, and emotionally affected by military training, even at this early stage of the deployment cycle. If there are problems, attempts must be made to resolve them before the additional stressors associated with deployment begin. Ideally, the military family has their life constructed so that if the service member has to deploy, the family is able to continue life as usual. Throughout the deployment cycle, continual assessment and reassessment of the readiness of service and family members must be conducted. Tracking how clients change throughout the deployment cycle will help to improve the treatment process and can inform future research.

Extended separation associated with deployment and the impact on the family unit should be a primary feature of a predeployment briefing. The way in which the family unit reacts and reorganizes to accommodate the service member's absence must be addressed. This introduces the concept of ambiguous loss, especially how it affects relationships during deployment. An activity where family members discuss their challenges in an open supportive setting is an effective technique. Fostering supportive networks within the family and among military

peers can reduce feelings of loss, isolation, and distress (Wilcox, 2010). Building this supportive network of people in similar circumstances is an important component of resilience to life challenges.

Stage 2: Mobilization

In Stage 2, military personnel have received a warning order for a potential deployment, which can occur within 72 hours to 12 months. The mobilization stage begins as soon as military personnel receive warning orders alerting for a possible deployment and lasts until the service member deploys. "Mobilization" has different meanings for the different departments of the military. A mobilized National Guard or Reserve unit may spend the months preceding a deployment at a mobilization training center, geographically separated from their family. On the other hand, as the deployment draws nearer, active duty service personnel will spend longer hours at work or in the field preparing for the deployment, and thus less time at home with family. It is in Stage 2 where the military "helps" to distance military personnel from their family and ease the service member's transition into the deployment by building unit cohesion.

While military personnel are spending more time training, the family should be preparing to take on the roles of the service member. If the activities from Stage 1 are complete, this process will be much easier and will reduce disorganization and scrambling to complete preparations for the deployment. Family members are often anxiously anticipating the loss of their service member and may be upset about spending less time with their service member before the deployment. Such circumstances can lead to enduring relationship conflict that can set the stage for further complications as the deployment progresses.

From a clinical perspective, military personnel must be provided with opportunities to bond with their unit members to increase unit cohesion and ease their transition. Ideally, family members should receive a call or visit by a chaplain, professional, or family advocacy worker to ensure preparations are being made for the transition to take over the service members' roles at home; each family member can choose a role or be assigned a role. To set up a support network for the family members, the unit members and their families can organize common times to socialize together; the larger the support system, the easier the transition—creating a transitional community eases anticipatory grief. Ensuring a strong social support network and mentorship opportunities helps to prepare military families and increases resiliency.

It is common to have feelings of frustration, fear, or anxiety associated with the anticipated absence of the service member. Family members may be in denial of the upcoming deployment. Emotions tend to be repressed, avoided, minimized, and denied. Forums for open communication between supportive networks and the family unit during these times can create enduring stability. Weekly family meetings provide the opportunity to vocalize issues related to the deployment, and promote harmony during this stage.

Stage 3: Deployment

Stage 3 of the deployment cycle refers to the time in which military personnel deploy to their mission location, typically overseas and often in or near a combat zone. This stage begins as soon as military personnel leave for the deployment. This transitional period typically lasts for several weeks or longer.

In Stage 3, military personnel are adjusting to the new work environment. They may be able to occasionally communicate with family members at home for a limited amount of time. It is critical that communication is positive and supportive, for both military personnel and family members—negative communication can lead to added distress. While military personnel are adjusting to the new work environment, the family members are adjusting to the new home environment—an environment without the service member physically present.

From a clinical perspective, the professional must ensure that both service and family members are adjusting to their respective environments and circumstances. This is most easily accomplished by facilitating routine communication between service members and their families via e-mails, texts, phone calls, or televideoconferencing, which can develop and strengthen the family unit throughout the deployment. However, military personnel often know that even with the best preparatory plans, problems will arise. Thus, in the event of a communication failure, a backup plan should be made. As previously highlighted, having an established support network during this initial period is essential. This is especially true for the spouse/partner, who may be on a military installation or otherwise away from parents, siblings, and friends. Stage 3 is also a time for service and family members to grieve—they don't have to appear brave and strong anymore. It is important to encourage service and family members to express their feelings in a supportive atmosphere.

Nothing creates relational difficulties more than absence, and when the specters of life's stressors are added to the mix, partners may begin to question their love and commitment. Infidelity by both partners occurs too often and may lead to divorce. As uncomfortable as it may be, the professional must address the possibility of infidelity and what the family will do if it occurs. If partners can reaffirm their commitment to each other, ideally each week or at least on a monthly basis during the deployment, then risk is minimized. Nevertheless, the possibility of infidelity has to be addressed openly. Few plan to be unfaithful; however, absence has relational challenges—infidelity and falling out of love being the two more prominent threats.

Stage 4: Sustainment

The bulk of the deployment occurs in Stage 4, which begins after the initial weeks of deployment and lasts until the weeks before homecoming. During this stage, military personnel perform their assigned missions, and have ideally learned to cope with their new environment and circumstances. They may be going to in-theater counseling or may be seeing a chaplain to discuss challenges. Additionally, military personnel may be seeking informal, but beneficial, support from military peers and military leaders (Wilcox, 2010).

The spouse/partner has ideally established a new routine and new independence—taking on new sources of social support for help with the deployment and establishing new patterns. They may be making new decisions independently and may begin using "my" language—my house, my car, my children, my dog, and my money. Ideally, both the service and family members have gained resilience and have feelings of confidence and control.

From a clinical perspective, it is important to know that much can happen in Stage 4. Ensuring that service and family members are adjusting to the deployment is critical. Often, the challenges related to balancing life's issues are highlighted during communicating back home. Issues may include illnesses, family problems, bad grades, civil matters, transportation problems, financial difficulties, and employment challenges. Although being secretive or deliberately shielding issues from each other is not recommended, service and family members must soften their responses and explanatory narratives, speaking with a calm voice. Communicating confidence that the concerns are being addressed with due diligence is extremely important. Family members must also recognize that their service member can often do very little to resolve problems at home while they are deployed. Spouses/partners may overreact to circumstances discussed over the phone, which can leave their deployed service member with a sense of powerlessness that can lead to hasty, unsafe, or inappropriate decisions. Overall, it is necessary to encourage and build resiliency among service and family members.

Stage 5: Redeployment

Stage 5 of the deployment cycle takes place the weeks before homecoming. At this point, military personnel are busy transferring forces, material, and people to support other operations and incoming units. There is often little communication back home during this stage due to the number of tasks that need to be completed before the redeployment. It is important for military psychologists to counsel family members on these circumstances and ways to cope.

At home, family members are experiencing a surge of energy, both positive and negative. On the positive side, they are excited that the service member will soon be home and the family is often busy planning homecoming activities. On the negative side, there may be worry of injury or death in the final weeks. For some, there is high anxiety associated with anticipating the invisible wounds of war or from having to reestablish their relationship after a long separation.

From a clinical perspective, the aim is to reduce the anxiety and distress that family members are experiencing. The military experience, both the theater of war and extended separation, can change individuals; military personnel likely have experienced and witnessed events that defy description and explanation. Both the spouse/partner and the service member have changed significantly over the course of the deployment. Often, there is an uncomfortable readjustment and redefining of relationships. Relationships have to be renegotiated with the realization that they may not return to predeployment status. There are anticipatory excitement, tension, and expectations. In preparing for the postdeployment stage, military families should be aware of the importance of not pushing for information or explanations, and should be reminded not to withdraw or isolate.

Stage 6: Postdeployment

Once military personnel return to their home installation, they begin Stage 6. This stage varies greatly between individuals based on their experiences and resources and typically lasts 3–6 months, but can last a year or longer. Stage 6 is often an exciting but stressful time. It is often marked with a joyous homecoming and "honeymoon period." As the honeymoon period ends, adjustment and transition challenges emerge. Challenges range from readjusting to life at home to fitting into the new family routine, redefining family roles, and dealing with unexpected challenges.

Military personnel have been away from the family for long periods of time and will need to reintegrate back into home and family life. It is important to keep in mind that time has passed, things have changed, and life will not be the same as it was in predeployment, especially if this was the first deployment. It is difficult for the nonmilitary spouse/partner who has been in charge of the household during the deployment, as well as for the service member who needs to reestablish their role in the family. The family will need to learn to work together as a team again and will have to take action and plan together to get back on track.

For military personnel, it is a time to slow down and adjust back to the normal, predeployment training routine. This stage is also filled with postdeployment briefings, trainings, assessments, evaluations, and counseling to facilitate successful reintegration of military personnel. For National Guard and Reserve members, this stage may be a little more complicated. National Guard and Reserve members will need to secure employment, ensure their benefits are active, and readjust to the civilian environment. In all returning military personnel, it will be important to watch for risky behaviors including increased alcohol use, drug use, violence, suicidal behavior, or isolation. All military personnel and family members should monitor their peers for adjustment problems.

Stage 6 is where clinicians focus their attention on emerging relational or family patterns. As new family patterns emerge, skill building must be focused on open communication. Perhaps the greatest risk for families and relationships is isolation and distancing from each other due to noncommunication. Energy that is required to rebuild relationships, families, and open communication patterns is taxing and time consuming. There may be a distancing that occurs from one another, which happens insidiously. Communication is diminished and intimacy may become conflicted. Families in this stage tend to not present for treatment and thus suffer in silence. Military psychologists need to be creative to reach these at risk families.

SWITCHING THE FOCUS FROM POSTDEPLOYMENT

Research typically tends to focus on the more than 2 million deployments that have taken place since September 11, 2001, and the associated challenges that impact military personnel in Stage 6, the postdeployment stage. Despite a focus on postdeployment reintegration, military personnel and their families experience transition challenges not only during postdeployment but also earlier in the deployment cycle. Focusing on transitional challenges early in the deployment cycle has the potential to mitigate later challenges. Stage 6 is primarily when problems that have been progressing throughout the deployment cycle have the most visibility, often due to the reintegration of the family unit. Moreover, focusing on postdeployment, where problems have emerged, only allows intervention using secondary and tertiary levels of prevention, after risk behaviors are already established and problems have emerged. That is, primary prevention strategies, which occur before the onset of risk behaviors and injury, are not implemented.

A key to the deployment cycle is to prepare for the later stages early—prevention is key. Beginning at Stage 1, military psychologists should help military personnel and their families plan how to thrive and survive as they transition through the stages. Part of this planning should include ways to comprehensively address the unique and multifaceted protective and risk factors faced across the deployment cycle.

Military psychologists do not work alone—they have the other support services and resources. Utilizing community resources that serve military populations is a feasible option if clients are hesitant to seek services from military sources, arising from concerns of stigma. Community resources will also be beneficial for those who are transitioning into the civilian environment after separating from the military. This chapter is not a treatment guide, but rather a way to conceptualize how military clients transition through the deployment cycle. Although it is most common to see military clients in postdeployment, having an understanding of their deployment cycle stage will help to more appropriately address their challenges and help prevent postdeployment problems.

ACKNOWLEDGMENT

The authors would like to thank Gunnery Sergeant Kevin J. Williams Jr. for his contributions to this chapter.

References

Peebles-Kleiger, M. J., & Kleiger, J. H. (1994). Re-integration stress for Desert Storm families: Wartime deployments and family trauma. *Journal of Traumatic Stress, 7*(2), 173–194.

Pincus, S., Leiner, B., Black, N., & Ward Singh, T. (2011). The impact of deployment on military families and children. In M. K. Lenhart (Ed.), *Textbooks of military medicine: Combat and operational behavioral health* (pp. 487–499). Washington, DC: Office of the Surgeon General.

Wilcox, S. (2010). Social relationships and PTSD symptomatology in combat veterans. *Psychological Trauma: Theory, Research, Practice, and Policy, 2*(3), 175–182. doi:10.1037/a0019062

64 AGING VETERANS

Avron Spiro III and Michele J. Karel

DEMOGRAPHICS

In 2010 there were 23 million US veterans, about 10% of the population over age 18. Nearly three-quarters of them served during wartime, most in Vietnam (34%), with smaller numbers serving in Korea (11%) or World War II (9%) (Department of Veterans Affairs [DVA], 2010). Veterans are older than the US population; their median age is 64 versus 49 for the US population, and 64% are aged 55 and older. Veterans, especially older ones, are much more likely to be men (92%), to be Caucasian (85%), to have at least a high-school diploma or GED (95%), and to be married (70%), compared to non-veterans (DVA, 2010). Further demographic information on veterans can be found online at www.va.gov/vetdata, where the VA's recent national survey provides information also on Active Duty as well as demobilized Guard and Reserve troops (DVA, 2010).

A LIFE SPAN PERSPECTIVE ON MILITARY SERVICE

In the study of aging, the role of military service is often overlooked, and its effects on health and well-being are seldom considered. While military service can have positive effects (e.g., occupational and leadership training, access to VA educational and health benefits), the focus is often on negative outcomes (e.g., traumatic exposure, increased risk of PTSD or other mental disorders, physical injuries or disabilities). While the immediate effects of these negative experiences are often considered, many effects (positive as well as negative) can take years to manifest. These long-term effects can occur in multiple domains including physical and mental health, social and occupational functioning, and family and marital well-being.

A focus on aging veterans leads to a consideration of the long-term effects of military service, linking active duty (from enlistment until discharge) with the rest of life as a veteran. The effects of military service, especially among those who were deployed to a warzone, can be positive or negative, and immediate or delayed. One viewpoint useful for considering these effects is the life span perspective, which is based on the following principles (e.g., Spiro, Schnurr, & Aldwin, 1997):

- **Development and aging are lifelong processes.** Experiences early in life such as military service can have both short- and long-term effects; sometimes the latter are unrecognized until later in life.
- **The effects of military service occur within an historical context.** For example, whether military service occurs during war- or peacetime has an impact on the frequency and severity of exposure to various events such as warzone deployment or combat. Behaviors learned during service can have lifelong effects, for example, the negative consequences of smoking on health are well known, as are the positive consequences of learning discipline and leadership on later social and economic achievement.
- **Timing matters.** Whether one enters the military at a younger or older age has effects on the start of postmilitary marital and occupational careers.
- **There is a good deal of variability among people.** Not all people exposed to a given event will have similar outcomes. For example, while some respond negatively to warzone experiences and develop physical or mental conditions, others may have positive outcomes such as wisdom or posttraumatic growth and draw strength across their life.

HEALTH ISSUES CONFRONTING OLDER VETERANS

Mental Health

In general, epidemiological studies of mental health show that older adults have lower rates of current and lifetime mental disorders than younger adults. A similar age-related trend is found among veterans using VA services. For example, in FY2011, the prevalence of mental health or substance abuse disorders was 35% among those aged 35–64, 21% in those 65–74, and 12% in those 75+ (Institute of Medicine [IOM], 2012).

Trauma

- In the United States, about 80% of the adult population has experienced at least one potentially traumatic event during their lifetime.
- Trauma prevalence varies by age and gender. In a national study conducted in Australia, the lifetime prevalence of trauma was about 40% for women and 70% for men. Men showed a linear increase with age, but women an inverse-U shape, largely due to the experience of combat by older men.
- According to the 2010 National Survey of Veterans (DVA, 2010), a third of veterans (34%) served in combat or a warzone and reported exposure to dead/dying/wounded.
- In the VA Normative Aging Study, about 75% of the World War II and Korean veterans reported lifetime exposure to potentially traumatic events, including combat.

Posttraumatic Stress Disorder (PTSD)

- PTSD is generally less prevalent in elders than in younger adults, but is higher among subgroups of the elderly (e.g., combat veterans, Holocaust or disaster survivors; victims of maltreatment or interpersonal violence [IPV]) (Cook, Kaloupek, & Spiro, in press).
- Rates of PTSD vary across cohorts of veterans, with Vietnam and OEF/OIF/OND veterans having higher rates of PTSD than veterans of prior wars. PTSD is more likely to occur in deployed than in nondeployed troops, and is higher among those who served in Vietnam (Magruder & Yeager, 2009).
- Partial PTSD, in which the full set of symptom criteria as required by the American Psychiatric Association's *Diagnostic and Statistical Manual* are not met, should also be considered.
- There are several possible trajectories of PTSD in the elderly: de novo (in reaction to a recent trauma), chronic, or delayed onset/reactivated (from trauma in earlier life).

Other Mental Disorders

- As is the case for PTSD, other mental disorders (e.g., mood, substance use, and other anxiety disorders) also present various lifetime trajectories; risk factors and prognosis may vary across the life course, and their comorbidity with PTSD is not uncommon.
- Approximately 35–38% of veterans aged 35–64 in 2011 had at least one mental health or substance use diagnosis; of these, approximately half had more than one additional mental health or substance use diagnosis (IOM, 2012).
- Older adults with PTSD are likely to have other comorbid Axis I disorders (anxiety or depression) and higher rates of suicidal ideation than those without PTSD (Pietrzak, Goldstein, Southwick, & Grant, 2011).
- The presence of PTSD may be associated with an increased risk of dementia.

Suicide

Veterans are at higher risk for suicide than nonveterans, except for those aged 65+ (Kaplan, McFarland, Huguet, & Valenstein, 2012). Predictors and characteristics of veteran suicide vary by age group. Older veterans often have health problems and depression, rarely have previous suicide attempts, and are most likely to commit suicide with a firearm (84%) compared to veterans in other age groups. Acute alcohol use is rare among older veterans who completed suicide, but present about one-third of the time for younger and middle-aged veterans (Kaplan et al., 2012).

Late-onset stress symptomatology (LOSS)

Among aging combat veterans, changes in social roles (e.g., retirement), physical and cognitive declines, and bereavement are normative events in later life, and can serve as triggers for reconsideration of one's earlier life experiences, such as military service, leading to increased reminiscences about wartime experiences. "Late-onset stress symptomatology" is a condition with some similarity to PTSD, especially the intrusion/reexperiencing criterion, but without the numbing, avoidance, or arousal that are also part of PTSD (Davison et al., 2006). LOSS may be better dealt with by psychoeducation rather than therapy, to help place these returning thoughts of earlier stress and trauma from deployment into a context of life review and meaning-making. LOSS may also provide an opportunity for veterans to engage in a process of meaning-making, acceptance, and growth later in life.

Diversity Issues

- Older women veterans likely experienced different types of warzone stressors and exposures than did men. For example, nurses who served in Vietnam reported high levels of exposure to dead and dying troops and civilians.
- Possible history of military sexual trauma (MST) and interpersonal violence (IPV) are important to address among all aging veterans, particularly women veterans. In general, women with PTSD are more likely than men with PTSD to endorse a history of sexual assault or intimate partner violence (Pietrzak et al., 2011).
- Racial and ethnic minority veterans may have a higher likelihood of developing PTSD given trauma (Cook et al., in press).

Physical Health

Health

- Most (72%) veterans reported good or better self-rated health (DVA, 2010). Those with PTSD or other mental disorders had worse self-rated health and quality of life.
- 64% of veterans have ever smoked; but most (69%) have quit (DVA, 2010). The long-term consequences on various chronic diseases of aging due to smoking and other negative health behaviors that may be initiated during service should not be ignored.
- Older adults with higher levels of PTSD symptoms have higher risk of mortality and of various diseases (i.e., arterial, lower gastrointestinal, dermatologic, musculoskeletal) (Pietrzak, Goldstein, Southwick, & Grant, 2012).

- Older veterans with PTSD report little social support, poor self-rated health, more at-risk drinking, and greater use of tobacco and have more suicidal ideation (Durai et al., 2011).

Health Care Access/Utilization

- About 28% of veterans are enrolled in VA health care (DVA, 2010).
- 21% of veterans have applied for disability compensation for physical or mental conditions, and about three-quarters of them have received a disability rating (DVA, 2010).
- 13% of veterans have no health insurance (DVA, 2010); most of these are under 65 and not yet eligible for Medicare.

Mental Health Assessment and Care

VA Health Care System as a Model

Older adults with mental health conditions are less likely to receive mental health services and, when they do, are less likely to receive them from specialists. Most mental health care for older adults occurs in primary care settings; numerous studies demonstrate that integrated, interdisciplinary, collaborative primary care models are more effective for older adults. In the VA health care system, mental health care is integrated into many primary and geriatric care settings, as well as in mental health specialty settings (Karlin & Zeiss, 2010).

Considerations in Assessment and Care of Older Veterans

Mental health screening

Older veterans often present with somatic complaints rather than mood symptoms, and may not reveal underlying mental health symptoms without specific questioning. Screening for symptoms of depression, anxiety, PTSD, and alcohol and other substance misuse is important to guide further evaluation and treatment planning. It is critical to ask older veterans about suicidal thoughts or plans and to follow up with further evaluation and safety planning as indicated.

Assessment of cognition and dementia

Cognitive screening is not advised for all older, asymptomatic veterans. However, it is important to attend to behavioral signs of dementia or collateral reports of memory or other functional concerns and, if present, to seek further evaluation. Brief cognitive assessments, such as the Saint Louis University Mental Status Examination (SLUMS) (http://www.stlouis.va.gov/GRECC/SLUMS_English.pdf) or Montreal Cognitive Assessment (http://www.mocatest.org/) can be administered. If cognitive impairment is indicated, the veteran should be referred for further evaluation.

Decision making and functional capacity assessment

Moderate to severe cognitive impairment and/or psychiatric illness can lead the older veteran to have compromised abilities to make medical decisions, manage finances, live independently, or drive safely. Vulnerable elders are at risk for various forms of elder abuse and exploitation, which must also be evaluated in the clinical setting. The American Psychological Association's Office on Aging website has important resources to support psychologists in evaluating capacity, as well as dementia, in the older adult (http://www.apa.org/pi/aging/).

Evaluation of the caregiver

Family members may be more active collaborators in the care and treatment of older veterans. Of course, capable veterans must consent to including family members in their care. Family and other caregivers (e.g., friends, neighbors) play critical roles in helping disabled veterans live at home; they are critical partners in care of the veteran and often need psychoeducation, skill building, and support themselves. Tools for caregiver evaluation and intervention are available online (http://www.caregiver.va.gov/; http://www.apa.org/pi/about/publications/caregivers/).

Interdisciplinary approach

Older veterans often have complex medical comorbidities, and may take multiple medications that can cause or exacerbate psychiatric or cognitive symptoms and functional impairments. To inform assessment and treatment, coordinated care is required to sort out complex presenting problems and plan a collaborative approach to care.

Evidence-based psychological interventions

Many evidence-based psychological treatments are effective with older adults, including those to treat depression, generalized anxiety disorder, PTSD, alcohol misuse, disruptive behaviors in dementia, family caregiver distress, and a range of behavioral health concerns including insomnia and chronic pain (e.g., Karel, Gatz, & Smyer, 2012).

Draw on strengths and veteran identity

Older veterans can draw on a lifetime of experience, including military service, to help them cope with late-life challenges. Values of hard work, commitment, courage, caring for family, and helping fellow veterans can be very helpful in managing mental distress and functional changes later in life. Older veterans do very well in group therapy settings, where they can share struggles, strategies for coping, and mutual support. For many older veterans, the experience of "talking about their feelings" is a new one, and can be experienced as a great relief and potential for growth in late life.

FUTURE TRENDS

The aging veterans of the future will differ in numerous ways from those of today, and military and VA psychologists will be challenged to consider the life span implications of military service on different generations of veterans. Since the initiation of the all-volunteer force in 1973, the military has become more diverse, including more women and minorities. Often, in addition to active duty troops, Guard and Reserve troops are deployed, and likely to experience increased injuries and other sequelae of warzone deployment. The average age of deployed troops is now older than was the case during Vietnam, and this may pose challenges to psychologists and other practitioners serving Active Duty troops. Deployment

exposures common in today's conflicts and their associated impacts (e.g., IEDs' roles in traumatic brain injury) may bring new challenges to understanding and caring for today's veterans in the future.

The issues that confront tomorrow's aging veterans, as well as the strengths they bring to bear given an increased focus on training resilience, likely will differ from those of today's veteran population. Tomorrow's veterans may bring higher rates of mental health symptoms and substance use disorders with them as they age, and some have suggested that troops who have participated in the current conflicts may be aging more rapidly than would be expected, perhaps as a result of the many stressors associated with repeated deployments. Aging veterans also will continue to face normative issues for relating to their own and their family's aging (e.g., care-giving, dementia, advance care planning). Helping those who have served our country prepare for optimal aging in their middle and later years is an important component of care now and into the future, and may require that military and VA psychologists expand their training in the emerging field of geropsychology.

References

Cook, J. M., Kaloupek, D., & Spiro, A. (in press). Trauma in older adults. In T. M. Keane, P. A. Resick, & M. J. Friedman (Eds.), *Handbook of PTSD: Science and practice—A comprehensive handbook* (2nd ed.). New York, NY: Guilford.

Davison, E. H., Pless, A. P., Gugliucci, M. R., King, L. A., King, D. W., Salgado, D. M.,... Bachrach, P. (2006). Late-life emergence of early-life trauma: The phenomenon of late-onset stress symptomatology among aging combat veterans. *Research on Aging, 28*(1), 84–114.

Department of Veterans Affairs. (2010, October). *National survey of veterans, active duty service members, demobilized National Guard and Reserve members, family members, and surviving spouses.* Washington, DC: Department of Veterans Affairs. Retrieved from http://www.va.gov/vetdata/docs/SurveysAndStudies/NVSSurveyFinalWeightedReport.pdf

Durai, U. N. B., Chopra, M. P., Coakley, E., Llorente, M. D., Kirchner, J. E., Cook, J. M., &

Levkoff, S. E. (2011). Exposure to trauma and posttraumatic stress disorder symptoms in older veterans attending primary care: Comorbid conditions and self-rated health status. *Journal of the American Geriatrics Society, 59*(6), 1087–1092.

Institute of Medicine. (2012). *The mental health and substance use workforce for older adults: In whose hands?* Washington, DC: National Academies Press.

Kaplan, M. S., McFarland, B. H., Huguet, N., & Valenstein, M. (2012). Suicide risk and precipitating circumstances among young, middle-aged, and older male veterans. *American Journal of Public Health, 102* (Supplement 1), S131–S137.

Karel, M. J., Gatz, M., & Smyer, M. A. (2012). Aging and mental health in the decade ahead: What psychologists need to know. *American Psychologist, 67*(3), 184–198.

Karlin, B. E., & Zeiss, A. M. (2010). Transforming mental health care for older veterans in the Veterans Health Administration. *Generations, 34*(2), 74–83.

Magruder, K. M., & Yeager, D. E. (2009). The prevalence of PTSD across war eras and the effect of deployment on PTSD: A systematic review and meta-analysis. *Psychiatric Annals, 39*(8), 778–788.

Pietrzak, R. H., Goldstein, R. B., Southwick, S. M., & Grant, B. F. (2011). Prevalence and Axis I comorbidity of full and partial posttraumatic stress disorder in the United States: Results from wave 2 of the national epidemiologic survey on alcohol and related conditions. *Journal of Anxiety Disorders, 25*(3), 456–465.

Pietrzak, R. H., Goldstein, R. B., Southwick, S. M., & Grant, B. F. (2012). Physical health conditions associated with posttraumatic stress disorder in U.S. older adults: Results from wave 2 of the national epidemiologic survey on alcohol and related conditions. *Journal of the American Geriatrics Society, 60*(2), 296–303.

Spiro, A., III, Schnurr, P. P., & Aldwin, C. M. (1997). A life-span perspective on the effects of military service. *Journal of Geriatric Psychiatry, 30*(1), 91–128.

65 SPIRITUAL RESILIENCY IN THE MILITARY SETTING

William Sean Lee and Willie G. Barnes

HISTORICAL OVERVIEW OF THE MILITARY CHAPLAINCY

Since 1775 local clergy who have volunteered to serve in the military have provided for the religious and spiritual needs of the individuals in America's military. The purpose of the military Chaplain is to ensure the right of free exercise of every service member to practice their religious faith according to the dictates of their own beliefs. This right is guaranteed to every military member by the First Amendment to the US Constitution.

While Chaplains perform worship services and minister according to the beliefs and practices of their particular religious group, they are also required to ensure the provisions of religious support that meets the religious requirement of every combatant as indicated in Department of Defense regulations, Joint Publications 1-05 (Department of Defense, 2006). Chaplains additionally serve on the unit commander's personal staff to advise him or her concerning the impact of religion, morals, morale, and ethics on the unit and its mission (*JP-1-05*, 2006).

It is usually the military Chaplain that is the first point of referral for support in the military community. Chaplains are trusted agents for support within the military for many service members. This is for two reasons: (1) the military Chaplain is an actual military member and is viewed as "one of us" by military members; and (2) military Chaplains are granted the right of privileged communications by virtue of law and regulation, and due to their endorsement for military service by their recognized religious organization or denominations. These two qualifications of Chaplains provide them with considerable trust and significant respect in the military culture and community. This trust and respect provides credibility for military Chaplains that often results in them being the first resource utilized as referral for support of military members and their families in need. It also results in the Chaplain being a trusted "gatekeeper" to enable successful facilitation of referrals to mental health care providers.

THE ROLE OF THE MILITARY CHAPLAIN

In the military, Chaplains are embedded with the service members as part of their units. As already described, this relationship has historically been the norm for the United States military Chaplains. Chaplains train, eat, sleep, recreate, deploy into harm's way, and return as "comrades-in-arms" with their fellow service members. The only difference, but the key essential difference between the military service member and the Chaplain, is that the Chaplain is a noncombatant. Under the Geneva Convention for Amelioration of the Wounded and Sick, Article 28, Military Chaplains are granted the special status of "noncombatant" (Geneva Convention, 1929). Due to their noncombatant status, US Military Chaplains are forbidden to carry or use weapons. The military Chaplain is the only noncombatant personnel in the entire military system. As a countermeasure, every military Chaplain has a Chaplain's Assistant,

who is a noncommissioned officer and who is thus a combatant. Chaplains in all military components have a Chaplain's Assistant as part of their two-person team. The Chaplain's Assistant provides direct support for the military Chaplain. However, in the combat setting, the Chaplain's Assistant's essential task is to keep their chaplain alive, since the Chaplains are noncombatants.

These defining characteristics of military Chaplains as both military members and as noncombatants uniquely equip them as initial points for religious support and for referral and support when individual warriors or units are in crisis. Consequently, the Chaplain is usually not seen primarily as a religious representative, but a safe and trusted spiritual presence to turn to for support, counsel, comfort, and guidance. Chaplains encourage those who have a particular religious tradition to actively practice their faith. While Chaplains perform religious acts according to their own faith traditions, they provide for the religious support of all military members and their families.

Chaplains recognize that while not everyone has a religious tradition, the military values the importance of caring for the spirituality of every military member. Therefore, Chaplains are trained and responsible for the spiritual well-being of military members (See AR 165-1, *Chaplain Activities in the United States Army Sec. 1*). The goal of spiritual resiliency for military members is to continue to find meaning and purpose in life while fulfilling the requirement of military service. In lectures on spiritual resiliency at Yellow Ribbon Deployment training, the first author calls these markers "Life Animators" (Lee, 2005–2012). Life animators for each individual are those relationships, activities, locales, beliefs, and values that make life worth living. Chaplains encourage and assist military members in identifying their life animators and coping skills through a spiritual life review and inventory counseling to enable spiritual resilience throughout the deployment cycle.

In the military context, spiritual resilience is critically important to the combatant before, during, and after deployments. The experience of deployments and war must be placed in a larger context of meaning for individuals to survive the conflicts and have the opportunity for personal psychospiritual growth. Military members are trained to fight, defend, and kill if needed when put in combat situations, often needing to make these decisions within a split-second. Many service members have made multiple deployments to combat zones, thus compounding the stress on spiritual resilience capacity and capability of the individual combatant. These combatants must be able to frame their actions and experiences in combat within a larger context of meaning and values to effectively readjust to their loved ones and the larger society (Colonel Michael Gaffney, personal communication, January 14, 2011).

The most poignant and compelling example of the value of spirituality and spiritual resilience in the military context is that of the United States Prisoner of War (POW) in the Vietnam War. The longest held POW was US Navy pilot Everett Alvarez Jr., who was in captivity in North Vietnam for over 8 years. He credited faith in God and faith in the United States as the keys to his survival and that of most POWs (Alvarez, 1991).

TAXONOMY OF SPIRITUAL DIAGNOSIS

Diagnosing spiritual issues that negatively impact the service member's therapeutic process can be challenging due to the inherent subjectivity of spirituality. Asking a few simple questions to determine the religious history and preference of the individual can be helpful. Such information can provide an objective framework for understanding the psychospiritual orientation of the military member. Subsequently, this objective framework can provide structure as service members relate and integrate their military experience into their personal narrative and definition of self. As part of the interdisciplinary care team, integrating the spiritual dimension for enhanced coping, comfort, and meaning is the key task of the military Chaplain.

Military Chaplains are the appropriate caregivers to ask the following questions:

• Do you have a religious preference?
• Are you currently practicing your religious faith?

• What is most meaningful to you in life and in your faith tradition?
• What religious activities are most meaningful, helpful, or comforting to you?

Mental health care providers can also ask these questions, provided they refer the service member to a military Chaplain or religious professional if issues are deemed to be beyond their level of education, experience, and expertise. While there is regrettable and continuing stigma related to accessing mental health care, there is almost no stigma attached related to seeking the support and help of a military Chaplain. Often it is validation and endorsement of mental health care by the Chaplain that results in a military member accessing a mental health provider. The *Taxonomy of Spiritual Diagnosis* (1978) was developed to provide both etiology and defining characteristics allowing for patient diagnosis. The categories of diagnosis were determined to be: spiritual concerns, spiritual distress, and spiritual despair.

Spiritual Concerns

A diagnosis of spiritual concerns would be considered for someone who might evidence an inner conflict about beliefs, questioned the credibility of their usual faith system and practice, discouragement, mild anxiety, bewilderment, existential ambiguity, or verbalize a struggle to integrate their actions in military service with their belief system. Patients (military service members) who are able and actively cope with their spiritual problems without additional resources are reflecting spiritual concerns rather than the more crisis-like condition of spiritual distress.

Spiritual Distress

Spiritual distress results from more significant challenges to the service member's faith, values, and belief system. These challenges overwhelm the individual's spiritual resources, subsequently requiring additional resources to process the distress and return to a state of existential equilibrium. Often, spiritual distress presents with a broad range of emotional and psychosomatic systems. These can include, but are not limited to: intense angst, concentrated anger or rage, lack of spiritual trust, guilt or shame, crying, depression, grief, disturbed sleep, and generalized insecurity or fear. Individuals in spiritual distress often will continue to engage their concerns and be receptive to therapeutic interventions because they recognize the need.

Spiritual Despair

A military member diagnosed with spiritual despair evidences a lack of will to live, and has generally lost faith in themselves, treatment, and their religious and/or spiritual systems. The service member indicates through their subjective and objective responses that they have lost hope, have a sense of meaninglessness about life, no longer value their faith for coping, comfort, or meaning, feel abandoned and exhausted, no longer wish to practice their faith, and are passive or resistant to the treatment process. Military members in spiritual despair have been effectively overwhelmed by their spiritual issues and are no longer responsive to therapeutic interventions to integrate spirituality for coping, comfort, or meaning.

Spiritual Contentment

In 1996 the taxonomy for spiritual diagnosis was modified by expansion to include an additional, initial baseline category designated *spiritual contentment* and created by the first author while serving for Spiritual Care and Bereavement at the Hospice of Baltimore, Maryland (Lee, 1996a). This category allows for the possibility of self-resilience among those facing traumatic events resulting in personal growth through successful integration of religious and/or spiritual resources for coping, comfort, and meaning. While a significant portion of veterans that have faced combat deployments report issues that include characteristics of spiritual concerns, many are able to demonstrate significant spiritual resilience through internal processing and adequate self-care. These military members ultimately report posttraumatic growth (Tedeschi & Calhoun, 2004).

SUGGESTED THERAPEUTIC TOOLS AND INTERVENTIONS TO PROMOTE SPIRITUAL HEALTH

There are some simple, yet effective therapeutic tools and interventions utilized by military Chaplains that can also be easily used by mental health providers to promote spiritual health and resiliency among military personnel.

- Spiritual Life Review. Have the military member draw a line across a piece of paper. Ask the service member to list above the line those days in their life when they felt more alive or closest to the God of their faith if they are from a religious tradition. Ask them to describe what made that day so memorable. Note and list from those days the particular relationships, activities, and locales that provided the animation for the military member. Then list below the line those days when they felt least alive or furthest from the God of their faith if they are from a religious tradition. Ask them to describe how they managed to survive those days. Note and list from those days the particular coping skills that enable the service member to survive. From these above and below line experiences a list of life animators (above the line) and coping skills (below the line) can be identified as spiritual resilience resources for the military member to utilize in the present situation (Adapted from Lee, 1996b).
- The Taxonomy of Spiritual Diagnosis Spiritual Life Review (Lee, 1996a) described earlier can be explained and taught to the military member as a self-care tool to access when they might benefit from additional support or therapeutic relationship to process memories or experiences or current concerns to maintain a state of spiritual resilience.
- Military service members often find it difficult to identify and verbalize their spiritual and existential struggles with deployment cycle experiences and their associated emotions. A very effective tool related to spiritual resiliency and used

by some military Chaplains is the "Five Statements of Life Inventory." Explain to the service member that most of life can be framed in response to where or with whom we feel the need to say the following five statements: (1) Forgive me; (2) I forgive you; (3) Thank you; (4) I love you; and (5) Good-bye. Often this will provide the individual with a simple, yet clear path for looking at behaviors, relationships, life animators, and expectations.

Since 2001 over 2 million US Military members have been deployed to Afghanistan and Iraq. Multiple tours, prolonged separations, and the stress associated with every deployment have presented innumerable challenges for military members and their families. Spiritually resilient warriors and their families are vital to the effective and success of our military sustainment and efforts. The Chaplain and the Chaplain's Assistant are essential to military service members and their unit cohesion. They help promote spiritual resilience and support the faith traditions and experiences of military service members. They may serve as an important resource to psychologists and other mental health professionals to refer service members to when spiritual or religious issues are central to the service member's care.

References

Alvarez, E., Jr. (1991). *Chained eagle: The heroic story of the first American shot down over North Vietnam*. New York, NY: Dell.

Department of the Army, Headquarters. (2004). AR 165–1, *Chaplain activities in the United States Army* (Washington, D.C.), 6-7, 8.

Department of Defense. (2006). Joint Publications (JP 1-05). *Chaplains in joint operations*. Chaplaincy section.

Gaffney, M. (2011, January 14). *Behavioral health summit lecture*, Towson University, Towson, Maryland.

Geneva Convention. (1929). *Amelioration of the wounded and sick, Article 28*.

Lee, W. S. (1996a) *Spiritual diagnosis* (modified). Towson, MD: Hospice of Baltimore.

Lee, W. S. (1996b). *Spiritual life review*. Towson, MD: Hospice of Baltimore.

Lee, W. S. (2005–2012). *Spiritual resiliency for pre and post-deployment*. Presentations made during Maryland Army National Guard Yellow Ribbon Deployment Cycle Support Events.

National Committee for the Classification of Nursing Diagnosis. (1978). *Spiritual diagnosis taxonomy*. In Proceedings of the National Committee for the Classification of Nursing Diagnosis, Glendale, CA.

Tedeschi, R. G., & Calhoun, L. (2004, April 1). Posttraumatic growth: A new perspective on psychotraumatology. *Psychiatric Times*. Retrieved from www.psychiatrictimes.com/ptsd/content/article/10168/54661

66 POSTTRAUMATIC GROWTH

Richard G. Tedeschi

Although it is clear that traumatic experiences can produce various symptoms of anxiety, depression, and other difficulties, researchers and clinicians have recognized that positive outcomes are also possible. Tedeschi and Calhoun (1996) introduced the term *posttraumatic growth* to describe both a process of development of these outcomes, and the outcomes themselves. They stated that posttraumatic growth develops as a result of the struggle with traumatic life events. Some similar terms that appear in the literature are *stress-related growth*, *perceived benefits*, and *adversarial growth*, but posttraumatic growth is the most commonly used term. *Benefit-finding* has also been used to refer to the reports of people who report positive outcomes from adversity, but this term has been seen mostly in literature pertaining to experiencing illness and can refer to outcomes that, while beneficial (e.g., better health behaviors), do not necessarily represent significant personal transformations. Posttraumatic growth is different from *resilience*, since resilience is an ability to be minimally affected by trauma, while posttraumatic growth is the result of a struggle to come to terms with major events.

Posttraumatic growth has been reported by people who have experienced various events, including bereavement, illness, natural disaster, criminal victimization, and war. In many of these circumstances, people have not voluntarily put themselves in the path of trauma, while in others people are quite aware of the risk to which they are exposing themselves. Despite the variety of events that may act as a catalyst for posttraumatic growth, and whether people are voluntarily exposing themselves to potentially traumatic circumstances or not, similar psychological processes seem to be involved in these transformative experiences.

DOMAINS OF POSTTRAUMATIC GROWTH

There are five types of positive outcomes in posttraumatic growth, as measured by the factors found on the Posttraumatic Growth Inventory (Tedeschi & Calhoun, 1996). These five factors are:

- *Appreciation of Life*
- *Personal Strength*
- *Relationships with Others*
- *New Possibilities*
- *Spiritual Change*

The factor of the inventory that is termed Appreciation of Life involves a recognition of how time alive is precious and should not be squandered. The Personal Strength factor involves an awareness of how capable the person has been in coping with the traumatic event, or somehow surviving it. The factor that is labeled Relationships with Others measures the degree to which people have changed their views of others and how they behave toward them, for example, showing more compassion toward them, or being more emotionally expressive. The New Possibilities dimension represents a greater awareness of options for living that may not have been available if the traumatic event had not occurred. Spiritual Change represents the ways in which religious or spiritual aspects of the person's life have been strengthened in the aftermath of trauma. The particular ways that people experience posttraumatic growth vary along these dimensions. It is rare that a person reports posttraumatic growth on every factor.

POSTTRAUMATIC GROWTH PROCESS

The process of growth as most recently conceptualized (Calhoun, Cann, & Tedeschi, 2010) first involves the characteristics of the person before the crisis situation occurs. Personality characteristics, such as extraversion or openness to experience, may influence the likelihood of subsequent growth, and gender may also affect the possibility of growth, with women showing a somewhat greater tendency to report posttraumatic growth. When a traumatic event or events are experienced by an individual, that person's *assumptive world* is challenged or reconsidered. The assumptive world is the set of core beliefs a person has about how the world works, and the life they expect to lead. It is so fundamental that people hardly question it, until events call what they believe into question. Traumas make people wonder who they are, what will happen to their lives, and what sort of world this really is.

The process of shattering of the assumptive world has been described by Tedeschi and Calhoun (1996) as a psychologically seismic event, which, like an earthquake, shatters the core beliefs on which people have depended for basic understanding of identity and of their life narrative or implicit autobiography. A challenge or disruption of the assumptive world, which can also be associated with the disruption of important goals or of one's life narrative, is likely to produce repetitive, intrusive thoughts as the person attempts to grasp the fact that the event has happened and the implications for living. For persons whose assumptive world provides a context for, and a full understanding of the event, there is no challenge to core beliefs, and there will be little or no posttraumatic growth produced in the aftermath of the event.

With time, after the challenging of the core beliefs by traumatic events the repeated intrusive thinking about what has happened, and how to comprehend it, can become more reflective, deliberate, and focused on making sense of events. This more deliberate form of rumination is particularly useful in the development of posttraumatic growth. It may become possible for a trauma survivor to engage in this form of rumination as their emotional distress comes under some control.

It is also important to recognize that people respond to traumatic events within a sociocultural context. People are exposed to ideas about and models for posttraumatic growth in the larger culture and in their close relationships. The presence of supportive others, particularly those who maintain support for as long as it is requested or needed, can play an important role in how a person copes with trauma and also the degree to which posttraumatic growth is encouraged. Culturally based ideas about trauma response and growth may influence the types of posttraumatic growth a person is likely to report. For example, in collectivist cultures where disclosures about one's successes or positive qualities are considered socially inappropriate, trauma survivors may be less likely to report posttraumatic growth that reflects personal strength.

When a trauma survivor can reconstruct an effective core beliefs system or assumptive world through the processes of rumination and disclosure, they may also come to construct a revised life narrative. They may see that their life path has changed, that their future is different from what they expected, because the trauma has interrupted the pathway for living they had implicitly created. A more profoundly understood appreciation for how to live life well, and a sense of meaning may also result, and that may be described as wisdom.

CHARACTERISTICS OF REPORTS OF POSTTRAUMATIC GROWTH

A variety of studies have shown that posttraumatic growth is a common experience in the aftermath of trauma. For most samples, the rates of posttraumatic growth are in the range of one-half to two-thirds. Although there are relatively few studies of posttraumatic growth over time, there seems to be a good deal of stability in these changes. However, some people may at first report posttraumatic growth in the context of a coping mechanism that allows them to feel less anxious about their experiences, and later this is consolidated into a more permanent change in life perspective and behavior (Zoellner & Maercker, 2006). Reports of posttraumatic growth from people undergoing these significant traumas do not imply that the horrors of their experiences have been forgotten or denied, but that posttraumatic growth is recognized along with the negative aspects of their experiences.

Combat and Posttraumatic Growth

Although combat can be so traumatic as to produce high numbers of veterans with posttraumatic stress disorder, depression, anxiety, and substance abuse, there is also evidence that a significant proportion of persons with combat experience report posttraumatic growth. An early indication of this outcome came from a study of American prisoners of war held in Vietnam (Sledge, Boydstun, & Rabe, 1980). These prisoners reported upon repatriating that their extended incarceration had helped them achieve gains in personal character. Similar reports have been obtained in other studies of American POWs, from Israeli POWs, and from combat veterans involved in a variety of conflicts. A recent study of operation OEF/OIF veterans has shown that over 70% report some aspect of posttraumatic growth (Pietrzak et al., 2010). The implications of these reports of posttraumatic growth are beginning to be understood. Reports of growth have been reported to be associated with more altruistic community involvement and less suicidal ideation. At the same time those reporting growth have sometimes reported elevated levels of distress, posttraumatic stress disorder, or depression as well. These relationships may be found because posttraumatic growth is only possible if events have caused significant enough distress to challenge the assumptive

world. Factors that seem to be associated with the development of posttraumatic growth in combat veterans include perceived social support, especially among combat unit members, and active coping strategies.

FACILITATING POSTTRAUMATIC GROWTH

Since the development of the concept of posttraumatic growth, there have been attempts made to describe a process by which it might be facilitated in trauma survivors. Tedeschi and Calhoun (2006) make a point that posttraumatic growth is to a great degree a product of *expert companionship*. For military service members, these expert companions may be unit members or others with military experience who are seen as understanding what this service and combat are like. They may be professionals who are seen as understanding what the experience of trauma and its aftermath are like. But whether they are friends, family, chaplains, or other professionals, they are people who can listen for extended periods to stories that involve fear, guilt, shame, and confusion. Expert companions cannot prescribe posttraumatic growth. They facilitate it through a kind of listening that is like a patient mentoring process, so that the pieces of the puzzle of traumatic experience are laid out in front of the trauma survivor and his or her companion in a way that they can be fashioned into a new life narrative. In doing this, there is a reconsideration of core beliefs, appreciation of paradox (e.g., that in loss there may be gain, that admission of vulnerability can be a strength), and experiments with new ways of living. Military unit leadership may have an effect on posttraumatic growth, since unit support has been found to be associated with it. Unit leaders who do not focus on distinctions based on rank, but relate to their team and the unit as a whole as expert companions may offer perspectives and ways of leading that indicate that all are valued and need to be supported. Out of such alliances within the unit, the connections of expert companionship may be more possible.

Strategies for psychologically preparing soldiers for the rigors of combat have been implemented in recent years, and a more expansive Comprehensive Soldier Fitness program of preparation for combat and the return from deployment is currently being implemented. This program includes a component of posttraumatic growth, as described by Tedeschi and McNally (2011). They outlined five components in this program that are derived from the model of posttraumatic growth process described above, and overlap with standard trauma treatment approaches. The first component is a psychoeducational focus on the process of the shattering of the core beliefs or assumptive world and the basic physiological and psychological responses that are normal reactions to the experience of combat, that must be understood as precursors to posttraumatic growth rather than merely negative symptoms to be conquered. The second component involves learning methods of anxiety reduction, and regulation of intrusive thoughts and images of trauma in order to set the stage for more deliberate processes of thinking that can allow for reconstruction of core beliefs. The third component in this process involves encouragement of disclosures by the trauma survivor about their traumatic experiences and what it is like to live in the aftermath of these experiences. These conversations lead to a fourth component that is at the heart of the posttraumatic growth process: a reconfiguration of shattered belief systems, and revision of life narratives in a way that includes aspects of the five elements of posttraumatic growth. This aspect of the process may be part applied existential philosophy, and part cognitive or narrative therapy, as new ways of looking at trauma and finding meaning in surviving it are encouraged. Scurfield (2006) offers ways that negative perspectives on combat trauma can be considered in a way that involves more self-compassion and meaning, and can allow for posttraumatic growth. For example, the aftermath of combat can be viewed as bonus time or a second chance. "Missions" can be encouraged that allow for a new sense of meaning, future orientation, and recognition of the benefits of trauma for oneself and for others. It is important that the expert companions who offer such perspectives are sensitive to the timing of their statements when working with trauma survivors, who are sometimes open to such perspectives, and sometimes not.

The new perspectives may allow the survivor of trauma to see that despite horror, fear, or shame, redemption is possible and that life that is meaningful can be lived in the aftermath of trauma. In the fifth component of the posttraumatic growth process, there is an appreciation of oneself as a classic hero. This kind of hero is one who has survived enormous challenges, ones that ordinary people may not appreciate, and then returns to the community with an enhanced appreciation of life and sense of purpose. This perspective provides a guide for action and preparation for future challenges, enhancing psychological resilience with core beliefs that can do a better job of incorporating the facts of past and future experiences. The five factors of posttraumatic growth outcomes can yield life principles that allow trauma survivors to better understand what it means to be human, to appreciate an essential connection to others, and to be more empathic and altruistic and act as an agent of social change.

References

Calhoun, L., Cann, A., & Tedeschi, R. (2010). The posttraumatic growth model: Sociocultural considerations. In T. Weiss & R. Berger (Eds.), *Posttraumatic growth and culturally competent practice: Lessons learned from around the globe* (pp. 1–14). Hoboken, NJ: Wiley.

Pietrzak, R., Goldstein, M., Malley, J., Rivers, A., Johnson, D., Morgan, C., & Southwick, S. M. (2010). Posttraumatic growth in veterans of Operations Enduring Freedom and Iraqi Freedom. *Journal of Affective Disorders, 126,* 230–235.

Scurfield, R. M. (2006). *War trauma: Lessons unlearned from Vietnam to Iraq.* New York: Algora.

Sledge, W., Boydstun, J., & Rabe, A. (1980). Self-concept changes related to war captivity. *Archives of General Psychiatry, 37,* 430–443.

Tedeschi, R., & Calhoun, L. (1996). The Posttraumatic Growth Inventory: Measuring the positive legacy of trauma. *Journal of Traumatic Stress, 9,* 455–472.

Tedeschi, R. G., & Calhoun, L. G. (2006). Expert companions: Posttraumatic growth in clinical practice. In L. G. Calhoun & R. G. Tedeschi (Eds.), *Handbook of posttraumatic growth:*

Research and practice (pp. 291–310). Mahwah, NJ: Erlbaum.

Tedeschi, R. G., & McNally, R. J. (2011). Can we facilitate posttraumatic growth in combat veterans? *American Psychologist, 66*, 19–24.

Zoellner, T., & Maercker, A. (2006). Posttraumatic growth in clinical psychology: A critical review and introduction of a two component model. *Clinical Psychology Review, 26*, 626–653.

67 WAYS TO BOLSTER RESILIENCE ACROSS THE DEPLOYMENT CYCLE

Donald Meichenbaum

Since the terrorist attacks of September 11, 2001, over 2 million individuals have been deployed to Iraq and Afghanistan, with nearly 800,000 who have been deployed multiple times. There is a linear relationship between the number of fire fights (direct combat experiences), the number of deployments, and the severity of psychiatric symptoms returning service members experience. The rate of posttraumatic stress disorder (PTSD) and related psychiatric disorders among veterans who have served in recent combat, however, is only in the range of 10 to 18% (Nash, Krantz, Stein, Westphal, & Litz, 2011). This means that somewhere between 80 to 90% of returning service members are impacted by their combat experiences, but most evidence some level of resilience. Such evidence of resilience is not unique to military personnel.

Bonanno (2004) documented that the upper level of post-trauma disorders following traumatic victimizing experiences is approximately 30%. In fact, resilience is the normative response to trauma experiences, whether the traumatic events are a natural disaster or due to accidents, illness, losses, or intentional human design—terrorist attacks, childhood sexual abuse, rape, domestic violence, and the like (Meichenbaum, 2006).

Such evidence of resilience, or the ability to "bounce back," the ability to continue forward and maintain equilibrium in the face of chronic adversity, the ability to live with ongoing fear and uncertainty, and the ability to adapt to the difficult and challenging life experiences is more the rule than the exception, more common than rare. Moreover, resilience is not a sign of exceptional strength, but a fundamental feature of normal coping, or what Masten (2001) characterizes as "ordinary magic."

Research has continually demonstrated from the time of World War I that veterans, as a group, resume normal lives and most (70%+) appraise the impact of their military service on their present lives as "mainly positive" and "highly important." The majority of military spouses believe that deployment had strengthened their marriage, contributed to the development of new skills, as well as to a sense of independence and self-reliance. The children of military families are also typically resilient, even after experiencing significant trauma and family deaths (Sheppard, Malatras, & Israel, 2010; Wiens & Boss, 2006).

For example, studies of aviators who were shot down, imprisoned, and tortured for years by the North Vietnamese indicated that some 61% reported that the imprisonment had produced favorable changes, increasing their self-confidence, and teaching them to value the truly important things in life (Adler, Bliese, & Castro, 2011; Meichenbaum, 2011).

Military organizations have been proactive and effective in putting into place a whole host of intervention programs designed to bolster resilience across the entire deployment cycle. Table 67.1, which was adapted from Pincus, House, Christensen, and Adler (2001), provides an illustrative (not exhaustive) list of resilience-bolstering interventions that have been developed. These programs may be implemented at the universal (primary), selected (secondary), or indicated (tertiary) levels, which, respectively, provide services for all service members and their families; for those identified as being in need or at high risk; and for those requiring more immediate and comprehensive services. These intervention programs require organizational policies and support that are designed to reduce risk, as in the case of sexual harassment and sexual assaults, reporting practices, and mental health services for victimized service members, or programs that are designed to provide stress inoculation training before deployment (Meichenbaum 2006; 2009). Whealin, Ruzek, and Southwick (2008) highlighted that such preparatory universal intervention programs should:

1. make future potential stressors more predictable so that when they occur, exposed individuals will perceive themselves to be more in control and more self-efficacious;
2. encourage more positive cognitive appraisals of potential stressful events by providing practice and mastery training;
3. teach emotion-regulation, stress management, and social problem-solving skills.

Each of these military-based interventions incorporates the "building blocks," or factors that research indicates are the prerequisites of resilience. In considering these components, it is important to keep in mind that post-trauma stress and resilience can coexist. Positive and negative emotions may co-occur, operating side-by-side following exposure to traumatic events. In fact, service members may be resilient in one domain of their lives, but not in other domains, or at one time in their lives and not at other times. Resilience is a dynamic, fluid process that develops over time, and its expression may be a slow developmental progression. There are multiple pathways to resilience, with no single dominant factor, or "magic bullet" that determines it. Rather resilience-engendering activities need to be practiced and replenished on a regular basis, so that such coping responses become automatic and incorporated into one's repertoire. Moreover, one can think of not only resilient individuals, but also resilient families, communities, and resilient combat units. Strong unit cohesion that nurtures trust and high morale within the unit, a sense of connection and belongingness and perceived support, and units that have competent and concerned military leadership that instills confidence provide protective factors that promote resilience within their unit.

Besides unit cohesion, some other resilience-engendering factors include social supports ("band of brothers/sisters," peer and family supports); stress management techniques and proactive, as compared to avoidant, coping style; cognitive flexibility and an optimistic future orientation; 3:1 ratio of positive to negative emotions; and having a resilient-oriented mind-set. Meichenbaum (2012) has enumerated specific ways to bolster resilient behaviors in six domains of life—physical, interpersonal, emotional, cognitive, behavioral, and spiritual.

The US Army Comprehensive Soldier Fitness (CSF) Program (see *American Psychologist*, January 2011—Volume 66, Number 1, and http://www.army.mil/csf/resources.html) trains master resilience trainers, who are deployed in large organizational units on ways that service members can bolster personal strengths, control negative emotions, adopt a resilient mind-set and enhance relationships with loved ones. A major feature of CSF is the way it provides service members with individualized feedback, using the Global Assessment Tool (GAT) that can guide self-paced training

... having a life

115 - hyper startle

131 TBI (mild)

140 Suicidal thoughts

155 Posttraumatic growth (PTG)

Common Post Deployment Problems?

9 expecations (be realistic)   31 - it takes 3-6 months to get
   • intimacy - may take time   used to your old life
19 • Anger - excessive anger 20   59 - not the same people
29 • Sleep   all changed
39 • financial issues
49 • breakups / divorce
61 • children change
85 • Substance use / gambling
   • Self-medicate 94
97 • Grief
105- living w/ too

TABLE 67.1. Stages of Deployment and Illustrative Resilience-Bolstering Interventions

Stage	Tasks	Possible Interventions
PREDEPLOYMENT The notification of deployment to the point of departure.	Service member and family preparation, accompanying responsibilities and reactions	1. Military training program 2. Comprehensive Soldier Fitness Program (CFP) 3. Battlemind Programs (War Resiliency Programs) for service members and spouses. 4. Family Readiness campaigns that establish both patterns of communication and service members' ongoing presence in the family. 5. Family Organizational Plans
SUSTAINMENT From the end of the first month through to the final month of deployment.	Handle a variety of deployments and home-front stressors.	1. Mental Health Advisory Teams (MHAT) 2. Trauma Risk Management Programs (TRIM) 3. Small Unit After-Action Reviews 4. Battlemind Debriefing 5. Combat and Operational Stress, First Aid and Control (COSC), (PIE's interventions, Proximity, Immediacy, Expectation) 6. Navy and Marine Corps Combat Intervention Programs 7. Bereavement groups, Memorial and ceremonial services 8. Provide Work-rest cycles, Sleep management, Substance abuse programs, R&R
REDEPLOYMENT Month preceding homecoming to home arrival	Initial readjustment, altered routines, altered family responsibilities, communication.	1. Prepare for reintegration stressors. 2. Information about resources and services. 3. Address barriers to help-seeking. 4. Educate about Web-based resources (e.g., Military One Source)
POSTDEPLOYMENT Arrival home to 6 months.	Renegotiate roles, Establish new routines. Cope with injuries, losses and postcombat reactions.	1. Yellow Ribbon Reintegration Program. 2. Coming Together around Military Families (CTAMF). 3. Families Overcoming under Stress Combat Injury (FOCUS-CI) 4. Sesame Street Workshop—Talk, Listen, Connect. Also bibliotherapy for children. 5. Army couples' expressive writing project. 6. Evidence-based treatment programs for PTSD, substance abuse, couples therapy. 7. Telehealth programs.

(Stage Model Adapted from Pincus et al., 2001)

modules. While initial results of the CSF have been promising, a more complete evaluation of this $120 million initiative is now underway (Nash et al., 2011).

At the military family level for Active Duty service members, Wiens and Boss (2006) have enumerated a number of protective factors including access to comprehensive health care, educational services, legal assistance, consistent employment, and a host of on-base and online organizations that have been specifically created to provide support to families, as well as high levels of community supports. For the families of National Guard members, because of their geographical dispersement, additional out-reach intervention programs have been established such as the Yellow Ribbon program, Military One Source, (http://www.yellowribbon.mil; http://www.militaryonesource.com), and the like. The importance of such intervention efforts is underscored by the findings that married service members are three times more likely than single service members to meet diagnostic criteria for PTSD and 2.7

times more likely to be clinically depressed. Deployed soldiers report that home-front stressors are a major contributor to their levels of stress when deployed (Adler et al., 2011). Resilience-engendering programs need to take these risk and protective factors into account and reduce the barriers such as stigma associated with help-seeking, as well as practical barriers (transportation, child care, easier access) (Meichenbaum, 2009).

RECOMMENDATIONS

Finally, it is worth highlighting that the present assessment approach for returning service members is designed to identify the self-reported presence of psychiatric symptoms on the postdeployment health assessment (PDHA) and postdeployment health reassessment (PDHRA). Given that the normative reaction to deployment is resilience, it would be useful to systematically and routinely assess for what "signs of resilience" returning service members and their families evidence. For example, see the Posttraumatic Growth Inventory (http://cust.cf.apa.org/ptgi/index.cfm). In what ways has the exposure to combat and related deployment activities actually strengthened individuals, families and contributed to their growth? There is a need to convey an explicit message to all service members and their families that as a result of deployment they are likely to become more resilient. That which gets assessed, usually gets attended to and highlighted (Meichenbaum, 2011, 2012).

References

Adler, A. B., Bliese, P. D., & Castro, C. A. (Eds.). (2011). *Deployment psychology: Evidence-based strategies to promote mental health in the military.* Washington, DC: American Psychological Association.

Bonanno, G. A. (2004). Loss, trauma, and human resilience: Have we underestimated the human capacity to thrive after extremely aversive events. *American Psychologist, 59,* 20–28.

Masten, A. S. (2001). Ordinary magic: Resilience-processes in development. *American Psychologist, 56,* 227–238.

Meichenbaum, D. (2006). Resilience and posttraumatic growth: A constructive narrative perspective. In L. G. Calhoun & R. G. Tedeschi (Eds.), *Handbook of posttraumatic growth* (pp. 355–368). Mahwah, NJ: Erlbaum.

Meichenbaum, D. (2009). Core psychotherapeutic tasks with returning soldiers: A case conceptualization approach. In S. Morgillo Freeman, B. A. Moore, & A. Freeman (Eds.), *Living and surviving in harm's way: A psychological treatment for pre- and post-deployment of military personnel* (pp. 193–210) New York, NY: Routledge.

Meichenbaum, D. (2011). Resiliency building as a means to prevent PTSD and related adjustment problems in military personnel. In B. Moore & W. E. Penk (Eds.), *Treating PTSD in military personnel* (pp. 325–344). New York, NY: Guilford Press.

Meichenbaum, D. (2012). *Roadmap to resilience: A guide for military, trauma victims, and their families.* Clearwater, FL: Institute.

Nash, W., Krantz, L., Stein, P., Westphal, R. J., & Litz, B. (2011). Comprehensive soldier fitness, battlemind, and the stress continuum model: Military organizational approaches to prevention. In J. I. Ruzek, P. P. Schnurr, J. J. Vasterling, & M. J. Friedman (Eds.), *Caring for veterans with deployment-related stress disorders: Iraq, Afghanistan, and beyond* (pp. 193–214). Washington, DC: American Psychological Association.

Pincus, S. H., House, R., Christenson, J., & Adler, L. E. (2001). The emotional cycle of deployment: A military family perspective. *U.S. Army Medical Department Journal, 4/5/6,* 15–23.

Sheppard, S. C., Malatras, J. W., & Israel, A. C. (2010). The impact of deployment on U.S. Military families. *American Psychologist, 65,* 599–609.

Whealin, J. M., Ruzek, J. I., & Southwick, S. (2008). Cognitive-behavioral theory and preparation for professionals at risk for trauma exposure. *Trauma, Violence, and Abuse, 9,* 100–113.

Wiens, T. W., & Boss, P. (2006). Maintaining family resiliency before, during, and after military separation. In C. A. Castro, A. D. Adler, & C. A. Britt (Eds.), *Military life: The psychology of serving in peace and combat* (Vol. 3, pp. 13–38). Bridgeport, CT: Praeger Security International.

PART V
Resources

68 COMMON MILITARY ABBREVIATIONS

Bret A. Moore

AAFES	Army and Air Force Exchange Service
ABN	airborne
AC	Active Component; alternating current
ACOS	assistant chief of staff
ACR	armored cavalry regiment (Army); assign channel reassignment
AD	active duty; advanced deployability; air defense; automatic distribution; priority add-on
AE	aeromedical evacuation; assault echelon; attenuation equalizer
AFB	Air Force base
AFSOUTH	Allied Forces, South (NATO)
AG	adjutant general (Army)
AGR	Active Guard and Reserve
ALCON	all concerned
AMEDD	Army Medical Department
AMEDDCS US	Army Medical Department Center and School
AO	action officer; administration officer; air officer; area of operations; aviation ordnance person
AOI	area of interest
AOR	area of responsibility
ARFOR	Army forces
ARNG	Army National Guard
BAH	basic allowance for housing
BAS	basic allowance for subsistence; battalion aid station
BCT	brigade combat team
BDE	brigade
BRAC	base realignment and closure

BSB	brigade support battalion
BSCT	behavioral science consultation team
C2	command and control
CA	chaplain assistant; civil administration; civil affairs; combat assessment; coordinating altitude;
CAB	combat aviation brigade
CAC	common access card; current actions center
CBRNE	chemical, biological, radiological, nuclear, and high-yield explosives
CBTZ	combat zone
CDR	commander; continuous data recording
CID	criminal investigation division
CMD	command
COA	course of action
COGARD	Coast Guard
COMM	communications
CONUS	continental United States
COS	chief of staff; chief of station; critical occupational specialty
COSR	combat and operational stress reactions
CP	check point; collection point; command post; contact point; control point; counterproliferation
DD	Department of Defense (form); destroyer (Navy ship)
D-day	unnamed day on which operations commence or are scheduled to commence
DEERS	Defense Enrollment Eligibility Reporting System

DEFCON	defense readiness condition
DEPMEDS	deployable medical systems
DEPORD	deployment order
DFAS	Defense Finance and Accounting Service
DNBI	disease and nonbattle injury
DOB	date of birth
DOD	Department of Defense
DODDS	Department of Defense Dependent Schools
DOR	date of rank
DOS	date of separation
DTS	Defense Travel System
DVA	Department of Veterans Affairs
EER	enlisted employee review
ENL	enlisted
EOD	explosive ordnance disposal
EXORD	execute order
FHP	force health protection
FM	field manual (Army)
FMF	Fleet Marine Force
FOB	forward operating base; forward operations base
FORSCOM	United States Army Forces Command
FRAGORD	fragmentary order
FY	fiscal year
G-1	Army or Marine Corps component manpower or personnel staff officer (Army division or higher staff, Marine Corps brigade or higher staff)
G-2	Army Deputy Chief of Staff for Intelligence; Army or Marine Corps component intelligence staff officer (Army division or higher staff, Marine Corps brigade or higher staff)
G-2X	Army counterintelligence and human intelligence staff element
G-3	Army or Marine Corps component operations staff officer (Army division or higher staff, Marine Corps brigade or higher staff); assistant chief of staff, operations
G-4	Army or Marine Corps component logistics staff officer (Army division or higher staff, Marine Corps brigade or higher staff); Assistant Chief of Staff for Logistics
G-5	assistant chief of staff, plans
G-6	Army or Marine Corps component command, control, communications, and computer systems staff officer; assistant chief of staff for communications; signal staff officer (Army)
G-7	Army component information operations staff officer; assistant chief of staff, information engagement; information operations staff officer (ARFOR)
GUARD US	National Guard and Air Guard
GWOT	global war on terror
HHC	headquarters and headquarters company
HHD	headquarters and headquarters detachment
HHQ	higher headquarters
HVI	high-value individual
HVT	high-value target
IED	improvised explosive device
IG	inspector general
JA	judge advocate
JAG	judge advocate general
LAV	light armored vehicle
LES	leave and earnings statement
LNO	liaison officer
MARDIV	Marine division
MARFOR	Marine Corps forces
MASCAL	mass casualty
MASH	mobile Army surgical hospital
MEDCOM	medical command; US Army Medical Command
MEDEVAC	medical evacuation
MEPRS	Military Entrance Processing and Reporting System
MILPERS	military personnel
MILVAN	military van (container)
MOU	memorandum of understanding
MP	military police (Army and Marine); multinational publication
MRE	meal, ready to eat
N-1	Navy component manpower or personnel staff officer
N-2	Director of Naval Intelligence; Navy component intelligence staff officer
N-3	Navy component operations staff officer
N-4	Navy component logistics staff officer

N-5	Navy component plans staff officer
N-6	Navy component communications staff officer
NATO	North Atlantic Treaty Organization
NAVFOR	Navy forces
NAVMED	Navy Medical; Navy medicine
NBC	nuclear, biological, and chemical
NDAA	National Defense Authorization Act
NIPRNET	Nonsecure Internet Protocol Router Network
OCONUS	outside the continental United States
OEF	Operation ENDURING FREEDOM
OPSEC	operations security
PAO	public affairs office; public affairs officer
PCS	permanent change of station
PHS	Public Health Service
PLT	platoon
PMOS	primary military occupational specialty
POC	point of contact
POTUS	President of the United States
POV	privately owned vehicle
POW	prisoner of war
PPE	personal protective equipment
PPP	power projection platform
PROFIS	professional officer filler information system
QRF	quick reaction force; quick response force
RDO	request for deployment order
RECON	reconnaissance
ROE	rules of engagement
RPG	rocket propelled grenade
RTB	return to base
RTD	returned to duty
S-1	battalion or brigade manpower and personnel staff officer (Marine Corps battalion or regiment)
S-2	battalion or brigade intelligence staff officer (Army; Marine Corps battalion or regiment)
S-3	battalion or brigade operations staff officer (Army; Marine Corps battalion or regiment)
S-4	battalion or brigade logistics staff officer (Army; Marine Corps battalion or regiment)

SAPR	sexual assault prevention and response
SARC	sexual assault response coordinator
Seabee	Navy construction engineer
SECAF	Secretary of the Air Force
SECARMY	Secretary of the Army
SecDef	Secretary of Defense
SECNAV	Secretary of the Navy
SERE	survival, evasion, resistance, and escape
SF	security force; security forces (Air Force or Navy) special forces; standard form
SIPRNET SECRET	Internet Protocol Router Network
SIR	serious incident report; specific information requirement
SITREP	situation report
SJA	staff judge advocate
SMART	special medical augmentation response team
SME	subject matter expert
SPRINT	special psychiatric rapid intervention team
SUBJ	subject
TD	temporary duty; theater distribution; tie down; timing distributor; total drift; transmit data
TDA	Table of Distribution and Allowance
TF	task force
TO&E	table of organization and equipment
TRADOC	United States Army Training and Doctrine Command
UAV	unmanned aerial vehicle
UMT	unit ministry team
USA	United States Army
USACHPPM	United States Army Center for Health Promotion and Preventive Medicine
USAF	United States Air Force
USCG	United States Coast Guard
USMC	United States Marine Corps
USN	United States Navy
USPHS	United States Public Health Service
UXO	unexploded explosive ordnance; unexploded ordnance

VA	Veterans Administration; victim advocate; vulnerability assessment	WMD	weapons of mass destruction
VBIED	vehicle-borne improvised explosive device	WRAIR	Walter Reed Army Institute of Research
VOL	volunteer	XO	executive officer
VTC	video teleconferencing	ZULU	time zone indicator for Universal Time
WARNORD	warning order		
WIA	wounded in action		

See http://www.dtic.mil/doctrine/dod_dictionary/for a comprehensive list of military acronyms and terms.

69   COMPARATIVE MILITARY RANKS

Bret A. Moore

Flag Rank Officers

Pay Grade	Army	Marine Corps	Navy/Coast Guard	Air Force
Special	General of the Armies	*none*	Admiral of the Navy	*none*
Special	General of the Army	*none*	Fleet Admiral	General of the Air Force
O-10	General	General	Admiral	General
O-9	Lt. General	Lt. General	Vice Admiral	Lt. General
O-8	Major General	Major General	Rear Admiral (upper half)	Major General
O-7	Brigadier General	Brigadier General	Rear Admiral (lower half)	Brigadier General

Commissioned Officers

Pay Grade	Army	Marine Corps	Navy/Coast Guard	Air Force
O-6	Colonel	Colonel	Captain	Colonel
O-5	Lt. Colonel	Lt. Colonel	Commander	Lt. Colonel
O-4	Major	Major	Lt. Commander	Major
O-3	Captain	Captain	Lieutenant	Captain
O-2	1st Lieutenant	1st Lieutenant	Lieutenant, JG	1st Lieutenant
O-1	2nd Lieutenant	2nd Lieutenant	Ensign	2nd Lieutenant

(Continued)

Warrant Officers

W-5	Chief Warrant Officer, Five	Chief Warrant Officer, Five	Chief Warrant Officer, Five	Chief Warrant Officer, Five (discontinued)
W-4	Chief Warrant Officer, Four	Chief Warrant Officer, Four	Chief Warrant Officer, Four	Chief Warrant Officer, Four (discontinued)
W-3	Chief Warrant Officer, Three	Chief Warrant Officer, Three	Chief Warrant Officer, Three	Chief Warrant Officer, Three (discontinued)
W-2	Chief Warrant Officer, Two	Chief Warrant Officer, Two	Chief Warrant Officer, Two	Chief Warrant Officer, Two (discontinued)
W-1	Warrant Officer, One	Warrant Officer, One	Warrant Officer, One (discontinued)	Warrant Officer, One (discontinued)

Enlisted Servicemen

E-9 (special)	Sergeant Major of the Army	Sergeant Major of the Marine Corps	Master Chief Petty Officer of the Navy/ Coast Guard	Chief Master Sergeant of the Air Force
E-9	Sergeant Major Command Sergeant Major	Master Gunnery Sergeant Sergeant Major	Master Chief Petty Officer Command Master Chief Petty Officer	Chief Master Sergeant Command Chief Master Sergeant
E-8	Master Sergeant First Sergeant	Master Sergeant First Sergeant	Senior Chief Petty Officer	Senior Master Sergeant
E-7	Sergeant First Class	Gunnery Sergeant	Chief Petty Officer	Master Sergeant
E-6	Staff Sergeant	Staff Sergeant	Petty Officer First Class	Technical Sergeant
E-5	Sergeant	Sergeant	Petty Officer Second Class	Staff Sergeant
E-4	Specialist/Corporal	Corporal	Petty Officer Third Class	Senior Airman
E-3	Private First Class	Lance Corporal	Seaman	Airman First Class
E-2	Private	Private First Class	Seaman Apprentice	Airman
E-1	Private	Private	Seaman Recruit	Airman Basic

http://en.wikipedia.org/wiki/Template:United_States_uniformed_services_comparative_ranks

INDEX

Printed in the USA/Agawam, MA
September 25, 2017

659243.005